CMP

Acknowledgement is made to P.H. Oosthuizen et al for the use of Figure 2 on p. 117, which appears on the front cover of this book.

Advanced Computational Methods in Heat Transfer

Vol 2: Natural and Forced Convection

Proceedings of the First International Conference, held in Portsmouth, UK, 17-20 July 1990.

Editors: L.C. Wrobel
C.A. Brebbia
A.J. Nowak

Computational Mechanics Publications
Southampton Boston

Co-published with

Springer-Verlag
Berlin Heidelberg New York London Paris Tokyo

L.C. Wrobel
Wessex Institute of Technology,
Computational Mechanics Institute,
Ashurst Lodge, Ashurst
Southampton SO4 2AA, UK

C.A. Brebbia,
Wessex Institute of Technology,
Computational Mechanics Institute,
Ashurst Lodge, Ashurst
Southampton SO4 2AA, UK

A.J. Nowak
Institute of Thermal Technology
Technical University of Gliwice
44-101 Gliwice
Konarskiego 22, Poland

British Library Cataloguing in Publication Data

Advanced computational methods in heat transfer
Vol 2. Natural and forced convection.
1. Heat transfer. Analysis. Applications of computer systems
I. Wrobel, L.C. (Luiz Carlos) II. Brebbia, C.A. (Carlos
Alberto) *1938-* III. Nowak, A.J.
536.200285

ISBN 1-85312-086-3
ISBN 1-85312-105-3 Set

ISBN 1-85312-086-3 Computational Mechanics Publications, Southampton
ISBN 0-945824-69-6 Computational Mechanics Publications, Boston, USA
ISBN 3-540-52878-4 Springer-Verlag Heidelberg Berlin New York London Paris Tokyo
ISBN 0-387-52877-6 Springer-Verlag New York Heidelberg Berlin London Paris Tokyo
        Set
ISBN 1-85312-105-3
ISBN 0-945824-97-1
ISBN 3-540-52879-2
ISBN 0-387-52879-2

Library of Congress Catalog Card Number 90 - 82731

CONTENTS

SECTION 1: NATURAL AND FORCED CONVECTION

SECTION 2: HEAT AND MASS TRANSFER

PREFACE

Heat transfer problems in industry are usually of a very complex nature, involving the coupling of different types of mechanisms like conduction, convection and radiation, and non-linear features such as temperature-dependent thermophysical properties, phase change and other phenomena. The solution of these problems requires the application of sophisticated numerical methods and extensive use of computer resources.

The present book and its two companion volumes contain edited versions of the papers presented at the First International Conference on Advanced Computational Methods in Heat Transfer, held in Portsmouth, England, in July 1990. The main objective of the conference was to bring together scientists and engineers who are actively involved in developing numerical algorithms as well as in solving problems of industrial interest; to discuss the behaviour of such methods in these extreme conditions and to critically evaluate them by comparison with experiments or established benchmarks wherever possible. All papers have been reproduced directly from material submitted by the authors, and their content is a reflection of the authors' opinion and research work.

The editors would like to thank Prof. D.B. Spalding for his opening address, and all the distinguished scientists who accepted our invitation to deliver special lectures. We are also indebted to the secretarial staff of Computational Mechanics, in particular L. Newman, J.M. Croucher and J. Mackenzie, for the hard work which eventually led to a successful and fruitful meeting.

L.C. Wrobel, C.A. Brebbia, A.J. Nowak
Southampton, July 1990

SECTION 1: NATURAL AND FORCED CONVECTION

Combined Forced and Free Laminar Convection in the Inclined Tube by BEM

P. Skerget, I. Zagar, A. Alujevic
Faculty of Engineering, University of Maribor and Turboinstitut Ljubljana, Yugoslavia

ABSTRACT

This paper deals with numerical simulation of the combined forced and free convection in inclined tubes at low flow rates. The boundary-domain integral method is used for simulation of three-dimensional time dependent laminar fluid flow. The velocity-vorticity formulation for the mass and momentum equations is used, and buoyancy effect is applied with Boussinesq approximation.

The method is validated against finite differences method results for the problem of pure natural convection in differentially heated cylinders. The results are compared for different inclinations of cylinders, various Rayleigh number values and aspect ratio (length-to-radius).

A numerical investigation has been made of combined forced and free laminar convection of upward and downward flow in the isothermaly heated tube, when inclination varies with respect to the gravity vector. At low flow rates in laminar flow regimes buoyancy effects have considerable influence on fluid flow and heat transfer characteristics of the flow development. In many practical situations, the length of the tube is not long enough for the flow to become fully developed and the problem of mixed convection in the entrance region becomes important. The orientation of the tube can have a considerable influence on the velocity and temperature profiles and the associated heat transfer and friction factor in the tube. The numerical results of the development of velocity profiles and temperature fields are presented for different Ra^* number values, and mixed convection factor Φ. They are compared with experimental and finite difference results.

GOVERNING EQUATIONS

The partial differential equations set, governing the motion of viscous incompressible fluid is known as nonlinear Navier-Stokes equations expressing the basic conservation balance of mass, momentum, and energy

$$\frac{\partial v_j}{\partial x_j} = 0 \tag{1}$$

$$\frac{\partial v_i}{\partial t} + \frac{\partial v_j v_i}{\partial x_j} = \frac{1}{\rho_o}\frac{\partial \tau_{ij}}{\partial x_j} + f_i \quad ; \quad i,j = 1,2,3 \tag{2}$$

$$\frac{\partial T}{\partial t} + \frac{\partial v_j T}{\partial x_j} = a_o \frac{\partial^2 T}{\partial x_j \partial x_j} \tag{3}$$

given in indicial notation for a right-handed Cartesian coordinate system, where v_i is the instantaneous component of the velocity, τ_{ij} is the component of the stress tensor and T is the temperature. The material properties such as density ρ_o, and the diffusivity a_o are assumed to be constant values.

For the Newton fluid the stress tensor is expressed by the relation

$$\tau_{ij} = -p\delta_{ij} + \eta_o\left(\frac{\partial v_i}{\partial x_j} + \frac{\partial v_j}{\partial x_i}\right) \tag{4}$$

η_o is dynamic viscosity and δ_{ij} is the Kronecker delta function. The Navier-Stokes equations can be derived with the substitution of the equation (4) in eq.(2)

$$\frac{\partial v_i}{\partial t} + \frac{\partial v_j v_i}{\partial x_j} = -\frac{1}{\rho_o}\frac{\partial p}{\partial x_i} + \nu_o \frac{\partial^2 v_i}{\partial x_j \partial x_j} + f_i \quad ; \quad i,j = 1,2,3 \tag{5}$$

with $\nu_o = \eta_o/\rho_o$ kinematic viscosity. By employing the Boussinesq approximation the buoyancy effect can be included in eq. (5) as follows

$$\frac{\partial v_i}{\partial t} + \frac{\partial v_j v_i}{\partial x_j} = -\frac{1}{\rho_o}\frac{\partial p}{\partial x_i} + \nu_o \frac{\partial^2 v_i}{\partial x_j \partial x_j} + g_i[1 - \beta(T - T_0)] \tag{6}$$

or

$$\frac{\partial v_i}{\partial t} + \frac{\partial v_j v_i}{\partial x_j} = -\frac{1}{\rho_o}\frac{\partial P}{\partial x_i} + \nu_o \frac{\partial^2 v_i}{\partial x_j \partial x_j} - g_i\beta(T - T_0) \tag{7}$$

$P = (p - \rho_o g_j r_j)$ being the modified pressure and r_j is the position vector. The Navier-Stokes equations represent a closed system of equations for the determination of velocity, pressure and temperature fields, subject to appropriate initial and boundary conditions of velocity and temperature.

Velocity-Vorticity Formulation

Introducing the vorticity vector ω_i and the vector potential Ψ_i of the solenoidal velocity field by the relations [7,9]

$$\omega_i = \epsilon_{ijk}\frac{\partial v_k}{\partial x_j} \tag{8}$$

$$v_i = \epsilon_{ijk}\frac{\partial \Psi_k}{\partial x_j} \quad ; \quad \frac{\partial \Psi_j}{\partial x_j} = 0 \tag{9}$$

where ϵ_{ijk} is the permutation unit tensor that equals 1 if the subscripts ijk are in cyclic order 12312, or equals -1, if they are in anticyclic order 32132, and is zero otherwise, the computation of the flow is divided into the kinematics given by the vector Poisson's elliptic equation

$$\frac{\partial^2 \Psi_i}{\partial x_j \partial x_j} + \omega_i = 0 \tag{10}$$

and into the kinetics described by the vorticity transport equation

$$\frac{\partial \omega_i}{\partial t} + \frac{\partial v_j \omega_i}{\partial x_j} = \frac{\partial \omega_j v_i}{\partial x_j} + \nu_o \frac{\partial^2 \omega_i}{\partial x_j \partial x_j} - \beta \epsilon_{ijk}\frac{\partial g_k T}{\partial x_j} \tag{11}$$

which simplifies to a scalar vorticity equation for the plane case

$$\frac{\partial \omega}{\partial t} + \frac{\partial v_j \omega}{\partial x_j} = \nu_o \frac{\partial^2 \omega}{\partial x_j \partial x_j} - \beta \epsilon_{ij}\frac{\partial g_j T}{\partial x_i} \tag{12}$$

where ϵ_{ij} is now the permutation symbol ($\epsilon_{12} = +1, \epsilon_{21} = -1, \epsilon_{11} = \epsilon_{12} = 0$).

BOUNDARY-DOMAIN INTEGRAL EQUATIONS

The boundary domain integral statement for the flow kinematics can be derived from the vector elliptic eq.(10) applying Green's theorem for the vector functions and the elliptic fundamental solution u^*, resulting in the following statement written in the vector notation [6,10]

$$c(\xi)\vec{v}(\xi) + \int_\Gamma (\vec{\nabla}u^* \cdot \vec{n})\vec{v}\,d\Gamma = \int_\Gamma (\vec{\nabla}u^* \mathrm{x} \vec{n})\mathrm{x}\vec{v}\,d\Gamma + \int_\Omega \vec{\omega}\mathrm{x}\vec{\nabla}u^*\,d\Omega \tag{13}$$

The integral statement(13) represents three scalar equations and only two of them are independent. It is completly equivalent to the continuity equation and vorticity definition, expressing the kinematics of the laminar and turbulent incompressible flow in the integral form. Boundary velocity conditions are included in boundary integrals, while the domain integral gives the contribution of the vorticity field to the development of the velocity field. Equation enables the explicit computation of the velocity vector in the interior of the domain $(c(\xi) = 1)$.

When the unknowns are the boundary vorticity values or the tangential velocity component to the boundary, one has to use the tangential component of the vector eq.(13)

$$c(\xi)\,\vec{n}(\xi) \times \vec{v}(\xi) + \vec{n}(\xi) \times \int_{\Gamma}(\vec{\nabla}u^* \cdot \vec{n})\,\vec{v}\,d\Gamma \;=\; \vec{n}(\xi) \times \int_{\Gamma}(\vec{\nabla}u^* \times \vec{n}) \times \vec{v}\,d\Gamma$$
$$+\; \vec{n}(\xi) \times \int_{\Omega} \vec{\omega} \times \vec{\nabla}u^*\,d\Omega \quad (14)$$

or when the normal velocity to the boundary is unknown, the normal component of the vector equation(13) has to be used

$$c(\xi)\,\vec{n}(\xi) \cdot \vec{v}(\xi) + \vec{n}(\xi) \cdot \int_{\Gamma}(\vec{\nabla}u^* \cdot \vec{n})\,\vec{v}\,d\Gamma \;=\; \vec{n}(\xi) \cdot \int_{\Gamma}(\vec{\nabla}u^* \times \vec{n}) \times \vec{v}\,d\Gamma$$
$$+\; \vec{n}(\xi) \cdot \int_{\Omega} \vec{\omega} \times \vec{\nabla}u^*\,d\Omega \quad (15)$$

in order to perform appropriate implicit system of discretized equations. Unknown boundary vorticity values are expressed in the integral form eq.(14) within the domain integral, excluding a need to use approximate formulae for determining boundary vorticity values, which would bring some additional error into the numerical scheme.

Vector equation(13) can also be given in the component form for individual x, y, z directions

$$c(\xi)v_i(\xi) + \int_{\Gamma} v_i \frac{\partial u^*}{\partial n}\,d\Gamma = \int_{\Gamma} v_k\left(\frac{\partial u^*}{\partial x_k}n_i - \frac{\partial u^*}{\partial x_i}n_k\right)d\Gamma - \int_{\Gamma} v_j\left(\frac{\partial u^*}{\partial x_i}n_j - \frac{\partial u^*}{\partial x_j}n_i\right)d\Gamma$$
$$+ \int_{\Omega} \omega_j \frac{\partial u^*}{\partial x_k}\,d\Omega - \int_{\Omega} \omega_k \frac{\partial u^*}{\partial x_j}\,d\Omega \quad (16)$$

for the cyclic combination of the indexes$(ijkij = 12312)$.

Describing the laminar transport of the vorticity and temperature in the integral statement, one has to take into account that each instantaneous component of the vorticity vector and temperature obey a nonhomogenous parabolic equation [8]; the following boundary-domain integral formulations can be derived for the space and plane vorticity transfer and temperature transport

$$c(\xi)\omega_i(\xi, t_F) + v_o \int_{\Gamma}\int_{t_{F-1}}^{t_F} \omega_i \frac{\partial u^*}{\partial n}\,dt\,d\Gamma = v_o \int_{\Gamma}\int_{t_{F-1}}^{t_F} \frac{\partial \omega_i}{\partial n}u^*\,dt\,d\Gamma$$
$$- \int_{\Gamma}\int_{t_{F-1}}^{t_F}(\omega_i v_n - v_i \omega_n + \beta\epsilon_{ijk}n_j g_k T)u^*\,dt\,d\Gamma$$
$$+ \int_{\Omega}\int_{t_{F-1}}^{t_F}(\omega_i v_j - v_i\omega_j + \beta\epsilon_{ijk}g_k T)\frac{\partial u^*}{\partial x_j}\,dt\,d\Omega + \int_{\Omega} \omega_{i,F-1}\,u^*_{F-1}\,d\Omega \quad (17)$$

$$c(\xi)\omega(\xi,t_F) + \nu_o \int_\Gamma \int_{t_{F-1}}^{t_F} \omega \frac{\partial u^*}{\partial n} \, dt \, d\Gamma = \nu_o \int_\Gamma \int_{t_{F-1}}^{t_F} \frac{\partial \omega}{\partial n} u^* \, dt \, d\Gamma$$

$$- \int_\Gamma \int_{t_{F-1}}^{t_F} (\omega v_n + \beta \epsilon_{ij} n_i g_j T) u^* \, dt d\Gamma$$

$$+ \int_\Omega \int_{t_{F-1}}^{t_F} (\omega v_i + \beta \epsilon_{ij} g_j T) \frac{\partial u^*}{\partial x_i} \, dt d\Omega + \int_\Omega \omega_{F-1} \, u^*{}_{F-1} \, d\Omega \tag{18}$$

$$c(\xi)T(\xi,t_F) + a_o \int_\Gamma \int_{t_{F-1}}^{t_F} T \frac{\partial u^*}{\partial n} \, dt \, d\Gamma = a_o \int_\Gamma \int_{t_{F-1}}^{t_F} \frac{\partial T}{\partial n} u^* \, dt \, d\Gamma$$

$$- \int_\Gamma \int_{t_{F-1}}^{t_F} T \, v_n u^* \, dt \, d\Gamma - \int_\Omega \int_{t_{F-1}}^{t_F} T v_i \frac{\partial u^*}{\partial x_i} + \int_\Omega T_{F-1} \, u^*{}_{F-1} \, d\Omega \tag{19}$$

DISCRETIZED BOUNDARY DOMAIN INTEGRAL EQUATIONS

Searching for an approximate solution of the field functions the boundary Γ has to be discretized in E boundary elements with N_e boundary nodes, and the domain Ω into C internal cells with N_c points [2]. Values of the field variables are to be approximated in individual elements and internal cells by the use of interpolation polynomials $\{\Phi\}$ and $\{\phi\}$ with respect to boundary or domain, and by time interpolation polynomials $\{\Psi\}$.

One can develop the following discretized statements for the instantaneous temperature and vorticity equations

$$[H]\{T\}_F = [G]\{\frac{\partial T}{\partial n}\}_F - \frac{1}{a_o}([G]\{Tv_n\} - [D_i]\{Tv_i\}_F - [B]\{T\}_{F-1}) \tag{20}$$

$$[H]\{\omega_i\}_F = [G]\{\frac{\partial \omega_i}{\partial n}\}_F - \frac{1}{\nu_o}([G]\{\omega_i v_n - v_i \omega_n + \beta(g_k n_j - g_j n_k)T\}_F$$

$$-[D_j]\{\omega_i v_j - v_i \omega_j + \beta g_k T\}_F - [D_k]\{\omega_i v_k - v_i \omega_k - \beta g_j T\}_F - [B]\{\omega_i\}_{F-1}) \tag{21}$$

the eq.(20) and eq.(21) can formally be written in a compact form for a field function f as

$$[H]\{f\}_F = [G]\{\frac{\partial f}{\partial n}\}_F + \{F_N(f, v_i)\} \tag{22}$$

The implicit system of equations is first solved for boundary unknowns vorticity or temperature respectively and corresponding flux values. The next kinetic step is explicit evaluation of the function values in the domain. For large Peclet and Reynolds number values the nonlinearity incorporated in the term $\{F_N\}$ becomes more and more severe, making the numerical computation difficult. The under-relaxation point iterative method has to be taken in explicit computation of the new domain values. The nonlinearity can be accounted for by decreasing the under-relaxation factor, or the time step, or by a successive increasing of the Peclet and Reynolds number values.

TEST EXAMPLES

Free convection in cylinders

The free convection in inclined cylinders has been studied first. The BEM results are compared with FDM results, Bontoux [1]. The cylinder geometry is depicted in Fig. 1, where R is radius, L lenght and $A = L/R$ aspect ratio. The inclination angle γ is referred to the vertical axis. The two circular endwalls are kept at constant temperatures $T_h = 1$ and $T_c = 0$, while the side wall is assumed perfectly conducting.

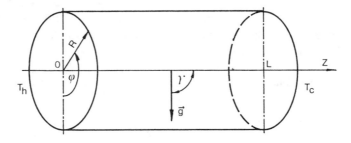

Fig. 1: Geometry of the cylinder

The linear 3-node triangular and 4-node quadrilateral boundary elements were used to model the boundary, while linear 8-node and 6-node brick internal cells are applied to discretise the domain. Mesh sizes $M = 5 \times 16 \times 9$ in radial, azimuthal and axial directions were used. The time step $\Delta t = t_F - t_{F-1} = 1s$, and the under-relaxation factor 0.1 were used. Free-convection motion numerical results for the $A = 5$ cylinder when the axis is inclined at an angle $\gamma = 180°$; $135°$; $90°$ with the gravity vector are presented for the Rayleigh number $Ra = 6250$. Velocity fields and isotherm contours for vertical, inclined and horizontal cylinders are depicted in Fig. 2, 3 and 4.

Fig. 2: Velocity fields and isotherm contours in vertical cylinder
$\gamma = 180^\circ$, $\phi = 0^\circ$ for $Ra = 6250$ at $t = 8$ and 40 s

Fig. 3: Velocity fields and isotherm contours in inclined cylinder
$\gamma = 135^o$, $\phi = 0^o$ for $Ra = 6250$ at $t = 8$ and $40\ s$

Fig. 4: Velocity fields and isotherm contours in horizontal cylinder
$\gamma = 90^o$, $\phi = 0^o$ for $Ra = 6250$ at $t = 8$ and 40 s

Mixed convection in entrace region of inclined tube

The inclined tube under consideration is shown schematically in Fig.5 together with boundary conditions. The fluid enters the tube with a uniform temperature $T_v = 0$ and fully developed laminar profile $v_v = 2\, v_o(1 - (r/R)^2)$, where v_o is average velocity of inlet flow. The tube wall temperature is held constant at uniform temperature $T_s = 1$. At the outlet of the tube fully developed flow and temperature profiles are assumed

$$\frac{\partial v_x}{\partial z} = \frac{\partial v_y}{\partial z} = \frac{\partial v_z}{\partial z} = \frac{\partial T}{\partial z} = 0$$

Due to the symmetry $(\omega_x,\, v_y,\, v_z,\, T)$ and asymmetry $(v_x,\, \omega_y,\, \omega_z)$ of the problem, the calculation is restricted to solution domain on one-half of the circular region, what gives a set of following boundary condition in $y - z$ plane

$$v_x = \omega_y = \omega_z = \frac{\partial T}{\partial n} = 0$$

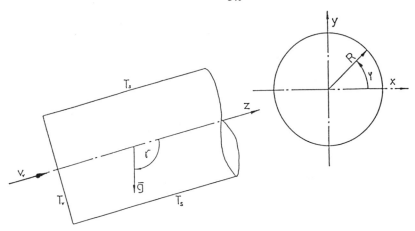

Fig. 5: Problem of mixed convection in entrance region of inclined tube

An examination of the dimensionless governing equations shows that the solution of the problem depends on three independent parameters: Prandtl number Pr, modified Rayleigh number Ra^* and mixed convection parameter Φ:

$$Pr = \frac{\nu}{a}, \qquad Ra^* = Ra\,\cos(\theta), \qquad \Phi = \frac{Ra}{Re}\,\sin(\theta)$$

where

$$Ra = \frac{8\,g\,\beta\,(T_s - T_v)\,R^3}{\nu\,a}, \qquad Re = \frac{2\,v_o\,R}{\nu}, \qquad \theta^o = \gamma^o - 90^o$$

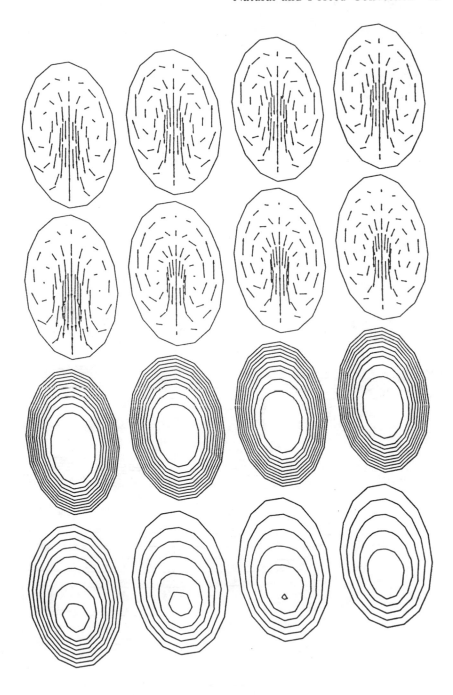

Fig. 6: Velocity fields and isotherm contours in the cross-sections
$(z/R = 1, 3, 5, 7)$ at $t = 2s$ and $t = 8s$, $(\Phi = 0, Ra^* = 1.25 \, 10^4)$

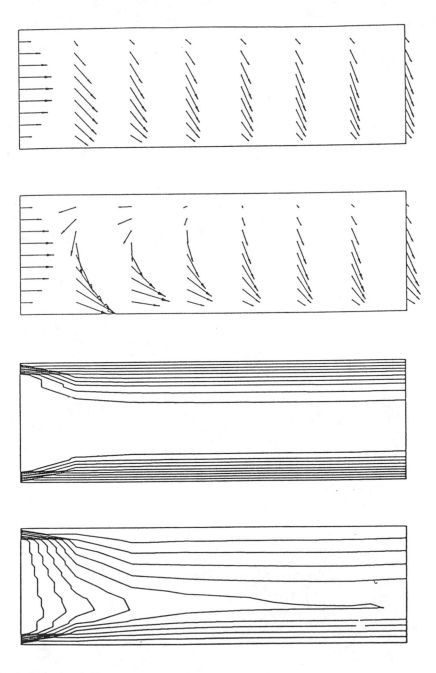

Fig. 7: Velocity fields and isotherm contours in symmetricaly plane ($\phi = 90^o$) at $t = 2s$ and $t = 8s$, ($\Phi = 0$, $Ra^* = 1.25\ 10^4$)

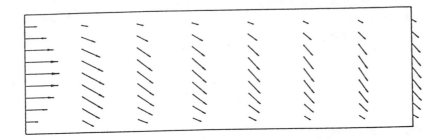

(a) Flow in horizontal tube ($\Phi = 0$, $Ra^* = 6250$)

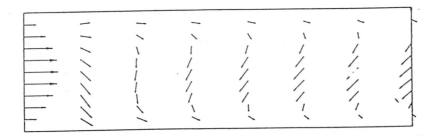

(b) Flow in inclined tube–upflow ($\Phi = 315$, $Ra^* = 6250$)

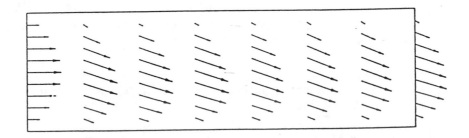

(b) Flow in inclined tube–downflow ($\Phi = -315$, $Ra^* = 6250$)

Fig. 8: Velocity fields in the symmetrical plane ($\phi = 90^o$) at time $t = 2s$ for different tube inclination

Values of dimensionless numbers are chosen in connection with model discretization. Mesh is described by $5 \times 8 \times 8$ nodal points in radial, circular and axial directions. The time step of the problem is $\Delta t = t_F - t_{F-1} = 1s$, and Prandtl number $Pr = 1$.

Developments of temperature and velocity field in the cross sections and the axial-symmetry plane of the horizontal tube ($\Phi = 0$ in $Ra^* = 1.25 \; 10^4$) are shown in Figs. 6 and 7. Velocity field from Fig. 6 displays typically buoyancy-induced secondary flow pattern, what agrees with results obtained from experiments [4] and finite differences method results [3]. The fluid rises along the heated wall and falls toward the center where it is carried out by the main flow. The consequence of the secondary flow pattern is the shift of the cold fluid core slightly to the lower tube wall, what is evident from the temperature field in cross and axial section of the tube. Recirculation region in the axial plane near the inlet is due to predominancy of the natural convection in this case. Influence of the tube inclination is studied at a little lower ratio of natural and forced convection. Results of velocity field for different inclination are shown in Fig. 8. Validation of results against experimental and finite differences method calculations is restricted only for quality comparison, due to coarse discretization in the analysis. This is particulary true for the upflow, where discretization cannot follow velocity field near the wall.

REFERENCES

[1] Bontoux,P., Smutek,C., Roux,B., Extremet,G.P., Schiroky,G.H., Hurford,A.C., Rosenberg,F.: *Finite Difference Solutions for Three Dimensional Buoyancy Driven Flows in Inclined Cylinder*, Vol.3, 3rd Int. Conf. on Num. Meth. for Nonlinear Problems, Dubrovnik. Pineridge Press, 1986.

[2] Brebbia,C.A., Telles,J.F.C., Wrobel,L.C.:*Boundary Element Methods-Theory and Applications*, Springer-Verlag, New York, 1984.

[3] Choudhury, D., Patankar, S.V.: *Combined Forced and Free Laminar Convection in the Entrance Region of the Inclined Isothermal Tube*, ASME Journal of Heat Transfer, Vol. 101, 1988.

[4] Huggenberger, M.: *Mixed convection heat transfer and pressure drop in channels with low fluid flow velocities*, PSI report, to appear.

[5] Skerget,P., Alujevic,A., Kuhn,G., Brebbia,C.A.:*Natural Convection Flow problems by BEM*, 9th Int. Conf. on Boundary Element Method, Stuttgart, Springer-Verlag, Berlin, 1987.

[6] Skerget,P., Alujevic,A., Zagar,I., Brebbia,C.A., Kuhn,G.:*Time Dependent Three Dimensional Laminar Isochoric Viscous Fluid Flow by BEM*, 10th Int. Conf. on BEM, Southampton, Springer-Verlag, Berlin, 1988.

[7] Skerget,P., Alujevic,A., Brebbia,C.A., Kuhn,G.:*Natural and Forced Convection Simulation Using the Velocity-Vorticity Approach*, Topics in Boundary Element Research (Ed. by Brebbia C.A.), Vol.5, Ch.4, Springer-Verlag , Berlin, 1989.

[8] Skerget,P., Kuhn,G., Alujevic,A., Brebbia,C.A.:*Time Depended Transport Problems by BEM*, Advances in Water Resources, Vol.12, No.1, 1989.

[9] Wu,J.C., Rizk,Y.M., Sankar,N.L.,:*Problems of Time Dependent Navier-Stokes Flow*, Developments in Boundary Element Methods, Vol.3, Chap.6, Elsevier Appl.Sci.Publ., 1984.

[10] Wu,J.C., Gulcat,U., Wang,C.M., Sankar,N.L.:*A Generalized Formulation for Unsteady Viscous Flow Problems*, Topics in Boundary Element Research (ed. Brebbia C.A.), Vol.5, Ch.3, Springer-Verlag, Berlin, 1989.

[11] Zagar,I.: *Three-Dimensional Transport Problems in Solids and Fluids by the Method of Boundary Elements*, M.Sc. Thesis, University of Maribor, 1989.

[12] Zagar,I., Skerget,P.: *Boundary Elements for Time Dependent 3-D Laminar Viscous Fluid Flow*, Mechanical Engineering Journal, Vol. 10-12, Ljubljana, 1989.

Flow and Heat Transfer in Bends of Two-Dimensional Corrugated Walls

R. Nassef, A.S. Mujumdar
E.T.S., Université de Quebec, and Chemical Engineering Department, McGill University, Montreal, Canada

Abstract

Developing laminar flow and heat transfer characteristics in a bend formed by corrugated walls was examined numerically using an orthogonal, boundary-fitted coordinate system (OBFC). The study was conducted for a fluid with Pr=0.7, and a Reynolds number ranging from 100 to 1500. On the basis of numerical experiments, a number of Nu-Re correlations for corrugated bends, and corrugated ducts with different corrugation angles are presented. The effects of curvature of the bend and of corrugation angle are also examined and discussed.

Introduction

Flow and heat transfer in corrugated ducts has been the subject of a number of exprimental and numerical investigations in recent years [1,2,3,4]. Such ducts which are frequently found in compact heat exchanger configurations, have been reported to produce dramatic improvement in the heat transfer rate. In this paper we examine numerically the developing laminar flow and heat transfer in two dimensional bends formed between two corrugated plates with a varying corrugation angle. These configurtions, which to our best knowledge have never been examined before in literature, are of potential practical importance for evaluation of the overall performance of heat exchangers utilizing such a novel configuration. The bends were considered as extensions of a straight corrugated duct with an entrance height H equal to the height of the corrugation element (Fig. 1). The inner wall of the bend consists of two elements of a corrugated surface idential to that used in the straight section of the duct. The upper wall uses larger elements, the size of which is selected so that both walls remain parallel and in phase with each other. Three types of corrugated surfaces are considered, with an angle $\alpha = 15°$, $30°$, and $45°$ (fig. 1).

The domain geometry was discretized in a 31x97 grid.   In addition, a plain duct section containing 31x12 grid points was added at the exit of the bend so that the solution domain contained a total of 31x109 nodes.   The finite difference method was used to solve the governing equations using the SIMPLE algorithm of Spalding and Patankar [9] and an orthogonal, boundary fitted coordinate system (OBFC).   Numerical solutions were obtained in the Reynolds number range of 100 to 1500 for a fixed Prandtl number of 0.7.   A parabolic velocity distribution was imposed at the inlet while outflow boundary conditions were assumed at the exit section.   These boundary conditions apply for a bend with a plain duct attached to both ends of the bend.   The results obtained in this study are, therefore, indicative of the performance of a stand alone bend, rather than one that form part of a corrugated duct.   For the purpose of evaluating the performance of the corrugated bend, the numerical results for the Nusselt number, and friction factor are compared with their corresponding values for a straight corrugated duct, as well as those for a plain, two dimensional bend.   The effect of curvature of the bend was examined by comparing the flow and heat transfer characteristics predicted for 4 bends with r_i= 25, 50, 75, and 100% of its previous value while maintaining the same entrance height as before (fig. 2).   The effect of the corrugation angle was examined by comparing the Nusselt number and friction factor in 5 bends with α=15°, 22.5°, 30°, 37.5°, and 45° respectively while maintaining the same entrance height H and corrugation span L (fig. 3).

Analysis and Mathematical Formulation:

Flow in corrugated bends combines some of the flow charateristics of plain bends and corrugated ducts.   In plain bends, the flow gives rise to centrifugal forces that prompt the formation of vortices in the flow field.   The fully developed flow in helical coils and curved rectangular ducts is charaterized by two secondary recirculation zones on the top and bottom halves of the duct section perpendicular to the main flow direction [5,6,7]. This secondary motion causes a large increase in the heat transfer rate as well as the frictional losses.   In a developing flow in bends between two semi-infinite surfaces, a recirculation zone forms next to the backward facing area of the inner wall and the flow path is diverted towards the outer wall.   This impingement action helps augment the local as well as the overall heat transfer coefficient.

In corrugated ducts the flow pattern bears a certain resemblance to the one just discussed with vortex formation in the backward facing zones on both walls, and an impingement action that brings about a large increase in both the heat transfer rate and the frictional losses.   In the numerical study conducted by Asako et al [1], the Nusselt number in a corrugated wall channel increased with the Reynolds and Prandtl numbers, as well as with the corrugation angle.   For an air flow in a corrugated duct with α= 30° and Re= 1000, the augmentation in heat transfer rate over that that can be attained in a plain duct was reported by Asako et al [1] to be in the order of 440%

Governing Equations

For a two dimensional steady laminar flow of an incompressible, constant property fluid with negligible viscous dissipation, the Navier Stokes equations and the energy equation can be written in terms of the orthogonal, boundary-fitted coordinate system as follows,

Mass Conservation:

$$\nabla_1 \ V_1 \ + \ \nabla_2 \ V_2 \ = \ 0 \tag{1}$$

Momentum Conservation:

$$\nabla_1 \ [V_1^2 \ - \ \frac{1}{Re} \ \frac{\partial V_1}{\partial X_1}] \ + \ \nabla_2 \ [V_1 V_2 \ - \ \frac{1}{Re} \ \frac{\partial V_1}{\partial X_2}] \ = \ - \ \frac{\partial P}{\partial X_1} \ + \ S_{v\,1} \tag{2}$$

Energy Conservation:

$$\nabla_1 \ [V_1 \ \Theta \ - \ \frac{1}{Re \ Pr} \ \frac{\partial \Theta}{\partial \overline{\overline{X}}_1}] \ + \ \nabla_2 \ [V_2 \ \Theta \ - \ \frac{1}{Re \ Pr} \ \frac{\partial \Theta}{\partial X_2}] \ = \ S_\Theta \tag{3}$$

The following non-dimensional variables are used in formulating the above equations,

$$X_1 = X = \frac{x}{H} \ , \ X_2 = Y = \frac{y}{H}, \ V_1 = U = \frac{u}{\overline{u}} \ , \ V_2 = \frac{v}{\overline{u}} \ , \quad \left.\vphantom{\begin{array}{c}1\\1\\1\end{array}}\right\}$$

$$P = \frac{p}{\rho \overline{u}^2} \ , \ \Theta = \frac{T - T_w}{T_i - T_w} \ , \ Re = \frac{\overline{u} H}{\nu} \ , \tag{4}$$

with $\overline{u} = \frac{\overset{\cdot}{m}}{\rho H}$.

The physical derivatives are defined as,

$$\nabla_j \ (\Psi) \ = \ \frac{\partial \Psi}{\partial X_j} \ + \ \frac{\Psi}{r_k} \ \text{ where } j \neq k \ , \ j=1,2 \tag{5}$$

r_k being the local radius of curvature of the k-th coordinate. S_v and S_Θ are source terms associated with the geometric form of each individual control volume. The Darcy friction factor and the local Nusselt number which are defined as:

$$f \ = \ - \ \frac{\Delta p}{L} \ . \ \frac{2H}{\rho \overline{u}^2 /2} \ , \ \text{ and} \tag{6}$$

$$Nu_x \ (\overline{x}) = \ \frac{[Q(\overline{x})/A].H}{(T_{bx} - T_w).k} \tag{7}$$

can be evaluated from the velocity and temperature distribution. The average Nusselt number was evaluated as,

$$Nu \ = \ \frac{1}{\Theta_b} \ \left[\int (\ \frac{\partial \Theta}{\partial Y} \ )_{Y=0} \ dX_B \ + \ \int (\ \frac{\partial \Theta}{\partial Y} \ )_{Y=1} \ dX_T \right] / \ \int (dX_B + dX_T) \tag{8}$$

Boundary Conditions:

Outflow boundary conditions were assumed at the exit section.   In order to ensure their applicability, a plain duct section was added at the end of the bend, the length of which was chosen so that no recirculation would occur as the flow exits from the solution domain.   The other boundary conditions are given below,

Inlet section

$$\left.\begin{array}{l} U = 6Y.(1\text{-}Y), \\ V = 0, \\ \Theta = 1. \end{array}\right\} \quad (9)$$

Walls

$$U = V = \Theta = 0. \tag{10}$$

Results and discussion:

Effect of Reynolds Number

The flow and heat transfer characteristics were examined numerically in 3 types of corrugated bends similar to the one outlined in figure 1 with corrugation angles of $15°$, $30°$, and $45°$ respectively and a Reynolds number varying between 100 and 1500.

Figure 4 displays the predicted streamline patterns in the 3 bends.   Formation of a separation zone can be observed on the rear facing sides of the corrugated walls.   The size of the recirculation bubbles is largest at $\alpha = 45°$, and decreases rapidly with α.   An increase in the Reynolds number also contributes to a reduction in size of the recirculation bubbles.   To better illustrate the effect of Re on the recirculation zone, the stream functions in a bend with a triangular roughness element (a configuration in which the recirculation bubble is more prominent) were computed at Re= 100 and 1000 (fig. 6).   The flow pattern is consistent with the one calculated for a corrugated bend with a marked reduction in the size of the separation zone at the higher Reynolds number.

Stream functions were also calculated for a developing flow in a corrugated straight duct section and compared with those previously obtained for a corrugated bend with the same α, L, and H.   Figures 4, and 5 show that the recirculation bubbles near the inner wall are of comparable size in these two configurations.   In corrugated bends the bubbles located near the top wall are much smaller in size.   This may be the result of the combined effect of the fluid undergoing a smaller deviation at the mid-section of the bend, and the centrifugal forces that propel the flow towards the upper wall.   The angle between the two wall sections at the mid-point is $45°$ larger in the case of the bend (fig. 1).   The larger angle causes a reduction of the adverse pressures gradient and the separation in the flow field.   The centrifugal forces tend to reduce recirculation near the upper wall while increasing it along the inner wall.

Figure 7 displays the local Nusselt number Nu_x (the mean value on the two walls) distribution in a corrugated wall bend with $\alpha=45°$ at Re= 1500. Also shown in this figure is the domain geometry and the recirculation bubbles along both walls. In the distribution, 4 peaks are clearly visible at locations that correspond to points of reattachment of the separation zone on both walls. Figure 8-a,b,c shows Nu as a function of Re for a straight corrugated duct and a corrugated wall bend at the same values of α, H, and L. Also shown are results obtained for a plain two dimensional bend with the same entrance height. For a corrugated bend, the Nusselt number values were higher than those obtained for a plain bend, but are well below those for a straight corrugated duct. In the log-log coordinates used in figure 8, the Nu-Re relation is linear in the range of high Re, and parabolic in the lower range. Only in the case of a corrugated duct with $\alpha=45°$, was the relation linear over the ntire Re range.

For corrugated ducts the trend indicated by our numerical predictions is consistent with the numerical results of Amano and Asako (ref. 1,2) a well as the experimental findings of O'Brien and Sparrow [3] (fig. 9). The large dicerepencies between results are understandable since in each case they correspond to a different set of flow or boundary conditions. The thermal boundary condition in Amano's work [2] is that of constant heat flux while a Dirichlet condition was assumed in both Asako [1] and O'Brien's [3] studies. The results of all the these studies are for a fully devloped flow. Further, O'Brien only considered turbulent flow (Re > 1500).

The following correlations represent the results of figure 8 for the Nusselt number in a developing laminar flow in two dimensional corrugated ducts and bends at Pr= 0.7,

$$Nu= C_1\, Re^m + C_2 Re^{-n} \tag{11}$$

The first term in the equation represent the linear portion of the Nu- Re curve, the second term makes a contribution that decrease rapidly with Re. Regression values of C_1, C_2, m, and n are listed below. These values may be used only over the range of flow and geometric parameters considered in this numerical study.

	Corrugated Duct				Corrugated Bend			
α	C_1	C_2	m	n	C_1	C_2	m	n
45°	1.357	0	0.5175	0	0.457	1367	0.6	1.5
30°	0.440	1000	0.575	1.2	0.5915	1112	0.5	1.3
15°	0.802	1100	0.421	1.3	0.730	735	0.427	1.2

The results given by equation 11 and the table above compare favorably with the correlation given by O'Brien for a turbulent, fully developed flow in a duct with a corrugation angle of 30°, which for air (Pr=0.7) becomes,

$$Nu = 0.3623\ Re^{0.614}.$$

The friction factor is found to be insensitive to variations in the Reynolds number, but it increased significantly with the corrugation angle. The values that were obtained numerically for the developing flow in ducts and bends at Pr=0.7 and Re= 100- 1500 are listed below,

	Corr. duct	Corr, bend
α	f	f
15°	0.673	0.732
30°	0.74	1.25
45°	1.09	1.375

Effect of curvature

The effect of curvature of the bend on the flow pattern was investigated by comparing the stream function computed for 4 corrugated bends with α=30°. The size of the corrugation elements on the inner wall were 25, 50, 75, and 100% of those used previously, while the entrance height H was maintained as before (fig. 2). The non- dimensional radius of curvature of the inner wall r_i^+ in these configurations is 1.1315, 2.263, 3.395, and 4.526 respectively. Figure 10 displays the stream functions computed for two bends with r_i^+ = 2.263 and 4.526. The increase in size of the separation zone at the lower r_i^+ resulting from higher centrifugal forces is clearly visible.
    Numerical results for the Nusselt number are plotted against the inner radius of curvature in figure 11. Also shown in this figure is a curve representing the friction factor variation with r_i^+. Both Nu and f increase with the curvature of the bend (a decreased r_i^+). The Nusselt number is observed to be much more sensitive to a variation in r_i^+, than is the friction factor.

Effect of corrugation angle

The flow and heat transfer characteristics were examined in five corrugated bends with a fixed entry height and corrugation angles of 15°,22.5°,30°,37.5°, and 45°. The Nusselt number and friction factor results at different corrugation angles are given in figure 12. While the friction factor shows a steady increase with the corrugation angle, the Nusselt number results maintaines a constant value up to α= 30°, then rise sharply with α.

Conclusions

The following conclusions emerge from the present numerical study,
    1.   The flow pattern in computed results bears much resemblance to that encountered in a corrugated duct but is modified by the centrifugal forces arising in the bend.
    2.   The Nusselt number values are lower than those reached in a straight corrugated duct with the same corrugation angle and entrance height, while the friction coefficient is higher. A new

correlation is given relating Nu to Re over parameter ranges studied.

3.   Both Nusselt number and friction factor increase with curvature of the bends. Nu being more sensitive to variations in r_i^+ than is f.

4.   The corrugation angle has an appreciable effect on the friction factor while its influence on the Nusselt number can be felt only for $\alpha \geq 30°$.

References

1.   Asako, Y., and Faghri, M., 1987,"Finite Volume Solutions for Laminar Flow and Heat Transfer in a Corrugated Duct", ASME J. Heat Transfer, vol. 109, pp. 627-634.

2.   Amano, R.S.,Bagherlee, A.,Smith, R.J., and Niess,T.G., 1987, "Turbulent Heat transfer in Corrugated Wall Channels with and Without Fins", ASME J. Heat Transfer, vol. 109, pp.62-63.

3.   O'Brien, J.E, and Sparrow, E.M, 1982, "Corrugated Duct Heat Transfer, Pressure Drop, and Flow Visualization", ASME J. Heat Transfer, vol. 104, pp. 410-416.

4.   Izumi, R., Yamachita, H., Kaga, S., and Miyazima, N., 1982, "Fluid Flow and Heat Transfer in Corrugated Wall Channels-Experimental Study for Many Bends", 19th National Heat Transfer Symposium of Japan, paper No. A 101.

5.   Cheng, K.C., and Akiyama, M., 1970, "Laminar Forced Convection Heat Transfer in Curved Rectangular Channels", Int. J. Heat Mass Trans.,vol. 13, pp. 471-490,

6.   Mori, T., And Nakayama, W., 1967, "Study on Forced Convective Heat Transfer in Curved Pipes", Int. J. Heat Mass TGrans., vol. 10, pp. 37- 59.

7.   Patankar , S.V., 1980, "Numerical Heat Trransfer and Fluid Flow", Hemisphere Pub. Co.

Nomenclature

A — surface area of the wall
f — friction factor, (eqn. 6),
H — entrance height, (fig. 1),
L — corrugation span of a corrugated bend (fig. 1),
L'— corrugation span of a straight corrugated duct (fig.1),
Nu_x— local Nusselt number (eqn. 7),
Nu— average Nusaselt number for the entire bend (eqn. 8),
P — non dimensional pressure, (eqn. 4),
r_i^+— non dimensional radius of curvature
    of the inner surface of the bend (fig. 1),
r_k— local radius of curvature,
Re— Reynolds number (eqn. 4),
T_{bx}— fluid bulk tepmperature at the station x,
T_i— inlet temperature,
T_w— wall temperature,
U,V— local velocities,
$\bar{u}$ — average entrance velocity,
x,y— curvi-linear coordinates,
X,Y— non-dimensional curvilinear coordinates,
$\bar{x}$ — horizontal distance along the bend (fig. 1),
X_B,X_T — curvilinear distance along the bottom, top walls,

Greek letters
α — corrugation angle (fig. 1),
θ — non dimensional temperature (eqn. 4),
ν — dynamic viscosity,

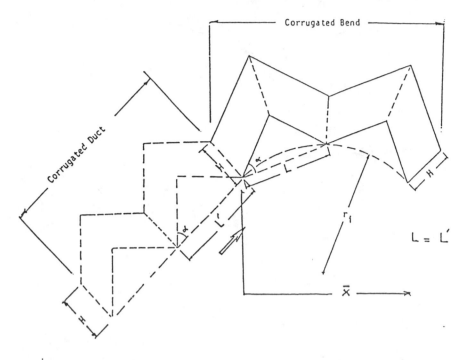

Fig. 1, Basic solution domain

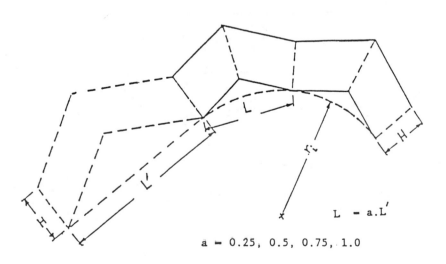

a ━ 0.25, 0.5, 0.75, 1.0

Fig. 2, Solution domain to study the effect of bends curvature.

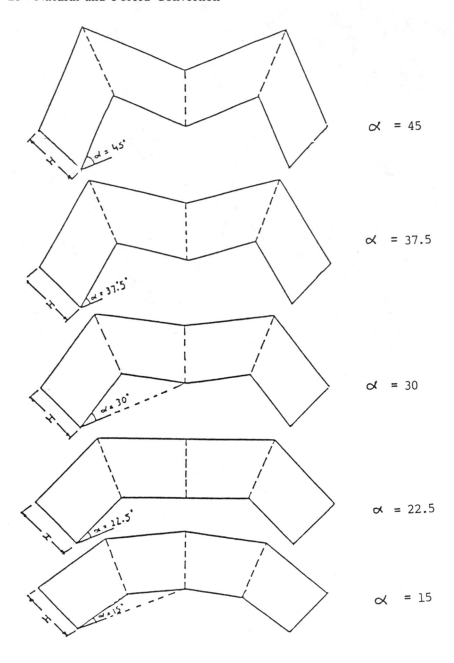

Fig. 3, Solution domain to study the effect

of corrugation angle.

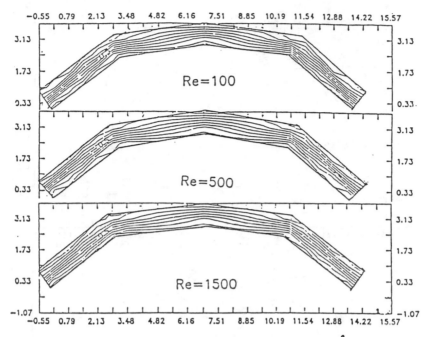

Fig. 4-A, Stream function in a corrugated bend, $\alpha = 15°$

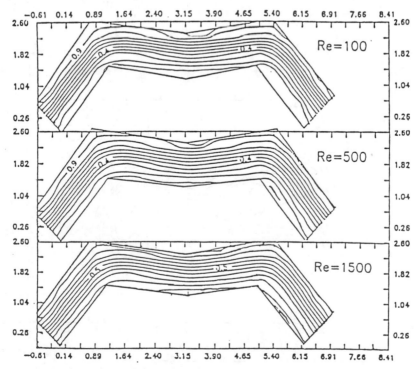

Fig. 4-B, Stream function in a corrugated bend, $\alpha = 30°$

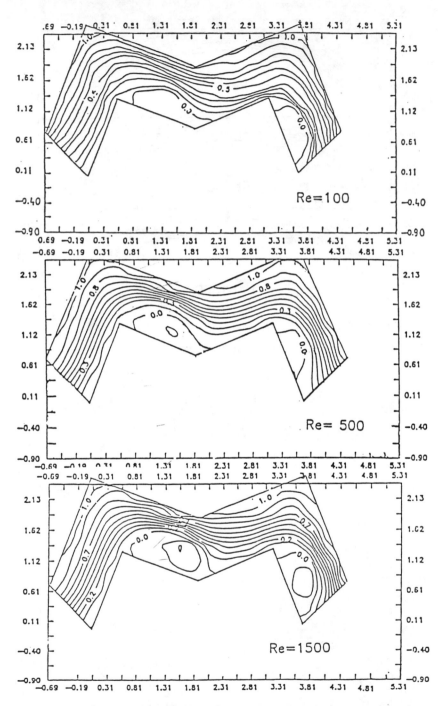

Fig. 4-C,   Flow pattern in a corrugated bend, $\alpha = 45°$

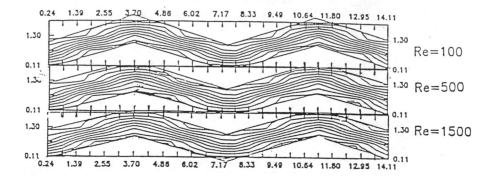

Fig. 5-A,   Flow pattern in a corrugated duct,   α= 15°

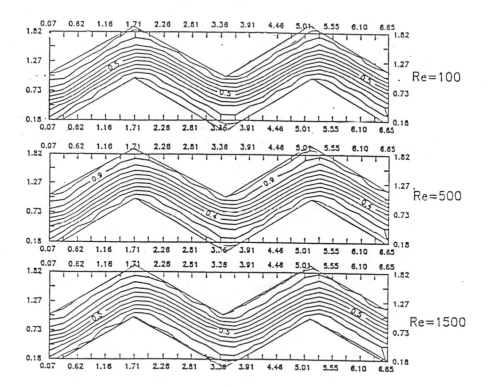

Fig. 5-B,   Flow pattern in a corrugated duct,   α= 30°

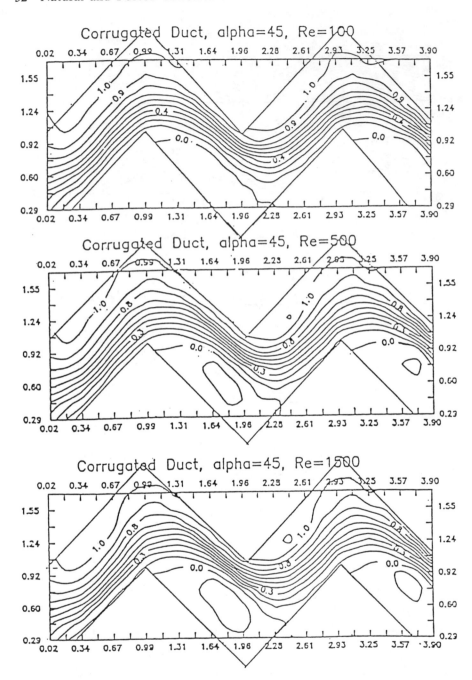

Fig. 5-G, Flow pattern in a corrugated duct, α= 45°

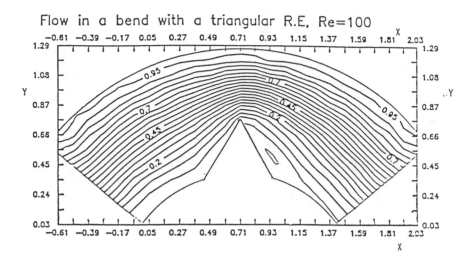

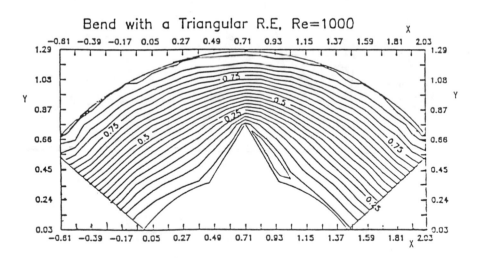

Fig. 6,

Flow pattern in a bend with a triangular roughness element

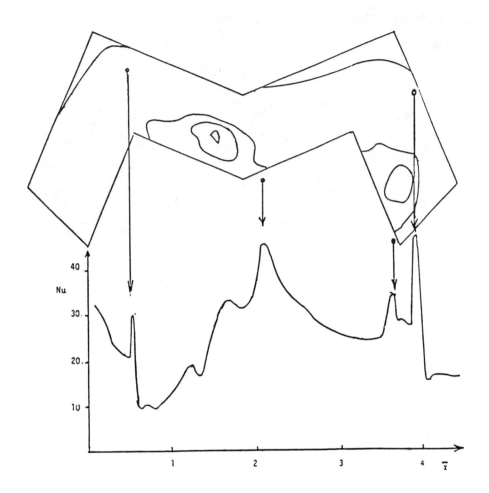

Fig. 7,

Local Nusselt number distribution in a corrugated bend,

$\alpha = 45°$,   Re=1500.

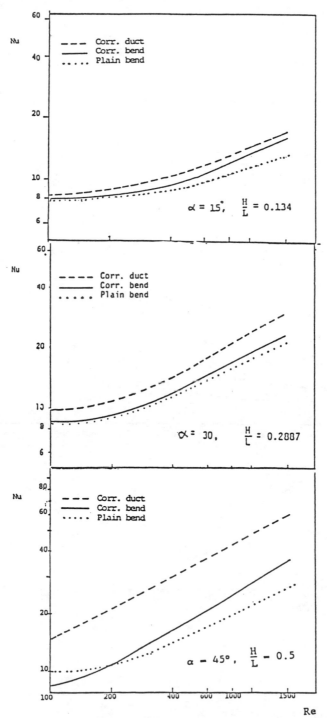

Fig. 8 : Average Nusselt number as a function of Re,

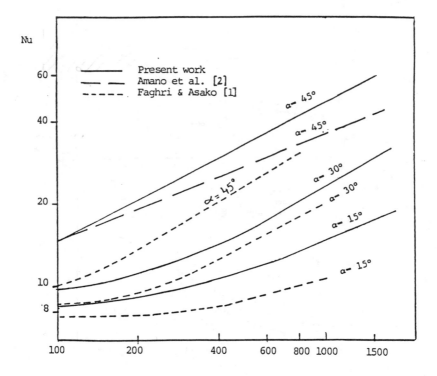

Fig. 9, Comparison of average Nusselt number in corrugated ducts with previous data

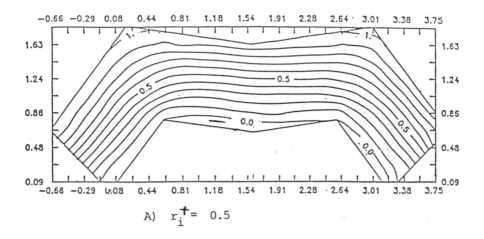

A) $r_i^+ = 0.5$

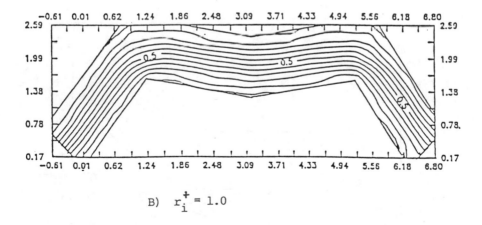

B) $r_i^+ = 1.0$

Fig. 10, Flow pattern in corrugated bends with different
curvatures, = 30 , Re = 1500

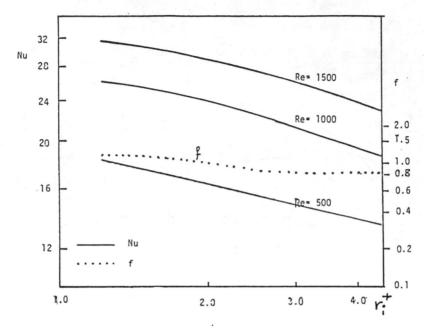

Fig. 11,   Effect of curvature "r_i^+" on Nu and f

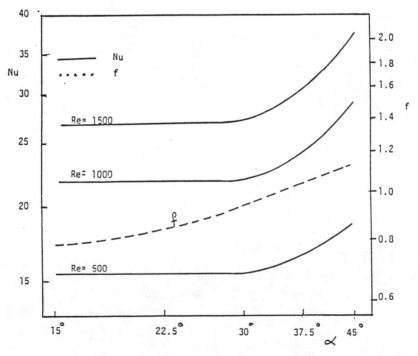

Fig. 12,   Effect of corrugation angle on Nu, and f

Effects of a Single Roughness Element on the Heat Transfer and Flow Situation in a Parallel Plate Duct (Numerical Calculation for a Porous-Type and a Solid-Type Element)

K. Ichimiya, H. Kondoh

Department of Mechanical Systems Engineering, Yamanashi University, Takeda-4, Kofu, Yamanashi 400, Japan

ABSTRACT

A numerical study is presented for the laminar flow and the heat transfer on the smooth heated surface opposite a single porous-type roughness element on the insulated wall in a parallel plate duct. The calculation was performed for the dimensionless height of an element H=0.357-0.642, the Reynolds number Re=100-1500 and the Darcy number Da=0.000025-0.025. The recirculation flow was generated behind the porous-type roughness element for low Darcy numbers and aided the increase of the local heat transfer. The characteristics were compared with the numerical results for an abrupt contraction and expansion flow due to a single solid-type roughness element in a flow passage.

INTRODUCTION

Turbulence promoters are utilized as a typical method of improving local heat transfer. Roughness elements are set on a heated wall [1,2], or twisted tape [3] is inserted into a flow passage. The temperature is almost in the middle range and the material is solid type. One of the authors has tried to improve the local heat transfer in a wide temperature range by setting the porous material in a flow passage. This is a method combined with the radiation shielding effect [4,5] in the high temperature range and the turbulence effect in the middle temperature range. Heat transfer augmentation and low flow resistance were obtained experimentally in turbulent flow region by inserting porous-type roughness elements [6].

In the present study, the effects of a porous-type turbulence promoter on the upper insulated wall on the local

heat transfer on the lower heated wall of a parallel plate duct
are examined fundamentally in a laminar flow. The characteristics of the heat transfer and the flow are obtained as basic
data for the practical use of several porous-type elements.

DESCRIPTION OF THE PROBLEM

Figure 1 shows a system which is composed of a porous-type or a
solid-type element on the insulated upper wall of a parallel
plate duct. The smooth lower wall is heated uniformly. Numerical analysis is performed in a normal temperature field with
thermal convection and conduction. The porous-type element
should be utilized in a wide temperature range. Consequently, a
porous-type element is set opposite the heated wall to apply
the element usefully in a high temperature field with thermal
radiation. Namely, a porous material absorbs the thermal
radiation and emits to the surroundings. The material is
considered to be a blend type which is made of alumina,
magnesium and silicone.

At the entrance of a parallel plate duct, the velocity of
the incompressible viscous fluid is assumed to be fully
developed in the laminar flow region. An element is set at a
distance 22.6 times the space of the flow passage from the
entrance. The fluid flows out from the exit at a distance 113
times the space after the element. A thermal situation is
almost developed at the exit since the dimensionless distance
$2(x/D_H)/(Re \cdot Pr)$ is 0.25 in a parallel plate duct with one-side
heating [7]. The distance from the element to the exit is 56.5
times the equivalent diameter D_H. Consequently, it is sufficient that the situation at the exit is assumed to be
developed thermally and hydraulically.

GOVERNING EQUATIONS AND BOUNDARY CONDITIONS

Governing equations (flow and energy) are expressed in
dimensionless form by introducing the stream function, the
vorticity and the following dimensionless factors.

$$u = \frac{\partial \psi}{\partial y} \quad v = -\frac{\partial \psi}{\partial x} \quad \omega = \frac{\partial v}{\partial x} - \frac{\partial u}{\partial y} \quad \Psi = \frac{\psi}{y_0 u_m} \quad \Omega = \frac{\omega y_0}{u_m} \quad \Theta = \frac{\lambda(T - T_0)}{q y_0} \tag{1}$$

In the Case of a Porous-Type Element
Dimensionless equations for the flow in a porous element are as
follows:

$$\frac{\partial^2 \Psi}{\partial X^2} + \frac{\partial^2 \Psi}{\partial Y^2} = -\Omega \tag{2}$$

$$\frac{\partial \Psi}{\partial Y} \frac{\partial \Omega}{\partial X} - \frac{\partial \Psi}{\partial X} \frac{\partial \Omega}{\partial Y} = \frac{2}{Re} \left(\frac{\partial^2 \Omega}{\partial X^2} + \frac{\partial^2 \Omega}{\partial Y^2} \right) - \frac{2}{Re} \frac{\Omega}{H^2 Da} \tag{3}$$

where X and Y are dimensionless coordinate x/y_0 and y/y_0, respectively, and Darcy number Da is defined as the ratio of permeability k and h^2. The flow outside a porous element is expressed by neglecting the last term of the right hand side of eq.(3).
The dimensionless energy equation is as follows:

$$\frac{\partial \Psi}{\partial Y}\frac{\partial \Theta}{\partial X} - \frac{\partial \Psi}{\partial X}\frac{\partial \Theta}{\partial Y} = \frac{2}{RePr}\left(\frac{\partial^2 \Theta}{\partial X^2} + \frac{\partial^2 \Theta}{\partial Y^2}\right) \tag{4}$$

Dimensonless boundary conditions for flow and temperature are as follows:
- at the entrance (X=0, $0 \leq Y \leq 1$)
  $U = -6(Y^2-Y)$, V=0, Θ=0
- on the upper wall ($0 \leq X \leq x_1/y_0$, Y=1)
  U=0, V=0, $\partial \Theta/\partial Y$=0
- on the lower wall ($0 \leq X \leq x_1/y_0$, Y=0)
  U=0, V=0, $\partial \Theta/\partial Y$=-1
- at the exit (X=x_1/y_0, $0 \leq Y \leq 1$)
  $U = -6(Y^2-Y)$, V=0, $\Theta = 2X/(Re \cdot Pr) - Y^4/2 + Y^3 - Y + 13/35$
- on the side surface of a porous element (X=x_0/y_0,
  X=$(xo+1)/y_0$, $(y_0-h)/y_0 \leq Y \leq 1$)
  $\partial U/\partial X|_p = \partial U/\partial X$, $(\tilde{\mu}/\mu)\partial V/\partial X|_p = \partial V/\partial X$

$\left.\begin{array}{c} \\ \\ \\ \\ \\ \\ \\ \\ \end{array}\right\} \tag{5}$

where the parameter $\tilde{\mu}/\mu$ is the ratio of the effective viscous coefficient in a porous element and the viscous coefficient of the fluid in a passage, and depends on the porosity ε. The present calculation is performed at ε=0.8 and $\tilde{\mu}/\mu$=1.2, obtained by Neal and Nader [8].

The boundary condition under the porous element is approximately evaluated by the following method. The flow rate passing through the porous element Qp is detertmined by equalizing the pressure drop through the element and the pressure differrence generated by the fluid under the element between Xo and Xo+1. As the result, Qp is expressed by the permeability k, the height h and the porosity ε.

$$Q_p = \int_{y_0-h}^{y_0} u\,dy = \frac{kh\varepsilon}{kh\varepsilon + (y_0-h)^3/12}Q \tag{6}$$

In eq.(6), as the permeability draws near to zero, k→ 0 , namely, Qp→ 0, the element approaches a solid element. On the other hand, as the height draws near to y_0, h→ y_0, namely, Qp→ Q, the porous element comes to occupy the whole cross section of the flow passage. The vertical velocity at the interface under the element is approximately zero. These boundary conditions are expressed in the following dimensionless form :
- under the element ($x_0/y_0 \leq X \leq (x_0+1)/y_0$, Y=$(y_0-h)/y_0$))

$$\frac{Q_p}{Q} = \int_{1-h/y_0}^{1} U\,dY = \frac{Da\varepsilon H^3}{Da\varepsilon H^3 + (1-H)^3/12} \tag{7}$$

$V=0$

In the Case of a Solid-Type Element
The governing equation for the flow is expressed by neglecting
the last term for the right hand side of eq.(3). The energy
equation is the same as eq.(4). Boundary conditions of the
upper and the lower walls of the flow passage are expressed in
eq.(5). Boundary conditions of a solid-type element are as
follows :
• on the side surface of the solid element ($X=x_0/y_0$,
  $X=(x_0+1)/y_0$)
    $U=0$, $V=0$, $\partial\Theta/\partial X=0$
• under the element ($x_0/y_0 \leq X \leq (x_0+1)/y_0$, $Y=(y_0-h)/y_0$)
    $U=0$, $V=0$, $\partial\Theta/\partial Y=0$

$$\left.\begin{array}{l} \\ \\ \\ \\ \end{array}\right\} \quad (8)$$

The local Nusselt number is expressed in eq.(9).

$$Nu = \frac{\alpha \cdot 2y_0}{\lambda} = \frac{2}{\Theta_w - \Theta_m} \tag{9}$$

NUMERICAL METHOD

Numerical calculation is carried out for Prandtl number Pr=0.7
and dimensionless width l/y_0=0.5 by changing the Darcy number
from 0.000025 to 0.025, the Reynolds number Re from 100 to 1500
, and the dimensionless height H from 0.354 to 0.642. Governing
equations are integrated and are transformed to the conservative form. Upwind differences are introduced into the finite
difference scheme. These equations are solved by the iteration
method [9]. If we denote the value of a quantity Φ which would
be computed in the (n-1)th iteration by $\Phi^{(n-1)}$ and the value
which would be computed in the normal way in the nth iteration
by Φ^n , then the value which is actually used in iteration n is
the computed form :

$\Phi = a \cdot \Phi^n + (1-a) \cdot \Phi^{(n-1)}$

Here, the coefficient a, which is termed the under-relaxation
parameter, stands for a number between 0 and 1. Grid space is
nonuniform and is close near the wall. The number of grids is
3300 maximum and 3219 minimum. The criterion for convergence is
chosen as follows :

$\text{Max}|\Phi^n - \Phi^{(n-1)}| < 8 \times 10^{-4}$

The quantity Φ corresponds to Ψ, Ω and Θ.

NUMERICAL RESULTS AND DISCUSSIONS

Flow situation
Figures 2 (a),(b),(c),(d),(e) and (f) show the dimensionless
stream function Ψ at H=0.5 and Re=1000 for various Darcy
numbers. The abscissa is the dimensionless distance at the
entrance and its scale is constructed to one-third comparing

with the ordinate. In the case of a large Darcy number (Fig. 2(a)), namely, high porosity, the stream lines are almost parallel, and a recirculating flow is not generated even after the porous-type element. However, at Da=0.01 (Fig. 2(b)), the stream lines begin to constrict from the edge of the element, and a recirculating flow whose scale is about four times the height of the flow passage appears after the element. As Da decreases from 0.0025 (Fig. 2(c)) to 0.001 (Fig. 2(d)), the recirculating flow, whose scale becomes larger, starts from the inside of the porous element and itself approaches the element. At Da=0.000025 (Fig. 2(e)), the characteristics of the flow are similar to those for a solid element (Fig. 2(f)). A small recirculating region is found at the upper edge of the element. This corresponds to the small separated region at the front of a solid element (Fig. 2(f)).

Figures 3 (a), (b), (c) and (d) represent the stream line for various Reynolds numbers at H=0.642 in the flow passage with a solid-type element. The scale of the recirculating flow after the roughness element is about $4.5y_0$ at Re=500 and $9.5y_0$ at Re=1500 along the flow direction. A small vorticity appears at X=28.6 and X=30.4 on the smooth heated surface for Re=800 and Re=1500, respectively. The position moves to the downstream region and the scale increases with increase of Re. These kinds of recirculating flows correspond to the flow after a sudden enlargement of the flow passage obtained by Armaly et al.[10] and affect the local heat transfer (Fig.6). However, for the porous-type element, this kind of small vorticity is not found on the smooth heated surface within the present conditions.

Local heat transfer
The isothermal lines are obtained in Fig. 4(a) for a porous-type element and in Fig. 4(b) for a solid-type element. These distributions are similar to each other. However, the temperature increase is milder for the porous-type element than for the solid-type element since the working fluid runs partly through the porous-type element. Temperature distribution around the solid-type element agrees with the results which Bunditkul and Yang [11] calculated for a contraction and enlargement flow.

Local Nusselt number on the smooth heated surface opposite a roughness element is shown in Fig. 5 for various Darcy numbers at H=0.5 and Re=1000. The distribution for the solid-type element is simultaneously represented in the figure. The local Nusselt number begins to increase at X=20 before the element, forms a peak under the element and decreases to the fully developed value. The rate of decrease of the Nusselt number increases with decrease of Da. The effective area for the augmentation of heat transfer is from X=20 to X=30 for Re=1000 and H=0.5.

Figure 6 shows the effect of H, namely, the rate at which a roughness element crosses the section of the flow passage. Generally, for the solid-type element, the local Nusselt number increases with increase of H because of an acceleration effect. However, for a porous-type element, the local heat transfer is not always improved, even for large H, since it depends on the Darcy number (Fig.7). Appropriate Darcy numbers should be selected for the improvement of heat transfer. For the solid-type element, the local Nusselt number decreases abruptly at X=31 for H=0.642 because of a small recirculating region on the smooth heated wall (Fig. 3(d)).

Figure 7 represents quantitatively the magnitude of heat transfer augmentation by the ratio of the maximum Nusselt number and the minimum Nusselt number before the element Numax/Nuin. The ratio does not depend on Re and tends to increase with H. In the case of the porous-type element, the flow rate which passes through the element increases for large H and Da (H=0.642 and Da=0.025) and the acceleration effect decreases. Consequently, though the local heat transfer is improved, the ratio Numax/Numin tends to decrease.

CONCLUSIONS

Effects of a single porous-type or solid-type roughness element opposite a smooth heated plate are analyzed numerically on the heat transfer and flow situation for the laminar flow in a parallel plate duct.
(1) For a porous-type roughness element, stream lines are almost parallel in the element at Da=0.025. It becomes clear how the recirculation flow behaves after the element as Da decreases.
(2) For a solid-type element, the heat transfer decreases locally at high value of H and Re since a small recirculating flow is generated on the heated surface.
(3) Heat transfer augmentation is performed at a high level with increase of H for a solid-type element. However, it depends on Da in a porous type element and is not always improved well at a high value of H.

Consequently, we should take into consideration the level of improvement and the effective area, and select the appropriate Darcy number and height of the roughness element.

ACKNOWLEDGEMENTS

The authors wish to thank to Mr. Takagi, I, Mr. Nebashi, T and Mr. Shinkai, S, who helped in the calculation.

REFERENCES

1. For example, Dipprey , D.F. and Sabersky, R.,Heat and Momentum Transfer in Smooth and Rough Tubes at Various

Prandtl Numbers, International Journal Heat and Mass Transfer,Vol.6,pp.329-353,1963.

2. For example, Webb, R.L., Eckert, E.R.G. and Goldstein, R.J.,Heat Transfer and Friction in Tubes with Repeated-Rib Roughness, International Journal Heat and Mass Transfer ,Vol.14,pp.1-11,1971.

3. For example, Date, A.W.,Prediction of Fully Developed Flow in a Tube Containing a Twisted Tape,International Journal Heat and Mass Transfer,Vol.17,pp.845-859,1974.

4. Hasegawa, S., Echigo, R., Ichimiya, K., and Kamiuto, K.,Augmentation on Heat Transfer by Thermal Radiation Shielding Plate Placed in a Duct, Proceedings of the 5th Int. Heat Transfer Conf.,Tokyo,Vol.1,pp. 108-112, 1974.

5. Echigo, R.,Effective Energy Conversion Method between Gas Enthalpy and Thermal Radiation and Application to Industrial Furnaces, Proceedings of the 7th Int.Heat Transfer Conf.,Munich,pp. 361-366,1982.

6. Ichimiya, K. and Mitsushiro, K.,Enhancement of the Heat Transfer of Wide Temperature Range in a Narrow Flow Passage (Effects of Porous-Type Turbulence Promoters in Normal Temperature Range),Proceedings of the 1st World Conf. on Experimental Heat Transfer, Fluid Mechanics and Thermo-dynamics,Yugoslavia,Vol.1,pp. 659-664,1988.

7. Kays,W.M.,Convective Heat and Mass Transfer,McGraw-Hill,New York,pp. 128,1966.

8. Neal, G. and Nader, W.,Practical Significance of Brinkman's Extension of Darcy' s Law (Coupled Parallel Flows within a Channel and a Bounding Porous Medium), Canadian Journal of Chemical Engineering,Vol.52,pp. 475-478,197 2.

9. Gosman, A.D., Pun, W.M., Runchal, A.K., Spalding, D.B. and Wolfstein, M.,Heat and Mass Transfer in Recirculating Flows,Academic Press,London,1969.

10. Armaly, B.F., Durst, Pereira, J.C.F. and Schonung, B., Experimental and Theoretical Investigation of Backward Facing Step Flow, Journal Fluid Mechanics,Vol.127,pp. 473-496,1983.

11. Bunditkul, S. and Yang, W.J .,Laminar Transport Phenomena in Constricted Parallel Duct, Letters in Heat and Mass Transfer,Vol.4,pp.249-260,1977.

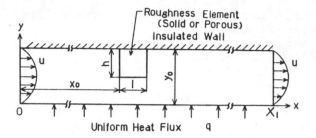

Figure 1 Coordinate system

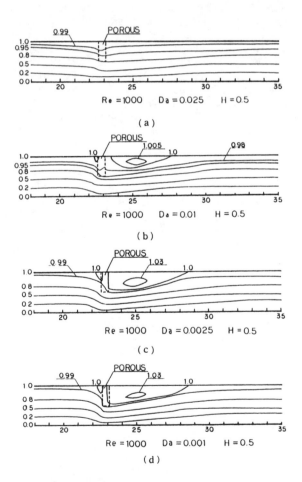

Figure 2 Dimensionless stream function
(1)   (a porous-type element)

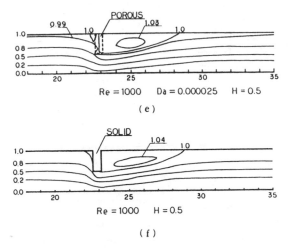

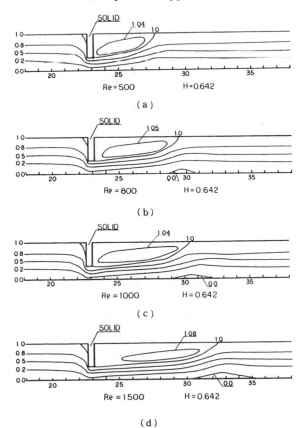

Figure 2 Dimensionless stream function
(2)   (a porous-type element)

Figure 3 Dimensionless stream function
(a solid-type element)

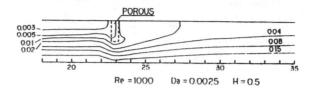

(a) A porous-type element

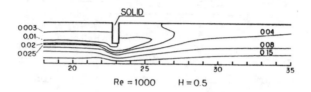

(b) A solid-type element

Figure 4 Dimensionless temperature distribution

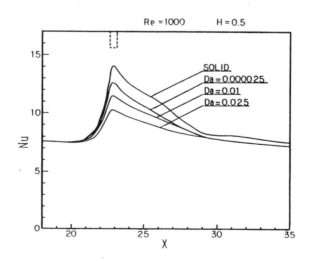

Figure 5 Local Nusselt number (effect of Da)

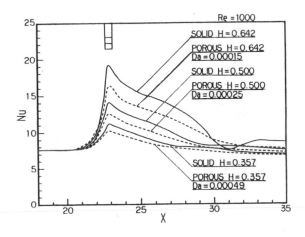

Figure 6 Local Nusselt number (effect of H)

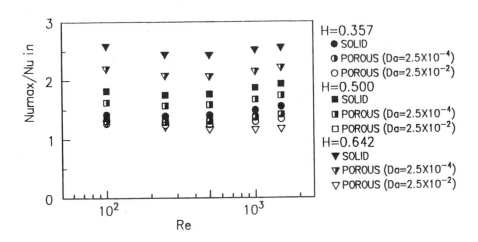

Figure 7 Maximum Nusselt number Numax/Nuin

A Combined Numerical and Experimental Investigation of Natural Convection in a Horizontal Concentric Annulus

M.W. Collins, J. Kaczynski(*), J. Stasiek(**)
Thermo-Fluids Engineering Research Centre, City University, London, EC1V 0HB, England

ABSTRACT

A numerical and experimental investigation has been carried out to extend existing knowledge of velocity and temperature distributions and local heat-transfer coefficients for natural convection within a horizontal cylindrical annulus. Results were obtained using water and air at atmospheric pressure. The diameter ratio D_o/D_i was 2.6, and the Rayleigh number based on the gap width varied from 3.10^5 to 1.10^7. A finite difference method was used to solve the governing constant - property equations numerically. The heat and fluid flow patterns in the annuli were visualised with the help of liquid crystals suspended as small tracer particles in the water. The numerical calculations are compared with experimental observations from the literature. Comparisons between the present numerical and experimental results under similar conditions show good agreement.

INTRODUCTION

The problem of accurately predicting heat transfer due to antural convection within enclosed spaces has recently received increased attention as reflected in the literature, but knowledge in this area is still rather limited. Applications have included nuclear reactor design, cooling of electronic equipment, aircraft cabin insulation, thermal storage systems, underground electric transmission cables and focusing solar collectors. The problem has been investigated experimentally by Lis [9] and Grigull and Hauf [5] among others [7] [8]. Flow patterns in air were observed by Bishop and Carley [1] and Kuehn and Goldstein [7]. Photographs showing the dependence of the flow patterns on the Grashof number and diameter ratio were obtained by Powe, Carley and Bishop [11]. The first determination of local heat-transfer coefficients in an annular geometry with air was made by Eckert and Soehngen [2] in 1948 using a Mach-Zehnder interferometer. Grigull and Hauf [5], Kuehn and Goldstein [7], and Kshmarov and Ivanov [8] used a similar technique to measure local heat transfer coefficients in air on the inner and outer cylinder for different diameter ratios. Several correlations have been proposed for experimental mean heat-transfer data for natural convection in horizontal annuli, with the results often given in terms of an equivalent conductivity. Some analytical solutions valid at small Rayleigh numbers have been obtained. More recently Jischke and Farschi [6] have used boundary - layer theory to analyse the flow in the laminar regime. Raithby and Hollands [12] have considered the high Rayleigh number boundary - layer limit with a conduction - layer model. Their empirical analysis correlates data quite well, but it gives no information about the flow and temperature fields. Numerical results were presented only for the laminar case up to a Rayleigh number (based on gap width L) of 10^5. An outer to inner diameter ratio of 2.6 was used in these solutions. As suggested by previous studies the diameter ratio was found to be significant [11], since the flow patterns were depended on it, especially when the flow regime underwent transition from laminar to turbulent.

In the present study, numerical solutions are presented for steady - state, two-dimensional natural convection in an annulus between two horizontal concentric, cylinders. The standard k-ε model is used, as proposed by Launder and Spalding [10] and developed by Faruk and Güceri [3]. The numerical calculations for air are compared with experimental results obtained using a Mach-Zehnder interferometer by Kuehn and Goldstein [7]. Two sets of experimental runs were made, one with air, the other using water. When water was used, the temperature and flow field are visualised with the help of thermochromic liquid crystals. The aim of the liquid crystals is to find the flow structures for the case fo two dimensional thermal convective flow between two concentric horizontal cylinders kept at different temperatures.

GOVERNING EQUATIONS

The two - equation (k-ε) turbulence model has been used here to predict the antural convective flow in an annulus for Rayleigh numbers above 10^4. Prediction of natural convection in a two dimensional cylindrical annuli requires the solution of coupled partial differential equations. In the laminar flow regime, these equations are obtained from the conservation of mass, momentum and energy. For turbulent flows, two additional equations, k* and ε* are considered [3] [4] [10]. Using the stream function - vorticity formulation, the five governing equations can all be represented by a single elliptic transport equation in the following form [4] [5]:

$$a_\varphi \left[\frac{\partial}{\partial r^*} \left(\varphi \frac{\partial \Psi^*}{\partial \theta} \right) - \frac{\partial}{\partial \theta} \left(\varphi \frac{\partial \Psi^*}{\partial r^*} \right) \right] - \left[\frac{\partial}{\partial r^*} \right.$$

$$\left. \left(b_\varphi r^* \frac{\partial (c_\varphi \varphi)}{\partial r^*} \right) + \left(\frac{b_\varphi}{r^*} \frac{\partial (c_\varphi \varphi)}{\partial \theta} \right) \right] = - r^* d_\varphi \qquad (1)$$

The laminar flow regime is described by three equations, φ representing the dependent variables ψ^*, ω^*, and T^*. For the turbulent flow regime, φ represents the time - averaged values of ψ^*, ω^*, T^*, k^* and ε^*, giving five equations to be solved. In generating these equations from equation (1), the multipliers a_φ, b_φ, c_φ and d_φ take appropriate expressions as summarised in Table 1.

Table 1. Coefficients for the flow equations [3]

Variable φ	a_φ	b_φ	c_φ	d_φ
Ψ^*	0	1	1	$- \omega^*$
ω^*	1	1	$1 + \mu_t^*$	$- Gr \left[\dfrac{\partial T^*}{\partial r^*} \sin\theta + \dfrac{1}{r^*} \dfrac{\partial T^*}{\partial \theta} \cos\theta \right]$
T^*	1	$\dfrac{1}{Pr} + \dfrac{\mu_t^*}{\sigma_T}$	1	0
k^*	1	$1 + \dfrac{\mu_t^*}{\sigma_k}$	1	S_k
ε^*	1	$1 + \dfrac{\mu_t^*}{\sigma_\varepsilon}$	1	S_ε

For the laminar regime, μ_t^* is set equal to zero in the expressions and only the first equations are considered. For the turbulent regime, even though two-dimensional mean motion is studied, the fluctuating components of velocities in all three dimensions were taken into account. In this case, σ_T, σ_k and

σ_ε (σ_T = 1.0; σ_k= 1.0; σ_ε = 1.3 [3] [4] are the turbulent Prandtl numbers for T, k and ε respectively.   The source terms S_k and S_ε for the k* and ε* equations for buoyancy driven recirculating flows in polar coordinates, are given by [3]:

$$S_k = -\mu_t^* \left\{ 2\left(\frac{\partial V_r^*}{\partial r^*}\right)^2 + 2\left(\frac{\partial V_\theta^*}{r^*\partial\theta}\right)^2 + 2\left(\frac{V_r^*}{r^*}\right) + \left[r^*\frac{\partial}{\partial r^*}\left(\frac{V_\theta^*}{r^*}\right)\right. \right.$$

$$\left. \left. + \frac{\partial V_r^*}{r^*\partial\theta}\right]^2 \right\} + Gr\,\frac{\mu_t^*}{\sigma_T}\left[\frac{\partial T^*}{r^*\partial\theta}\sin\theta - \frac{\partial T^*}{\partial r^*}\cos\theta\right] + \varepsilon^* \qquad (2)$$

$$S_\varepsilon = -C_1\,\mu_t^*\,\frac{\varepsilon^*}{k^*}\left\{2\left(\frac{\partial V_r^*}{\partial r^*}\right)^2 + 2\left(\frac{\partial V_\theta^*}{r^*\partial\theta}\right)^2 + 2\left(\frac{V_r^*}{r^*}\right)^2 \right.$$

$$\left. + \left[r\frac{\partial}{\partial r^*}\left(\frac{V_\theta^*}{r^*}\right) + \frac{\partial V_r^*}{r^*\partial\theta}\right]^2 \right\} + C_3\,Gr\,\frac{\mu_t^*}{\sigma_k}\frac{\varepsilon^*}{k^*}\left[\frac{\partial T^*}{r^*\partial\theta}\sin\theta\right.$$

$$\left. - \frac{\partial T^*}{\partial r^*}\cos\theta\right] + C_2\,\frac{\varepsilon^{*2}}{k^*}; \qquad (3)$$

where

$$k^* = kL^2/\nu^2, \quad \varepsilon^* = \varepsilon L^4/\nu^3, \quad V_r^* = LV_r/\nu, \quad V_\theta^* = LV_\theta/\nu. \qquad (4)$$

The effect of buoyancy on the creation of turbulent energy is taken into account by the second term in the expression for S_k. The empirical constants are taken as C_1 = 1.44 and C_2 = 1.92 [3] [4]. C_3 is put equal to C_1 by assuming similar contributions from buoyancy and gradient production terms, and a sensitivity study on C_3 is presented in [4].

The boundary conditions to be satisfied a solution to equiation (1) are given in Table 2 [4] [5]. For ψ^*, ω^* and T^*, the same boundary conditions are used for laminar and turbulent regimes; for the latter, of course, time-averaged quantities are considered instead.

Table 2. Summary of the boundary conditions [3]

Variable	Inner Cylinder	Symmetry Plane	Outer Cylinder
Ψ^*	$\Psi^* = 0$	$\Psi^* = 0$	$\Psi^* = 0$
ω^*	$\omega^* = -\dfrac{\Psi^*_p}{\Delta r^{*2}}$	$\omega^* = 0$	$\omega^* = -\dfrac{\Psi^*_p}{\Delta r^{*2}}$
T^*	$T^* = 1$	$\dfrac{\partial T^*}{\partial \theta} = 0$	$T^* = 0$
k^*	$k^* = 0$	$\dfrac{\partial k^*}{\partial \theta} = 0$	$k^* = 0$
ε^*	$\varepsilon^* = \dfrac{C_\mu^{0.75} k_p^{*1.5}}{\chi \, \Delta r^*}$	$\dfrac{\partial \varepsilon^*}{\partial \theta} = 0$	$\varepsilon^* = \dfrac{C_\mu^{0.75} k_p^{*1.5}}{\chi \, \Delta r^*}$

In table 2 for ε^* subscript p represents the nearest grid point from the wall, χ is the von Karman constant taken as 0.42, the empricial constant $C_\mu = 0.09$ and ψ^*_p is to be evaluated at Δr^* into the fluid.

Local Nusselt numbers, representing the ratio of the effective to the actual thermal conductivity, are defined for the ineer and outer cylinder surfaces respectively by:

$$N_i = r_i^* \ln (R) \left[\frac{\partial T^*}{\partial r^*} \right]_{r^* = r_i^*} \qquad (5)$$

$$N_o = - r_o^* \ln (R) \left[\frac{\partial T^*}{\partial r^*} \right]_{r^* = r_o^*} \qquad (6)$$

where

$$r_i^* = \frac{r_i}{L} \; ; \; r_o^* = \frac{r_o}{L} \; ; \; R = \frac{r_o^*}{r_i^*} \qquad (7)$$

c

Mean or average values of the Nusselt number at each surface may then be defined by:

$$\overline{N}_i = r_i^* \frac{\ell n \ (R)}{\pi} \int_o^\pi \left[\frac{\partial T^*}{\partial r^*} \right]_{r^* = r_i^*} d\theta \qquad (8)$$

$$\overline{N}_o = r_o^* \frac{\ell n \ (R)}{\pi} \int_o^\pi \left[\frac{\partial T^*}{\partial r^*} \right]_{r^* = r_o^*} d\theta \qquad (9)$$

Thus, a total Nusselt number $\overline{N}_T$ was defined simply as the arithmetic mean of $\overline{N}_i$ and $\overline{N}_o$. While a simple energy balance will immediately show that these two values should be equal, due to the numerical techniques involved, the values actually obtained differed slightly.

As is well known, the mean Nusselt number or equivalent conductivity is a function of the Grashot number, Prandtl number and diameter ratio, i.e.

$$\overline{Nu}_T = \frac{\overline{N}_i + \overline{N}_o}{2} = k_{eq} = f \ (Gr, \ Pr, \ R) = F \ (Ra, \ R) \qquad (10)$$

NUMERICAL RESULTS

The system of equations (1) together with the boundary conditions, Table 1 and 2 was solved numerically via a finite difference method, as presented by Faruk and Güceri [3, and Gosman et al [4] In general, 25 grid spacings in the radial direction and 41 in the anuglar direction were utilised. The Gauss-Seidel technique was selected as being relatively straightforward and simple to apply. However, it shoudl be noted that other methods, such as "line-by-line" or S.I.P. (Strongly Implicit Procedure), might be more advantages from the stand-point of computational time required.

The calculations were performed on a IBM PC AT digital computer. To obtain convergence, about 1500-2000 iterations were needed for the laminar and turbulent flow predictions presented here.

The convergence criteria needed for termination of the computation were preassigned as:

$$\left[\left(\varphi^{(N)} - \varphi^{(N-1)}\right) / \varphi^{(N)}\right]_{max} \leq \delta \qquad (11)$$

where

$$\delta \in < 0,001 \div 0.005 > \qquad (12)$$

$\varphi^{(N)}$ – values of function φ in N iteration step.

An under-relaxation factor of 0.5 was found to be satisfactory.

In order to assess the validity of the numerical results, test calculations were conducted for air and water, which could be directly compared with experimental results reported by Kuehn and Goldstein [7] The calculated local Nusselt number (for air) is shown in Fige 1.   In Fig. 2, the isotherms are presented for a high Rayleigh number flow, $Ra_L = 55400$, $Pr = 0.701$, along with the interferogram analogues available from [7] for comparison.   Fig. 3 shows a comparison of temperature distribution for different angular positions.

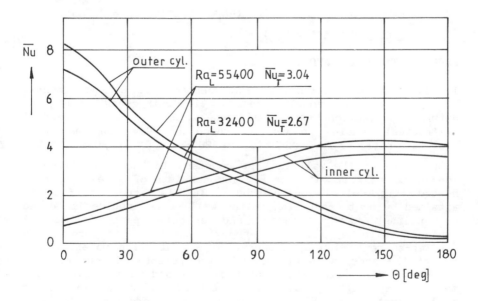

Fig. 1.   Distribution of local Nusselt numbers for the inner and outer cylinders for air.

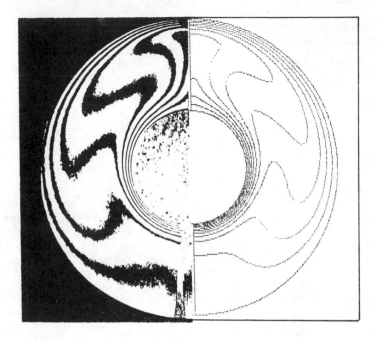

Fig. 2   Comparison of experimental and numerical isotherms
        [7].   Experimental (Ra$_L$ = 4.7 10 ; Pr = 4.7 10^4;

        L/D$_i$ = 0.8).   Numerical (Ra$_L$ = 5.5 10^4 Pr = 0.703;

        L/D$_i$ = 0.8 ΔT* = 0.1)

EXPERIMENTAL SUTDY AND CONCLUSION

The flow visualisation has been performed in an experimental
stand, the circular annulus cross-section being shown in Fig. 4.
Natural convection has been generated in the horizontal
cylindrical annulus so formed, which is 70mm long and with
diameter ratio D$_o$/D$_i$ ≈2.6 (D$_i$ = 30.8mm, D$_o$ = 80mm).   The
temperature of the wall is kept constant in these experiments.
The inner cylinder '1' (see Fig. 5) consists of an electrically
heated thick-walled copper tube.   The isothermal surface is
attained by heat conduction in the wall.   The annulus is formed
by an inner copper tube '1' and an outer perspex tube '2'
equipped with a further perspex tube '3' to provide a
counter-current flow of temperature – controlled water.   Each
end of the annulus is closed with the perspex windows of 10mm
thickness.   The temperatures of the inner and outer cylinder are
measured by eight thermocouples connected to an indicating
millivoltmeter.   Electric   power   was   provided   to   the
inner-cylinder resistor by a regulated D.C. power supply.   The

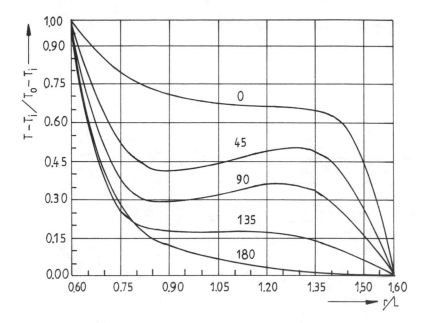

Fig. 3   Dimensionless radial temperature distribution in air
for Ra_L = 5.5 10^4; Pr = 0.706; L/D_i = 0.8)

power input was read from a wattmeter with an accuracy of
± 0.10W.   The cooling water was maintained as isothermal to
within ± 0.05° by a constant - temperature water bath.   A test
cell was built to accommodate liquids as well as air at
atmospheric pressure.   Temperature and flow distributions have
been visualised using thermochromic liquid crystals in the
microencapsulated form of about 30 ÷ 100 μm in diameter.   Fig. 5
shows the experimental arrangement for the visualisation study
of natural convection.   Flow structures were visualised using
photographic records of the motion of tracer particles
illuminated by a sheet of white light as follows.

Using a cylindrical lens and a slit, a plane sheet of light is
obtained with an adjustable width (1 to 2mm).   The flash was
triggered by a programmable impulse generator at prescribed time
sequences.   Usually 5 to 10 flashes were used to take one photo.
The flash sequences were also specially encoded to obtain
information about the direction of the flow.   To detect the
three-dimensional structures of the flow field monitored
photogrpahs were taken at differential 'X' horizontal cross
sections of the cylindrical annulus.

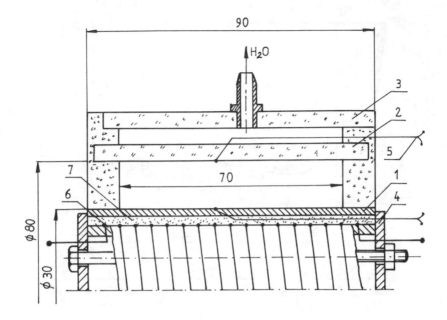

Fig. 4 Cross section of a cylindrical annulus:
1.   inner cylinder; 2.   outer cylinder; 3.   cooling cylinder;
4.   disk of insulating material; 5.   thermo-couples;
6.   electric heating element; 7.   quartz sand.

The average Nusselt numbers (equivalent conductivities) for air
are given in table 3.   The present results agree well with the
experimental correlation of Bishop [1], Kuehn and Goldstein [8]
and the conduction - boundary - layer model developed by Raithby
and Hollands [12].

At the Conference, results for water will be presented, together
with photographs of flow and thermal visualisation using liquid
crystals as tracer particles.

*   Polish Academy of Sciences, I.M.P., Gdansk, Poland.

** On leave of absence from Technical University of Gdansk,
     Poland.

Table 3 Experimental mean heat-transfer results for air.

ΔT	$\frac{1}{2}(T + T)$	Q	Ra_L	Pr	$\overline{Nu} \cong \overline{k}_{eq}$
K	K	W	−	−	−
13.46	304.59	0.428	$1.9\ 10^4$	0.703	2.56
20.67	308.58	0.757	$2.82\ 10^4$	0.703	2.91
37.58	317.80	1.489	$7.39\ 10^4$	0.701	3.19

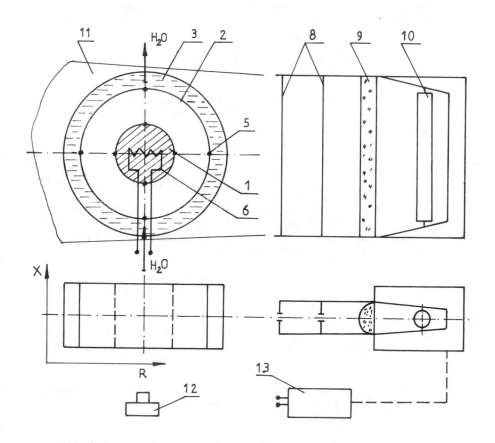

Fig. 5 Experimental stand for flow visualisation.
1.   inner cylinder; 2.   outer cylinder; 3.   cooling cylinder;
5.   thermocouples; 6.   electric heating element; 8 slits;
9.   cylindrical lens; 10 photoflash 11.   light sheet;
12.   camera; 13.   programmable impulse generator.

References

1.   Bishop E. H. and Carley C. T.   Photographic studies of natural convection between concentric cylinders Proc. Heat Transfer Fluid Mech. Inst. 1966 pp. 63-78.

2.   Eckert, E. R. G. and Soehngen E. E. Studies on heat transfer in laminar free convection with the Zehnder-Mach interferometer. Wright-Patterson AFB Tech. Rep. No. 5747, ATI-44580.   1948.

3.   Farouk B. and Güceri S. I.   Laminar and Turbulent Natural Convection in the Annulus Between Horizontal Concentric Cylinders.   J. Heat Transfer Vol. 104, 1982, pp 631 -636.

4.   Gosman A. D.,   Pun W. M.   Runchal A. K.,   Spalding D. B. Wolfshtein M.   Heat and Mass Transfer in Recirculating Flows.   Academic Press, London, 1969.

5.   Grigull U. and Hauf. W.   Natural convection in horizontal cylindrical annuli.   3rd Int. Heat Transfer Conf., Chicago, 1966 pp. 182-195.

6.   Jischke M. C. and Farschi M.   Boundary layer regime for laminar free convection between horizontal circular cylinders.   J. Heat Transfer Vol. 102, 1980 pp. 228-235.

7.   Kuehn T. H. and Goldstein R. J.   An experimental and theoretical study of natural convection in the annulus between horizontal concentric cylinders.   J. Fluid Mech. Vol. 74, 1976, pp. 695-719.

8.   Koshmorov Y. A. and Ivanov A. Y.   Experimental study of Heat Transfer through a Rarefied Gas Between Coaxial Cylinders. Heat Transfer, Sov. Res. Vol. 5, 1973, pp. 29-36.

9.   Lis J.   Experimental investigation of natural convection heat transfer in simple and obstructed horizontal annuli. 3rd Int. Heat Transfer Conf. Chicago, 1966, pp. 196-204.

10.   Launder B. E. and Spalding D. B.   The Numerical Computation of Turbulent Flows.   Computer Methods in Applied Mechanics and Engineering Vol. 3, 1974 pp. 269-289.

11.   Powe R. E., Carley C. T. and Bishop E. H. Free convective flow patterns in cylindrical annuli.   J. Heat Transfer Vol. 91, 1969 pp. 310-314.

17.   Raithby G. D. and Hollands K. G. T.   A general method of obtaining approximate solutions to laminar and turbulent free convection problems Adv. in Heat Transfer, Vol. 11, 1975, pp. 265-315.

Forced Convective Heat Transfer between Horizontal Flat Plates Heated Periodically from Below

M. Hasnaoui, E. Bilgen, P. Vasseur

Ecole polytechnique, Mechanical Engineering Department, C.P. 6079 St. A, Montréal, P.Q., H3C 3A7, Canada

ABSTRACT

Two-dimensional Navier-Stokes and energy equations are numerically solved for laminar air flow in a horizontal channel with localized heating from below. The relative strength of the forced flow and the buoyancy effects are examined for a wide range of Rayleigh (Ra) and Reynolds (Re) numbers. For a fixed geometry and a given Rayleigh number, a complicated solution structure is observed upon increasing the Reynolds number. For $Re = 0$, i.e. in the case of pure free convection, a steady symmetric flow pattern is obtained. This pattern becomes asymmetric for Re below a critical value but the rolls remain attached to the heating elements. Above the critical Re, the rolls are carried downstream with a time dependent velocity and the flow becomes periodic in time. For a sufficiently large Re the pure forced convection limit is recovered.

NOMENCLATURE

A	dimensionless length of the heating elements, L'_H/L'
B	aspect ratio , H'/L'
g	acceleration due to gravity
h	local heat transfer coefficient
$\bar{h}$	average heat transfer coefficient on upper wall
H'	channel height
L'	total length of calculation domain
Nu	mean Nusselt number, hH'/k
Pe	Peclet number, $RePr$

Pr Prandtl number, ν/α
Q_C dimensionless conduction
Ra Rayleigh number, $(g\beta\Delta T'H'^3)/(\nu\alpha)$
Re Reynolds number, $H'u_o'/\nu$
t' time
T dimensionless temperature of fluid, $(T'-T_C')/(T_H'-T_C')$
T' temperature of fluid
$\Delta T'$ temperature difference, $(T_H'-T_C')$
u_o average velocity of fluid
u,v dimensionless velocities in x and y directions, $(u',v')/u_o'$ or $(u',v')\,H'/\alpha$
u',v' velocities in x' and y' directions
x,y dimensionless Cartesian coordinates, $(x',y')/H'$
x',y' Cartesian coordinates
ω' vorticity

Symbols
α thermal diffusivity
β volumetric coefficient of thermal expansion
ν kinematic viscosity
ρ fluid density
τ dimensionless time, $t'u_o'/H'$ or $t'\alpha/H'^2$
Ψ dimensionless stream function, $\Psi'H'/u_o$ or Ψ'/α
Ω dimensionless vorticity, $\omega'H'/u_o$ or $\omega'H'^2/\alpha$

Subscripts
c critical
C cooled wall
H heated wall
loc local

Superscript
$'$ dimensional variables

INTRODUCTION

An understanding of the basic properties of convection in a small channel heated from below is important both from a theoretical and an application point of view. Compact heat exchangers and electronic equipment packages represent industrial situations in which channel-flow heat transfer is encountered. The flow in these channels is often laminar due to small dimensions and low velocities employed, and heat coefficients are characteristically low [1]. The existing literature in this domain has focused considerable attention on flow between two horizontal flat plates, where the lower and upper plates are respectively heated and cooled. Reviews of past studies on this subject have been presented by Osborne and Incropera [2], Ouazzani et al. [3] and others. A related problem of practical interest is when finite heat sources are located on the bottom surface of the horizontal channel. This situation may be encountered in electronic cooling applications where, in general, the wiring boards are not heated uniformly due to discrete spacing of heat dissipating components. Numerical and experimental results have been obtained by Kennedy and Zebib [4,5] for laminar convection through a channel with a single localized heat source. Forced convection in a horizontal rectangular channel with discrete heat sources has been considered by Incropera et al. [6]. However, the study was mostly concerned with a turbulent flow regime where the buoyancy effect is small. Recently, the mixed convection flow and heat transfer between parallel plates, with localized heat sources dissipating an equal and uniform heat flux over their length, have been studied by Tomimura and Fujii [7]. The temperature field was found to be considerably affected by the discrete heating because of the intermittent development of thermal boundary layers.

The present study considers the free convective effects caused by finite heat sources on the bottom of a horizontal channel through which a fluid is flowing because of an external pressure gradient. The relative strength of the free and forced convection is numerically examined for a wide range of Rayleigh and Reynolds numbers and for various heated segment lengths. The behaviour of steady state solutions and the development of sustained oscillatory solutions is also studied.

MATHEMATICAL FORMULATION

Consider the laminar flow of an incompressible, constant property, Boussinesq fluid through a channel bounded by two parallel horizontal plates. The physical model and coordinate system are shown in Fig. 1a. The upper surface is assumed to be isothermally cooled at T'_C, heat sources, at a constant temperature T'_H, cover a portion of the bottom surface while the rest of it is adiabatic. The channel has a height H' and the heating elements of length L'_H are regularly spaced at distances L' along the bottom surface. The third dimension of the channel (direction perpendicular to the figure) is large enough so that the flow and heat transfer are two-dimensional.

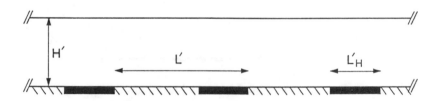

Fig. 1a Schematic of the physical problem

With these assumptions, the conservation equations for momentum and energy for mixed convection are:

$$\frac{\partial \Omega}{\partial \tau} + \frac{\partial u\Omega}{\partial x} + \frac{\partial v\Omega}{\partial y} = -\frac{Ra}{PeRe}\frac{\partial T}{\partial x} + \frac{1}{Re}\nabla^2\Omega \tag{1}$$

$$\frac{\partial T}{\partial \tau} + \frac{\partial uT}{\partial x} + \frac{\partial vT}{\partial y} = \frac{1}{Pe}\nabla^2 T \tag{2}$$

$$\nabla^2\Psi = -\Omega \tag{3}$$

$$u = -\frac{\partial \Psi}{\partial y} \qquad v = \frac{\partial \Psi}{\partial x} \tag{4}$$

The above equations are obtained using the nondimensional parameters defined in the nomenclature.

The boundary conditions are (see Fig. 1b)

For $y = 0$:

$$\left.\begin{array}{ll} \Psi = 0 & u = v = 0 \\ T = 1 & \text{for heated surfaces} \\ \dfrac{\partial T}{\partial y}\bigg|_{y=0} = 0 & \text{for adiabatic surfaces} \end{array}\right\} \qquad (5)$$

For $y = 1$:

$$T = 0 \quad \Psi = -1 \quad u = v = 0 \qquad (6)$$

For $x = 0, \; B$:

$$\textit{Periodic conditions for all } y \qquad (7)$$

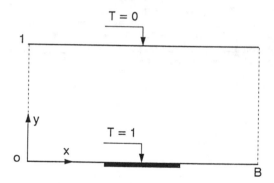

Fig. 1b Numerical domain and coordinate system

For natural convection, by taking $Pe = 1$ hence $Re^{-1} = Pr$, equations (1) and (2) may be modified. In this case, the scale factor for velocity, vorticity, stream-function and time will be

$$\left.\begin{array}{ll} (u, v) = (u', v')H'/\alpha' & \Psi = \Psi'/\alpha \\ \Omega = \omega' H'^2/\alpha & \tau = t'\alpha/H'^2 \end{array}\right\} \qquad (8)$$

The boundary conditions will be the same as (5) - (7) except $\Psi = 0$ on the upper wall.

From the above equations, boundary conditions and geometry, it appears that the present problem is governed by five dimensionless parameters: A, B, Ra, Re and Pr numbers.

Two quantities of practical interest are the local and overall Nusselt numbers. A local Nusselt number which is a direct measure of the local heat flux, is obtained as

$$Nu_{loc} = \frac{hH'}{k} = \frac{\partial T}{\partial y}\bigg|_{y=1} \qquad (9)$$

The mean Nusselt number is obtained as

$$Nu = \frac{\bar{h}H'}{k} = \frac{1}{BQ_C} \int_0^B \frac{\partial T}{\partial y}\bigg|_{y=1} dx \qquad (10)$$

This gives the net heat transfer rate leaving the system through the upper wall.

NUMERICAL METHOD

The conservation equations (1) to (3) describing the flow and heat transfer in this problem were solved numerically using a finite-difference discretization procedure. Central-difference formulas were used for all spatial derivative terms, in the Poisson, energy and vorticity equations. A modified alternate direction implicit procedure was adapted to obtain from equations (1) and (2) the vorticity and temperatures profiles. The finite-difference forms of the vorticity and energy equations were written in conservative form for the convective term in order to preserve the conservative property. Values of the stream function at all grid points were obtained with equation (3) via a successive over-relaxation method. Suitable values for the relaxation parameters were between 1.780 and 1.847. The velocities at all grid points were determined with the dimensionless form of equation (4) using updated values of the stream function. Variation by less than 10^{-4} over all grid points for the stream function was adopted as the convergence criterion.

For the present work, uniform mesh sizes were used for both x and y directions. As expected, it was found that the necessary number of grid lines depended on the Rayleigh number and the calculation domain. A grid of 41 × 21 for the aspect ratio $B = 2$ was found to model accurately the flow fields described in this study. Using a grid of 61 × 31 gave a variation in the order of 3 % for the average Nusselt number and 0.7 % for Ψ_{max} at $Ra = 10^5$. Also, the use of a calculation domain of $B = 4$ gave identical results in comparison with $B = 2$. Time step sizes of $\delta\tau = 0.001$ and $\delta\tau = 0.0004$ were tried for some cases. Both time step sizes gave identical results. Hence $\delta\tau = 0.001$ was used for the rest of the simulations.

RESULTS AND DISCUSSION

To verifiy the accuracy of the present numerical solution procedure, some test calculations were carried out for free convection and mixed convection between parallel plates fully heated from below $(A = 1)$. For $Re = 0$, a comparison of Nu with the experimental results of [8-9] has shown that the agreement was good. For $Re > 0$, the comparison with the numerical results of [3] was excellent.

Flow and temperature fields and heat transfer rates for various Ra, Re and A will be examined for $B = 2$ and $Pr = 0.72$ (air) in the following sections.

i) Pure free convection $(Re = 0)$

Figs. 3a - c illustrate the various modes of convection in the channel for given values of the governing parameters, namely $Ra = 5 \times 10^4$, and $A = 0.9$. The heat source on the bottom surface is indicated by a heavy line. Positive and negative values of Ψ in the figure title, correspond to clockwise and counter-clockwise circulations respectively. In all these graphs, the increments between adjacent isotherms and streamlines are $\Delta T = T_{max}/11$ and $\Delta \Psi = (\Psi_{max} - \Psi_{min})/11$. Figure 3a shows the streamlines (left) and isotherms (right) obtained by using the rest state ($\Psi = T = 0$) as initial conditions. The resulting flow pattern consists in four counter-rotative cells of approximately equal size. However, this solution is not unique and other modes of convection are possible, the final steady state achieved being a function of initial conditions. For instance Figs 3b and 3c show two possible modes of convection associated with one pair of cells. When dealing with two cells above a heated element, the direction of their rotation seems to be imposed a priori by the physics of the problem, the fluid being ascendant above the heated element as illustrated in Fig. 3b. However it is also possible, with appropriate initial conditions, to obtain a steady state convective pattern where the fluid motion above the heated element is not ascendant but rather descendant. Such reversed circulations illustrated in Fig. 3c have already been observed in the past by Robillard et al. [10] in the case of a porous cavity partially heated from below.

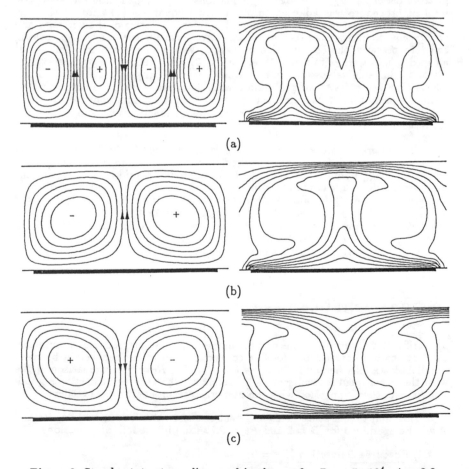

Figure 3 Steady state streamlines and isotherms for $Ra = 5 \times 10^4$, $A = 0.9$
a) $\Psi_{max} = 14.57$; $\Psi_{min} = -14.57$ b) $\Psi_{max} = 23.15$; $\Psi_{min} = -23.15$
c) $\Psi_{max} = 23.16$; $\Psi_{min} = -23.16$

At this stage it must be mentioned that, due to periodic boundary conditions imposed by eq. (7), any solution at $Re = 0$ will display a symmetry with respect to a vertical axis located at $x = 1$. The effect of Ra on the average Nu is depicted in Fig. 4a for various values of the dimensionless length of heat source A. The steady state flow pattern obtained for all the data presented in this figure consists of a pair of convective cells similar to that of Fig. 3b. All the values in Fig. 4 a were obtained by using the rest state as initial conditions, for the values of 0.25 $\leq A \leq 0.75$. The curve with $A = 1$, which corresponds to a channel fully heated from below, is also included for comparison. The results of Fig. 4a indicate that the overall heat transfer always increases with Ra and A. The variations of Nu_{loc} on the upper wall are shown in Fig. 4b for $Ra = 5 \times 10^5$ and various values of A. This local Nusselt number is a direct measure of amount of heat lost through the upper wall at any location. The variation of Nu_{loc} is symmetric with respect to the center of the heated segment where its value reaches a maximum. This behaviour is consistent with the flow and temperature patterns of Fig. 3b.

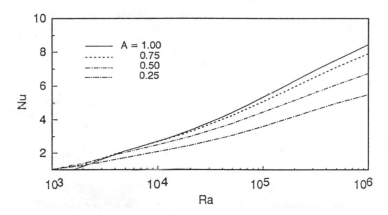

Figure 4 a)   Average heat transfer as a function of Rayleigh number and size of heated element

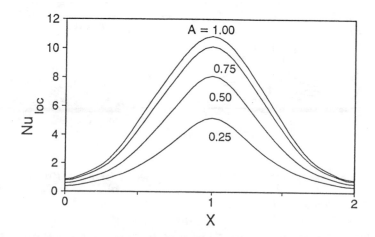

Figure 4 b)   Local heat transfer over the heated element as a function of the size of the element for $Ra = 5 \times 10^5$

ii) Mixed convection ($Re > 0$)

The effects of an externally-induced flow, on the pure natural convective heat transfer will be discussed next. To present the increasing influence of this forced flow (or relatively decreasing strength of the buoyancy), the streamlines and isotherms for $Ra = 10^5$ and $Re = 1$ and 6 are presented in Figs. 5 a-b. For $Re = 1$, the externally induced flow is very weak and the buoyancy effects dominate the convective flow and heat transfer. The resulting flow pattern, depicted in Fig. 5a, consists of two counter-rotating cells almost symmetric with respect to a vertical plane passing through the center of the heated element. This symmetry is destroyed as the strength of the forced flow is progressively enhanced. This is illustrated in Fig. 5b where the quasi-symmetric flow produced at $Re = 1$ is now skewed toward the downstream when the Reynolds number is increased to 6. The convective cells remain attached to the heating elements with their centers displaced downstream. The enhanced forced convective effects also change the strength and nature of the recirculation. The left convective cell, which is displaced above the heated element, is stronger than the right cell located above adiabatic segment of the channel. However, the values of Ψ_{max} and Ψ_{min} in the title of Figs. 5a and 5b indicate that the relative strength of two convective cells decreases as the intensity of the forced flow is enhanced. There exists a critical Reynolds number Re_c above which a steady state solution is not possible. In general, Re_c is a function of Ra, A and Pr.

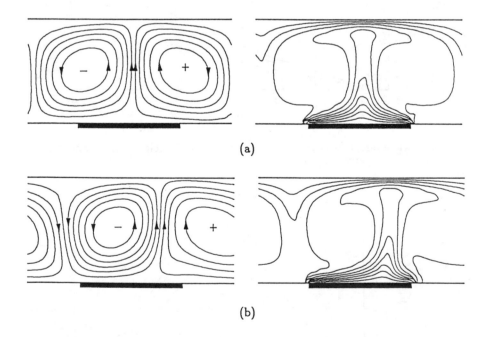

(a)

(b)

Figure 5 Steady-state streamlines and isotherms for $Ra = 10^5$, $A = 0.5$ a) $Re = 1$; $\Psi_{max} = 36.36$, $\Psi_{min} = -40.00$ b) $Re = 6$; $\Psi_{max} = 4.90$, $\Psi_{min} = -7.40$

Figure 6 presents Re_c as a function of Ra obtained for $A = 0.5$ and $Pr = 0.72$. With the actual numerical approach, it is difficult to determine precisely the threshold values of Re_c since the time periodicity at the onset of unsteady motion has a very long period. Thus, the lower curve in Fig. 6 represents an upper limit for the steady state region for which the maximum relative value of the stream function during two consecutive time steps was lower than 10^{-3}. Conversely, recurrence times as long as approximately 10 were obtained for the results defining the lower limit of the unsteady state region.

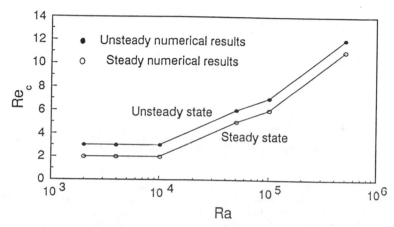

Figure 6   Critical Reynolds number versus Rayleigh number for $A = 0.5$

Figures 7 a-b show the Nusselt number and maximum stream function computed at every time step for the case of $Ra = 10^4$, $A = 0.5$ and $Re = 2, 3$ and 4. For $Re = 2$, the initial values of T, Ψ and Ω were taken as zero; for $Re = 3$ and 4, earlier solution results were used to initialize T, Ψ and Ω. Fig. 7 shows that for $Re = 2$, a steady state condition is reached in a dimensionless time unit of about $\tau = 0.8$. The resulting steady state profiles are similar to those presented in Fig. 5 with two asymmetric counter-rotating cells. As Re is increased up to 3, a low frequency oscillation is observed. To ensure a sustained oscillation, the time integration was continued up to $\tau = 20$. The basic frequency for $Re = 3$ is $f = 0.24$ and it increased to $f = 0.41$ for $Re = 4$.

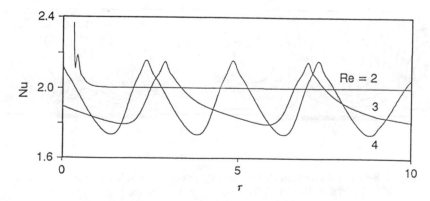

Figure 7 a   Nusselt number variation with dimensionless time τ for $Ra = 10^4$, $A = 0.5$ and various values of Re

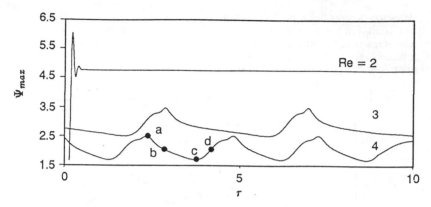

Figure 7 b   Maximum stream function variation with dimensionless time τ
for $Ra = 10^4$, $A = 0.5$ and various values of Re

The streamlines and isotherms at the time indicated with a, b, c and d in
Fig. 7b, during a single oscillation of the flow at $Re = 4$, are depicted in Figs 8
a-d. It is seen that, during one complete cycle, the direction of the cells passing
over the heated element alternates with time, being successively clockwise and
counterclockwise.

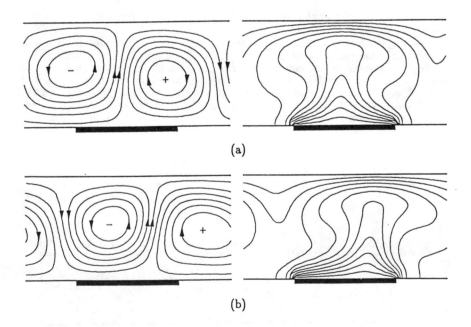

Figure 8   Stream function and temperature contours over one complete cycle
for $Ra = 10^4$, $A = 0.5$ and $Re = 4$. a, b are from Fig. 7 b.

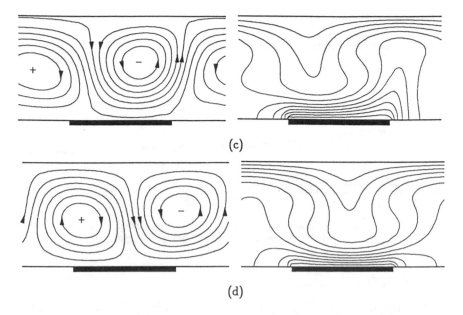

(c)

(d)

Figure 8   Stream function and temperature contours over one complete cycle
for $Ra = 10^4$, $A = 0.5$ and $Re = 4$. c, d are from Fig. 7 b.

The instantaneous Nusselt number is maximum for two approximately equal
cells with the flow ascendant over the heated element (compare Fig. 7a and 8a).
It reaches a minimum value when a single counterclockwise cell pass over the
heated element (Fig. 8c).

Finally, the influence of Re on the cyclic variation of the instantaneous cell
velocity V_c is illustrated in Fig. 9 for $Ra = 5 \times 10^4$, $A = 0.5$ and $Re = 5.4, 6$ and
10. The basic frequency for $Re = 5.4$ is $f = 0.215$ and it increases to $f = 0.408$
for $Re = 10$.

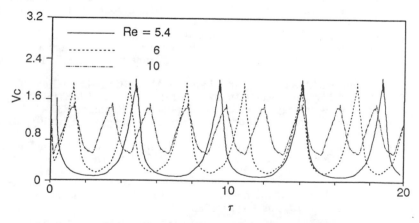

Figure 9   Instantaneous cell velocity V_c with dimensionless time τ for Ra
$= 5 \times 10^4$, $A = 0.5$ and $Re = 5.4, 6$ and 10.

CONCLUSION

1. In the case of pure convection ($Re = 0$) a pair of counter rotating cells is produced; the cells are symmetrically located above the heating elements. The overall heat transfer rate always increases with the Rayleigh number and the size of the heated elements.

2. In the case of mixed convection ($Re > 0$) there is, for a fixed geometry and a given Rayleigh number a critical Reynolds number below which the flow remains steady. Above that threshold value the flow is periodic and the cells are carried downstream with a time dependent velocity.

ACKNOWLEDGMENT

Financial supports by Natural Sciences and Engineering Research Council Canada and the FCAR Province of Quebec are acknowledged.

REFERENCES

1. Mixed Convection Heat Transfer, The Winter Annual Meeting of the American Society of Mechanical Engineering, Boston, HTD - Vol. 84, Edited by Prasad, V., Catton, I. and Cheng P., 1987.
2. Osborne, D.G. and Incropera, F., Laminar, Mixed Convection Heat Transfer for Flow between Horizontal Parallel Plates with Asymmetric Heating, International Journal of Heat and Mass Transfer, 28, 1: 207-217, 1985.
3. Ouazzani, M.T., Caltagirone, J.P., Meyer, G. and Mojtabi, A., Etude Numérique et expérimentale de la Convection Mixte entre deux Plans Horizontaux à Températures Différentes, International Journal of Heat and Mass Transfer, 32: 261-269, 1989.
4. Kennedy, K.J. and Zebib, A., Combined Forced and Free Convection between Parallel Plates, Proceedings, 7th International Heat Transfer Conference, Grigul et al. Editors, Hemisphere Publishing Corp., Washington D.C., 3: 447-451, 1982.
5. Kennedy, K.J. and Zebib, A., Combined Free and Forced Convection between Horizontal Parallel Plates, International Journal of Heat and Mass Transfer, Vol. 26, 3: 471-474, 1983.
6. Incropera, F.P., Kerby, J.S., Moffat, D.F. and Ramadhyani, S., Convection Heat Transfer from Discrete Heat Sources in a rectangular Channel, International Journal of Heat and Mass Transfer, Vol. 29, 7: 1051 - 1058, 1986.
7. Tomimura, T. and Fujii, M., Laminar Mixed Convection Heat Transfer between Parallel Plates with Localized Heat Sources, Proc. Int. Symp. on Cooling Technology for Electronic Equipment, Honolulu, 233 - 247, 1987.
8. Chandrasekhar, S., Hydrodynamic and Hydromagnetic Stability, Oxford, 1961.
9. Hollands, K.G.T., Raithby, G.D. and Konicek, L., Correlation Equations for Free Convection Heat Transfer in Horizontal Layers of Air and Water, International Journal of Heat and Mass Transfer, 18: 879 - 884, 1975.
10. Robillard, L., Wang, C.H. and Vasseur, P., Multiple Steady States in a Confined Porous Medium with Localized Heating from Below, Num. Heat Transfer, Vol. 13, pp. 91-110, 1988

Free Convection on a Horizontal Plate Subjected to a Mixed Thermal Boundary Condition

G. Malarvizhi, G. Ramanaiah

Department of Mathematics, Anna University, Madras 600 025, India

ABSTRACT

Similarity analysis of the problem of free convection on a heated horizontal plate is considered assuming that the plate is subjected to a mixed thermal boundary condition. It is shown that the mixed thermal boundary condition is characterised by a parameter ϕ lying between 0 and $\pi/2$ and the two extreme values correspond to the cases of prescribed plate temperature and prescribed surface heat flux. If one has to compute the heat transfer coefficient for various values of ϕ, there is no need to solve the boundary value problem every time; it is enough to solve a polynomial equation provided the solution is known for any particular value of ϕ.

INTRODUCTION

The problems of free, mixed and forced convection over a horizontal plate have been studied by several authors [1-7]. These studies assume that the plate is subjected to a prescribed temperature (Dirichlet's condition) or a prescribed surface heat flux (Neumann's condition). The convection problem when the plate is subjected to a mixed thermal boundary condition has not received sufficient attention. In this paper we consider the similarity analysis of the problem of free convection on a heated horizontal plate when it is subjected to a mixed thermal boundary condition

$$a_0(x) \ (T-T_e) - a_1(x) \ \frac{\partial T}{\partial y} = a_2(x) \quad \text{at } y=0 \tag{1}$$

where T is the temperature, T_e is the ambient temperature, x measures the distance along the plate from the leading edge and y measures the distance normal to it. The functions $a_o(x)$, $a_1(x)$ and $a_2(x)$ are prescribed functions permitting the similarity solution to the convection problem. It is assumed that these functions are all either positive or negative.

ANALYSIS

The free convection on the horizontal plate is governed by the boundary layer equations,

$$\frac{\partial u}{\partial x} + \frac{\partial v}{\partial y} = 0 \tag{2}$$

$$u \frac{\partial u}{\partial x} + v \frac{\partial u}{\partial y} = -\frac{1}{\rho} \frac{\partial p}{\partial x} + \nu \frac{\partial^2 u}{\partial y^2} \tag{3}$$

$$\frac{1}{\rho} \frac{\partial p}{\partial y} = g \beta (T - T_e) \tag{4}$$

$$u \frac{\partial T}{\partial x} + v \frac{\partial T}{\partial y} = \frac{\nu}{Pr} \frac{\partial^2 T}{\partial y^2} \tag{5}$$

and the boundary conditions,

$$u = v = 0 \text{ at } y=0 \tag{6}$$

$$u \to 0, \ p \to p_e, \ T \to T_e \text{ as } y \to \infty \tag{7}$$

besides   Equation (1), where u   and v   are velocity components in x and y directions, p is the pressure, p_e and ρ are the pressure and density of the ambient fluid.   The symbols ν, g, β and Pr denote the kinematic viscosity, gravitational acceleration, coefficient of   thermal expansion and Prandtl number respectively.

Consider

$$C = c_o x^\lambda, \ c_o > 0, \ G = \frac{g \beta C x^3}{\nu^2} \tag{8}$$

We shall determine the constants c_o and λ with the aid of Equation (1) in the manner explained later.   Introducing the similarity variables,

$$\eta = \frac{y}{x} G^{1/5}, \ f(\eta) = \frac{\psi}{\nu G^{1/5}} \tag{9}$$

$$F(\eta) = \frac{(p_e - p) x^2}{\rho \nu^2 G^{4/5}}, \ \theta(\eta) = \frac{T - T_e}{C}$$

where ψ is the stream function, we find

$$u = \frac{\partial \psi}{\partial y} = \frac{\nu G^{2/5}}{x} f' \tag{10}$$

$$v = - \frac{\partial \psi}{\partial x} = - \frac{\nu G^{1/5}}{5x} [(\lambda+3)f+(\lambda-2)\eta f'] \tag{11}$$

and hence Equations (3)-(7) reduce to

$$5f''' + (\lambda+3)ff'' + (2\lambda+1)(2F-f'^2) + (2-\lambda)\eta\theta = 0 \tag{12}$$

$$F' + \theta = 0 \tag{13}$$

$$5\theta'' + Pr[(\lambda+3)f\theta' - 5\lambda f'\theta]= 0 \tag{14}$$

$$f(o) = f'(o) = f'(\infty) = F(\infty) = \theta(\infty) = 0 \tag{15}$$

where the primes denote differentiation with respect to η. Equation (1) becomes

$$b_0 c_0 \theta(0) - b_1 c_0^{6/5} \theta'(0) = 1 \tag{16}$$

where

$$b_0 = \frac{a_0(x)}{a_1(x)} x^\lambda, \qquad b_1 = \left[\frac{g\beta}{\nu^2}\right]^{1/5} \frac{a_1(x)}{a_2(x)} x^{2(3\lambda-1)/5} \tag{17}$$

For the existence of similarity solution b_0 and b_1 must be constant. Setting

$$b_0 c_0 = \cos\phi, \quad b_1 c_0^{6/5} = \sin\phi, \quad 0 \le \phi \le \pi/2 \tag{18}$$

where c_0 is the positive root of the equation

$$b_1^2 c_0^{12/5} + b_0^2 c_0^2 - 1 = 0, \tag{19}$$

Equation (16) becomes

$$\theta(0) \cos\phi - \theta'(0) \sin\phi=1 \tag{20}$$

which takes the form $\theta(0)=1$ in the case of prescribed temperature ($a_1=0$, i.e., $\phi=0$) and $\theta'(0)=-1$ in the case of prescribed heat flux ($a_0=0$, i.e., $\phi=\pi/2$).

The local Nusselt number defined by

$$Nu_x = - \frac{x}{(T-T_e)} \frac{\partial T}{\partial y}\Big|_{y=0}$$

takes the dimensionless form

$$\frac{Nu_x}{G^{1/5}} = - \frac{\theta'(0)}{\theta(0)} \qquad (21)$$

We have reduced the convection problem to solving Equations (12)-(15) and (20) involving a parameter ϕ. If we know the solution for any particular value of ϕ, say $\phi = \phi_0$, then the solution for any other value of ϕ can be computed easily with the aid of the following two theorems.

Theorem 1.   The Equations (12)-(15) are invariant under the transformation group

$$\eta = A \eta_0, \quad f(\eta) = \frac{1}{A} f_0(\eta_0), \quad F(\eta) = \frac{1}{A^4} F_0(\eta_0),$$

$$\theta(\eta) = \frac{1}{A^5} \theta_0(\eta_0) \qquad (22)$$

where A is a positive constant.

Theorem 2.   If $\{f_0(\eta_0), F_0(\eta_0), \theta_0(\eta_0)\}$ is the solution of the convection problem for $\phi = \phi_0$, then $\{f(\eta), F(\eta), \theta(\eta)\}$ is the solution for any ϕ provided A is the positive root of the equation

$$A^6 - A \theta_0(0) \cos\phi + \theta_0'(0) \sin\phi = 0 \qquad (23)$$

The truth of these theorems can be verified by substitution in the relevant equations with the aid of Equation (22).

The significance of these theorems is evident if we note that

$$- \frac{\theta'(0)}{\theta(0)} = - \frac{\theta_0'(0)}{\theta_0(0)} \frac{1}{A} \qquad (24)$$

Recalling Equation (21), we observe that computation of heat transfer coefficient for various values of ϕ amounts to solving Equation (23) for A and then using Equation (24). For instance, in the case of prescribed heat flux ($\phi = \pi/2$), Equation (24) gives

$$\frac{1}{\theta(0)} = [-\theta_0'(0)]^{5/6} \qquad (25)$$

where θ_0 corresponds to the case of prescribed temperature.

DISCUSSION

The boundary value problem has been solved numerically for the case $\phi = 0$ and the solutions for other cases have been obtained using Theorem 2.   The numerical values of $-\theta'(0)/\theta(0)$, $f''(0)$ and $F(0)$ which represent heat transfer coefficient, dimensionless stress and pressure drop at the plate respectively are given in Table 1 for various values of ϕ and Pr=0.72.   These quantities have been normalized by their corresponding values for $\phi = 0$ as

$$N = -\frac{\theta'(0)/\theta(0)}{\theta_o'(0)} , \quad S = \frac{f''(0)}{f_o''(0)} , \quad P = \frac{F(0)}{F_o(0)} \tag{26}$$

which become with the aid of Equation (22),

$$N = \frac{1}{A} , \quad S = \frac{1}{A^3} , \quad P = \frac{1}{A^4} \tag{27}$$

Figure 1 shows the variation of N, S and P with ϕ.   For any given λ , as ϕ increases from 0 , N, S and P decrease from the value 1 to reach the minimum values N_c, S_c and P_c at $\phi = \phi_c$ and then increase to reach the maximum values N_m, S_m and P_m at $\phi = \pi/2$.   It can be seen that there exists a value $\phi = \phi_1$ such that $N \leq 1$, $S \leq 1$, $P < 1$ for $0 \leq \phi \leq \phi_1$ and $N > 1$, $S > 1$, $P > 1$ for $\phi_1 < \phi < \pi/2$. The maximum value A_c of A which makes N, S and P minimum is obtained from Equation (23) as the positive root of the Equation

$$A_c^{12} - A_c^2 - [\theta_o'(0)]^2 = 0 \tag{28}$$

while the corresponding value ϕ_c of ϕ is given by

$$\phi_c = \arctan[-\theta_o'(0)/A_c] \tag{29}$$

Likewise the value ϕ_1 for which A = 1 is given by

$$\phi_1 = 2 \arctan[-\theta_o'(0)] \tag{30}$$

The values of N_c, S_c, P_c and ϕ_c are given in Table 2 and the values of N_m, S_m, P_m and ϕ_1 are given in Table 3 for some selected values of λ .   The profiles of the dimensionless temperature θ , the streamwise velocity f' and the pressure drop F are shown in Figures 2-4 respectively.   The profiles for $\phi = \phi_1$ coincide with those for $\phi = 0$ since A=1 for $\phi = \phi_1$.

In the above analysis we have referred to 'the positive roots' of Equations (19), (23) and (28).   By Descartes rule of signs one can easily see that each of these equations possesses a unique positive root.

CONCLUSIONS

In this study it is shown that

1.   The problem of free convection with a mixed thermal boundary condition is characterised by a parameter ϕ lying between 0 and $\pi/2$ and $\phi=0$ corresponds to the case of prescribed temperature while $\phi=\pi/2$ corresponds to the case of prescribed heat flux.

2.   If one has to compute the heat transfer coefficient $-\theta'(0)/\theta(0)$ for various values of ϕ, then there is no need to solve the boundary value problem every time.   It is enough to solve Equation (23) and use Equation (24) with the known solution for the case $\phi=0$.

3. $-\theta'(0)/\theta(0)$ is maximum when the plate is subjected to a prescribed heat flux $(\phi=\pi/2)$ while it is minimum when it is subjected to the mixed boundary condition with $\phi=\phi_c$.

4. $-\theta'(0)/\theta(0)$ for the mixed boundary condition with $\phi=\phi_1$ coincides with that for the case of prescribed temperature $(\phi=0)$.

5.   The heat transfer coefficient for the prescribed temperature is related to that for prescribed heat flux as

$$-\theta'(0)\big|_{\phi=0} = \left[\frac{1}{\theta(0)}\big|_{\phi=\pi/2} \right]^{1.2} \tag{31}$$

We propose to extend the analysis presented here to the cases of vertical plate and other geometries in subsequent papers.

Table 1   The values of A, $-\theta'(0)/\theta(0)$, $f''(0)$ and $F(0)$

λ	ϕ	A	$-\theta'(0)/\theta(0)$	$f''(0)$	$F(0)$
	0	1.0000	0.3574	0.9784	1.7349
	$\pi/6$	1.0085	0.3544	0.9539	1.6772
0	$\pi/4$	0.9922	0.3602	1.0015	1.7898
	$\pi/3$	0.9615	0.3717	1.1006	2.0297
	$\pi/2$	0.8424	0.4243	1.6366	3.4449
	0	1.0000	0.4613	1.0369	1.5068
	$\pi/6$	1.0179	0.4532	0.9832	1.4037
1/3	$\pi/4$	1.0062	0.4585	1.0719	1.4701
	$\pi/3$	0.9807	0.4704	1.0992	1.6287
	$\pi/2$	0.8790	0.5248	1.5266	2.5239
	0	1.0000	0.7531	1.2379	1.0522
	$\pi/6$	1.0418	0.7229	1.0947	0.8931
2	$\pi/4$	1.0404	0.7239	1.0992	0.8981
	$\pi/3$	1.0258	0.7342	1.1468	0.9503
	$\pi/2$	0.9538	0.7896	1.4264	1.2711

Table 2   The values of N_c, S_c, P_c and ϕ_c

λ	N_c	S_c	P_c	ϕ_c
0	0.9883	0.9654	0.9542	19.45°
1/3	0.9815	0.9457	0.9282	24.36°
2	0.9589	0.8816	0.8453	35.84°

Table 3   The values of N_m, S_m, P_m and ϕ_1

λ	N_m	S_m	P_m	ϕ_1
0	1.1871	1.6728	1.9858	39.33°
1/3	1.1377	1.4724	1.6751	49.53°
2	1.0484	1.1525	1.2083	73.97°

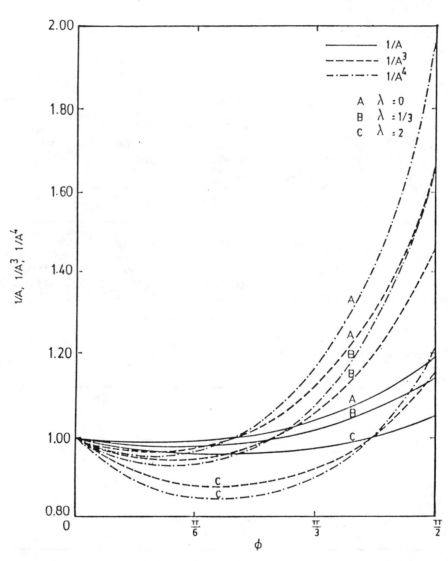

Figure 1   Plots of N, S and P versus ϕ

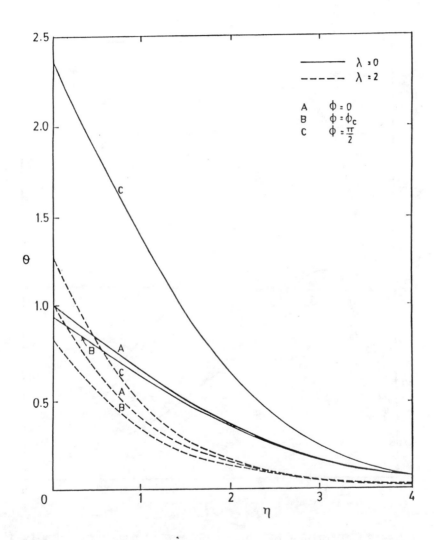

Figure 2   Profiles of dimensionless temperature θ

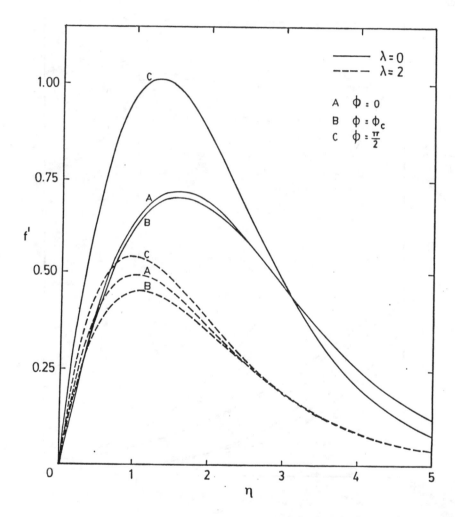

Figure 3   Profiles of dimensionless streamwise velocity f'

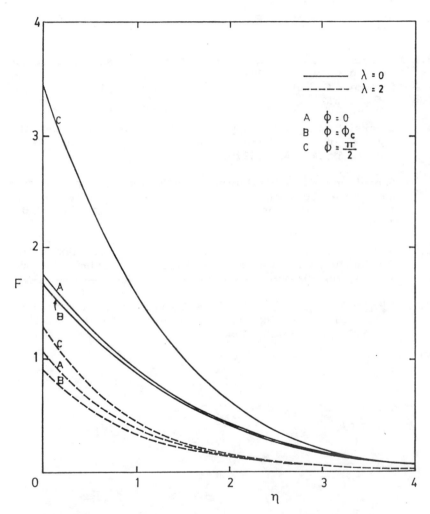

Figure 4   Profiles of dimensionless pressure drop F

REFERENCES

1.   Stewartson, K. On the Free Convection from a Horizontal
      Plate, ZAMP, Vol.9, pp.276-281, 1958.

2.   Gill, W.N., Zeh, D.W. and Casal, E.D. Free Convection
      on a Horizontal Plate, ZAMP, Vol.16, pp.539-541,
      1965.

3.   Clarke, J.F. and Riley, N. Natural Convection Induced
      in a gas by the Presence of a Hot Porous Horizontal
      Surface, Q. Jl. Mech. Appl. Math. Vol.28, pp.373-
      396, 1975.

4.   Clarke, J.F. and Riley, N. Free Convection and the
      Burning of a Horizontal Fuel Surface, J. Fluid Mech.
      Vol.74, pp.415-431, 1976.

5.   Schneider, W. A Similarity Solution for Combined Forced
      and Free Convection Flow over a Horizontal Plate,
      Int. J. Heat Mass Transfer, Vol.22, pp.1401-1406,
      1979.

6.   Merkin, J.H. and Ingham, D.B. Mixed Convection
      Similarity Solutions on a Horizontal Surface, ZAMP,
      Vol.38, pp.102-116, 1987.

7.   Lin, H.T. and Yu, W.S. Free Convection on a Horizontal
      Plate with Blowing and Suction, J. Heat Transfer,
      Vol.110, pp.793-796, 1988.

Transient Free Convection from a Vertical Flat Plate in a Non-Darcian Porous Medium with Inertia Effects

S. Haq(*), J.C. Mulligan(**)

(*) Center for Computer Science, Research Triangle Institute, RTP, North Carolina 27709, USA

(**) Department of Mechanical and Aerospace Engineering, North Carolina State University, Raleigh, NC 27695-7910, USA

ABSTRACT

Transient, buoyancy induced flow and heat transfer in a non-Darcian fluid-saturated porous medium adjacent to a suddenly heated semi-infinite vertical wall is analyzed. The partial differential equations governing non-Darcian, unsteady free convection under boundary layer assumptions are found to be of a singular parabolic type. These equations are solved in a semi-similar domain, using a successive relaxation method and the effects of inertia on the transient profiles of the temperature field and the local Nusselt number are evaluated. The results confirm, as in the case of Darcian convection, that during the initial stage before the effects of the leading edge are influential, heat transfer and flow phenomena in non-Darcian porous media are governed by transient one dimensional conduction. Solutions are presented for the transition from this initial conductive stage to a fully two-dimensional transient, which ultimately terminates in steady convection. It is shown that in a non-Darcian porous medium the transient development of the temperature field and local Nusselt number takes place monotonically and, unlike in homogenous media, neither exhibit an overshoot. Inertia effects are also shown to increase the time required to reach steady state, which results in a decrease in the Nusselt number.

INTRODUCTION

The subject of free convection in fluid saturated porous media is important in many environmental, geophysical and engineering applications. Examples include convection in geothermal reservoirs, in energy storage beds, in beds of fossil fuels, in packed bed chemical reactors, and in thermal insulations. Cheng [1] and Catton [2] have presented excellent surveys of buoyancy induced flows in porous media and described many studies of steady convection. Similarity solutions have been shown to exist for steady free convection from a vertical wall under Darcy's law [3-5].

The Darcian model of fluid flow is valid only for small Reynolds numbers, based upon pore size. At higher Reynolds numbers one may expect turbulence within the pores and microscopic inertial forces to play an important role [6]. Thus Darcy's flow model is essentially appropriate only in very low permeability porous media, while flow with high pore Reynolds numbers should be analyzed using a nonlinear Darcy-Forchheimer type model [7]. Several analyses of inertia effects in free convection in porous media near an isothermal heated vertical wall have been been made, using velocity squared terms in the momentum equation [8-11]. These studies indicated the heat transfer coefficient is decreased and the Darcian flow model, neglecting inertia effects, over predicts the heat transfer. All of these studies are restricted to steady state conditions.

The case of free convective flows which are transient, however, present additional difficulties for porous media, whether Darcian or non-Darcian. Classical problems involving homogenous media which are non-similar have been shown to exhibit an essential singularity in the boundary-layer formulation of the governing equations [12-14]. The singularity has been related to overshooting of the boundary layer thickness and the temperature field, and the creation of a transient pattern consisting of a one-dimensional conduction or diffusive regime followed by a two-dimensional convective approach to steady flow. In homogenous fluids the singularity occurs at the juncture of these flow regimes. Recently, Telionis [15] and Williams et al.[14] have reviewed literature on this topic.

Transient free convection in porous media has not received as much attention. Recently, some studies were carried out to analyze the transient problem under Darcy's law [5,16,17]. Most recently Chen, et al. [19] considered non-Darcian flow and have presented numerical solutions of free convection from an impulsively heated vertical plate using a volume averaged momentum equation which includes nonlinear convective and boundary terms. Their steady-state solutions agreed well with the results of previous steady state investigations. However the transient solution could not be verified, and the characteristic transient domains for the case of the isothermal wall temperature could not be identified. Such corroboration is essential because numerical methods applied to unsteady singular parabolic equations have generally presented numerical difficulties [14] and have even been questioned in the past for their false numerical diffusion property [12,13].

In this paper the transient free convection from an isothermal semi-infinite vertical wall into a high porosity medium with inertia effects is analyzed. As in the case of Darcian convection, the heat transfer is shown to exhibit a 2-D transient convective domain after an initial 1-D conductive domain. Furthermore, the heat transfer phenomenon in the 2-D transient domain is governed by a singular-parabolic type PDE before a steady state is reached. The transient temperature field and Nusselt number are obtained numerically for the full time domain. It is shown that neither an overshooting of the temperature, and hence undershooting of the Nusselt number, nor a discontinuity in either the temperature or Nusselt number occurs and the time require to reach a steady state is increased due to non-Darcian effects.

ANALYSIS

Consider a Cartesion coordinate system with the origin at the leading edge, the x-axis directed upwards along the wall and the y-axis normal to the wall, into a homogenous porous medium of permeability K. The permeability and mean pore size of the porous media are assumed to be sufficiently small to discount the presence of a macroscopically unstable and oscillatory motion which has been observed in beds of large diameter particles, large void spaces, and very non-uniform boundary regions. It is also assumed that the thermal properties of the solid and liquid phases, and the microscopic heat transfer coefficient are such that thermal equilibrium can be assumed. This should be the case in porous media involving relatively fine beds of materials encountered in most applications. Hence the characteristic response time at the microscopic level is assumed to be small compared compared to that at the macroscopic level and even in relatively rapid transient conditions, variations of the boundary layer characteristics occur over distances which are large compared to the pore size of the bed. The governing equations, for non-Darcian flow with inertia terms included in the momentum equations, under boundary layer assumptions are given as [18]

$$u_x + v_y = 0 \tag{1}$$

$$u + (K'/\nu)|u|u = -(K/\mu)(dp/dx - \rho g) \tag{2}$$

$$\sigma T_t + u T_x + v T_y = \alpha T_{yy} \tag{3}$$

where K' is the inertia coefficient, σ is the heat capacity ratio of the saturated porous medium, and α is the thermal diffusivity of the porous medium defined as $k/(\rho c_p)_{fluid}$, k being the volume weighted conductivity. ρ in equation (2) is the fluid density. The momentum boundary and convective terms including time derivative of the velocity have been shown to be negligible for natural convection in porous media of under small Darcy number [18]. The boundary layer assumptions, implying negligible streamwise conduction, are valid for flow under large Raleigh number [1-4,18]. The initial and boundary conditions are

$T = T_\infty$,   for $t \leq 0$ at $x \geq 0$, $y \geq 0$

$T = T_\infty$,   for $t > 0$ at $x \geq 0$, $y \to \infty$

$T = T_w$, $v = 0$, for $t > 0$ at $x \geq 0$, $y = 0$

$T = T_\infty$, $u = 0$, for $t > 0$ at $x = 0$, $y \geq 0$

Non-dimensional variables are now defined as $X = x/x^0$, $Y = y/y^0$, $\tau = t/t^0$, $U = u/u^0$, $V = u/u^0$, $\theta = \Delta T/\Delta T_w$ where

$$x^0 = K^2 g\beta\Delta T_w/\nu^2 \qquad y^0 = \sqrt{(\alpha K/\nu)} \qquad t^0 = \sigma K/\nu$$

$$u^0 = \nu x^0/K \qquad\qquad v^0 = \nu y^0/K \qquad \Delta T = T-T_\infty \qquad\qquad (4)$$

A non-dimensional heat transfer group is also defined as

$$Nu_x = [-(T_y)_{y=0}]x/\Delta T_w \quad \text{or} \quad Nu_x/\sqrt{Ra_x^*} = [-(\theta_Y)_{Y=0}]/\sqrt{X} \qquad (5)$$

where $Ra_x^* = xKg\beta\Delta T_w/(\nu\alpha)$ is known as modified Rayleigh number for a porous medium. Using the Boussinesq approximation and introducing the non-dimensional variables yields general equations in the format

$$U_X + V_Y = 0 \qquad\qquad (6)$$

$$U+(Gr')U^2 = \theta \qquad\qquad (7)$$

$$\theta_\tau + U\theta_X + V\theta_Y = \theta_{YY} \qquad\qquad (8)$$

$$
\begin{array}{lll}
\theta = 0, & \text{for } \tau \leq 0 \text{ at } X \geq 0, Y \geq 0 & (9) \\
\theta = 0, & \text{for } \tau > 0 \text{ at } X \geq 0, Y \to \infty & (10) \\
\theta = 1, V = 0, & \text{for } \tau > 0 \text{ at } X \geq 0, Y = 0 & (11) \\
\theta = 1, U = 0, & \text{for } \tau > 0 \text{ at } X = 0, Y \geq 0 & (12)
\end{array}
$$

where $Gr' = g\beta KK'(\Delta T_w)/\nu^2$ is a modified Grashof number expressing the relative importance of inertia effects as compared to viscous effects. It should be noted that if $Gr' = 0$ then the inertial effects do not exist and buoyancy is balanced by the viscous effect; that is, we recover Darcy's law.

Following the semi-similar approach presented by Haq and Mulligan [17], the above equations are transformed by introducing a finite-domain mapping of the unbounded τ and X variables, using semi-similar coordinates (ξ,η) and stream function ψ given as

$$\xi = 1-\exp(-\tau/X), \quad \eta = Y/\sqrt{(X\xi)}, \quad \psi = [\sqrt{(X\xi)}]f(\xi,\eta) \qquad (13)$$

The number of independent variables in equations (4)–(12) are then reduced from three to two and a semi-similar formulation of the equations is obtained. Equations (6)–(12) can be written as

$$\xi(1-\xi)[1 + \ln(1-\xi)U](\partial\theta/\partial\xi) = \theta_{\eta\eta} + C(\xi,\eta)\theta_\eta \qquad (14)$$

$$U + (Gr')U^2 = \theta(\xi,\eta) \qquad\qquad (15)$$

$$U(\xi,\eta) = f_\eta$$

where $C(\xi,\eta) = (1/2)[\xi+(1-\xi)\ln(1-\xi)]f + \xi(1-\xi)\ln(1-\xi)(\partial f/\partial\xi) + (1-\xi)(\eta/2)$

For $\xi = 0$, $\theta_{\eta\eta} + (\eta/2)\theta_\eta = 0$ (16)
   $\eta = 0$, $\theta = 1$
   $\eta \to \infty$, $\theta = 0$

For $\xi = 1$, $\theta_{\eta\eta} + (f/2)\theta_\eta = 0$ (17)
   $\eta = 0$, $\theta = 1$
   $\eta \to \infty$, $\theta = 0$

Equation (5) can be written as $Nu_x/\sqrt{Ra_x^*} = [-(\theta_\eta)_{\eta=0}]/\sqrt{\xi}$.

The singular parabolic nature of the non-Darcian problem is now evident. The change of the sign of the factor $[1 + \ln(1-\xi)U]$ in equation (14) from positive to negative causes the equation to become singular. In terms of non-dimensional variables, this coefficient can be written as

$$\gamma(X,Y,\tau) = [1 + \ln(1-\xi)U] = (1 - \tau U/X)$$ (18)

It is now clear that the inertial effects influence this factor implicitly, through a nonlinear $U(\theta)$ coupling, instead of the the linear coupling in Darcian problems, and this coupling is dependent upon Gr'. The direction of influence of the PDE equation (14), which is of a parabolic type, is altered by the change in sign of the coefficient. If ξ_c represents the value of ξ at which the coefficient changes sign then for the positive value of the coefficient, γ, the domain of influence of the PDE is $0 \le \xi \le \xi_c$ and equation (16) is considered as the initial condition. However for the negative value of the coefficient, the domain of influence of equation (14) is $\xi_c \le \xi \le 1$ and equation (17) becomes the initial condition.

The heat transfer at a point X^0 is characterized by 1-D conduction in the Y-direction until a signal from the leading edge reaches X^0. This leading edge effect propagates with a velocity field which results during 1-D conduction, as though the plate were infinite and conducting into a semi-infinite medium. At the instant the leading edge effect reaches X^0, the heat transfer process changes from that of 1-D conduction to a 2-D transient which finally approaches a 2-D steady state. The distance of penetration of this effect can be formulated as [20]

$$X_p = \int_0^\tau U(Y,\tau)\, d\tau$$ (19)

The governing equations for the initial 1-D conduction stage are

$$U + (Gr')U^2 = \theta   \text{and}   \theta_\tau = \theta_{YY}$$ (20)

where for $\tau=0$ or $Y=0$, $\theta = 1$ and as $Y \to \infty$, $\theta \to 0$. The solution of equations (20) is then

$$\theta = erfc(Y/(2\sqrt{\tau}))$$ (21)

$$U = [-1 + \sqrt{(1 + 4Gr'\theta)}]/(2Gr')$$ (22)

The penetration distance X_p at any time can then be determined by numerically solving equations (19) and (20). This numerical result then provides the critical time that is required for a leading edge signal at a specified value of Y to propagate a given distance along the plate. The value of the coefficient γ at this critical time can then be determined for a fixed location in X as a function of Y. Since θ has a maximum value of unity at Y=0, for Gr' $\geq$ 0 the maximum value of U also exists at the wall and is less than or equal to unity. Thus the maximum flow velocity is reduced due to inertial effects and therefore the critical time is higher for non-Darcian flow than for the Darcian flow. This implies that the maximum value of (X_p/τ) is $(X_p/\tau)_{crit} = U_{max}$, and the factor $\tau U/X$ of equation (18) at the critical point becomes equal to one. Therefore, at the critical point the value of γ is equal to 0 at Y = 0. Hence, γ=0 corresponds to the first arrival of a leading edge effect.

Thus, we conclude that the solution of the initial transient problem given by equations (21) and (22) is valid only for the initial domain given as

$$0 \leq \tau \leq (1/U_{max})X \quad \text{or} \quad 0 \leq \xi \leq 1-\exp(-1/U_{max}) \tag{23}$$

The temperature and velocity fields during this initial transient, in a transformed semi-similar domain, are

$$\theta = \text{erfc}\{(\eta/2)\surd(-\xi/(\ln(1-\xi)))\} \quad \text{for} \quad 0 \leq \xi \leq \xi_{crit} \tag{24}$$

$$U = [-1 + \surd(1 + 4Gr'\theta)]/(2Gr') \tag{25}$$

Analytical solutions are difficult to obtain for the time domain $\tau_{crit} < \tau < \tau_{ss}$, and hence no accurate predictions are available for either this regime, or the time required to reach steady state, τ_{ss}.

NUMERICAL SOLUTIONS

The semi-similar transformations defined by equations (13) map the infinite domain of the governing equations from $0 \leq \tau < \infty$, $0 \leq X < \infty$, and $0 \leq Y < \infty$ to $0 \leq \xi \leq 1$ and $0 \leq \eta < \infty$. Accordingly, the parabolic type initial conditions and boundary conditions are mapped to four conditions which are similar to those encountered with elliptic equations and therefore refered to here as "boundary" conditions. Convection phenomena in (ξ, η) are then mathematically described by equations (14)-(17), and even though the equations remain parabolic their representation is elliptic. These equations can now be solved numerically by an successive relaxation (SOR), a method generally used in solving elliptic type PDEs. A detailed analysis of this numerical method as applied in this context was first carried out by Wang [21]. More recently, the technique has been verified for the solution of singular parabolic Darcy-flow problems by Haq and Mulligan [17].

Equation (14) is discretized by a second order central difference, which yields

$$\theta(i,j) = \{ \, [\theta(i,j+1) + \theta(i,j-1)]/(\Delta\eta)^2 + C(i,j)[\theta(i,j+1) - \theta(i,j-1)]/(2\Delta\eta) -$$
$$\xi(i)[1-\xi(i)+A(i)U(i,j)][\theta(i+1,j)-\theta(i-1,j)]/(2\Delta\xi)\} \, / \, [2/(\Delta\eta)^2] \qquad (26)$$

$$U(i,j) = [-1 + \surd(1 + 4Gr' \, \theta(i,j))]/(2Gr') \qquad\qquad (27)$$
where
$$A(i) \quad = [1-\xi(i)]\ln[1-\xi(i)]$$

$$C(i,j) = (1/2)[\xi(i)+A(i)]f(i,j)+[1-\xi(i)]\eta(j)/2 + \xi(i)A(i)[f(i+1,j)-f(i-1,j)]/(2\Delta\xi)$$

$$f(i,j) = \int_0^{\eta(j)} U(\xi(i),\eta) \, d\eta \qquad\qquad (28)$$

$$\xi(i) = (i-1)\Delta\xi \quad \text{and} \quad \eta(j) = (j-1)\Delta\eta \quad \text{for} \quad i,j = 1,2,3,\dots$$

Equation (16) provides the boundary condition at $\xi=0$. It describes 1-D conduction heat transfer and can be represented analytically in terms of complimentary error functions. Alternatively, equation (24) also provides a boundary condition at $\xi=\xi_{crit}$, that is at the end of 1-D conduction domain. Equation (17) is another boundary condition for equation (14) and it describes steady-state convective heat transfer. This is a second order ordinary differential equation and hence can be solved numerically by using a 4th order Runge Kutta scheme. With all four "boundary" conditions known the finite difference equation (26) is solved iteratively using an SOR method. The initial guess is obtained for all i,j by linear interpolation between the solutions of equation (24) and (17). The f(i,j) is computed before each iteration using a corrected trapezoidal rule and the iterations are terminated after a convergence criterion, defined as $\max | \, \theta(i,j)^{n+1} - \theta(i,j)^n \, |$ less than a given Tolerance is satisfied, where n denotes the iteration level. While the number of iterations is dependent upon the initial guess, the final converged solution should be independent of the initial guess and mesh size.

RESULTS

The results presented are based on a mesh size of $\Delta\eta=0.1$, where the η domain was approximated by $0 \le \eta \le 10$ for Gr'=0 to $0 \le \eta \le 15$ for Gr'=10, and a convergence criterion of Tolerance=0.00001 was used. The number of mesh points in ξ coordinates varies with Gr' since the boundary at critical ξ depends on Gr'. Numerical experiments indicated that a relaxation factor, p = 0.6 (under relaxation) minimized the number of iterations of the successive relaxation scheme and ensured the stability and convergence of the results. Further mesh refinement was not found to alter the results significantly.

The analytical and numerical results are present in Table (1) and figure (1)-(3). Table (1) indicates that the inertia effects become significant for Gr'>0.1. It is shown that the maximum value of the velocity decreases due to the inertia losses and hence the critical time, that is the time when the solution departs from the initial 1-D domain, increases. Thus the effect of the convective terms in the energy equation is delayed and the heat transfer group at the critical time decreases with an increase in Gr'.

The temperature profiles plotted in figures (1 a-d) versus η for $Gr' = 0$, 0.1, 1, and 10 respectively indicate an increase in boundary layer thickness with increasing value of Gr'. Figures (1a), (1b) and (1c) indicate a local maxima in the slope of the temperature field at the wall (i.e θ_η at $\eta=0$), hence an overshoot after passing through a local minima is present in the value of the function $-\theta_\eta$ at $\eta=0$. Figure (1d), by contrast, for $Gr'=10$ indicates a monotonic behavior of the slope of the temperature field near the wall. The boundary layer thickness, however, is not affected by this behavior near the wall and increases monotonically between $\xi=0$ and $\xi=1$ as shown in figures (1). The temperature profile, plotted versus ξ in figure (2), goes through a local maxima near ξ_{crit} for small values of Gr'. However, the temperature remains higher at $\xi = 1$, as compared to the value at $\xi = 0.5$, for all stations of η.

Table 1. Effects of modified Grashof number on critical point and heat transfer group.

Gr'	U_{max}	$(t/X)_{crit}$	u_{crit}	Number of mesh in u	$Nu_X/!Ra_X^*$ Critical point	$Nu_X/!Ra_X^*$ Steady-state +	$Nu_X/!Ra_X^*$ Steady-state ++
0.00	1.000	1.000	0.632	37	0.564	0.444	0.444
0.01	0.990	1.010	0.636	37	0.561	0.442	0.442
0.10	0.916	1.092	0.664	34	0.540	0.430	0.430
1.00	0.618	1.618	0.802	20	0.444	0.366	0.366
10.00	0.270	3.702	0.975	25	0.293	0.251	0.251
100.00	0.095	10.512	0.9999	--	0.174	--	0.152

+ This study ,   ++ Reference [8]

The results presented in figure (3) indicate a monotonic transition of the heat transfer group from the 1-D conduction domain, through the 2-D domain of the essential singularity, and to steady-state, without a local minima or undershoot. This result is in contrast with the classical natural convection problem, where the essential singularity is associated with an undershoot in the heat transfer group. The steady-state value of the heat transfer group found in this study is in agreement with the published result as indicated in table (1). The fractional change in the heat transfer group, between the critical point and the steady-state, decreases and approaches 0 as Gr' increases. Thus for infinite Gr' we will have only a 1-D conduction domain, without convective heat transfer.

CONCLUSIONS

Transient, buoyancy induced non-Darcy flow in a non-Darcian fluid-saturated porous medium adjacent to an impulsively heated semi-infinite vertical wall has been analyzed and the heat transfer phenomena are shown to be initially governed by a parabolic PDE of 1-D conduction type. After a signal from the leading edge arrives at a location, and before steady-state is reached, the heat transfer phenomena in such a non-Darcian porous medium are governed by a

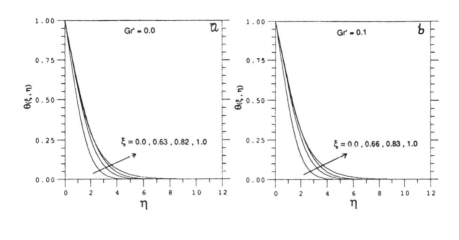

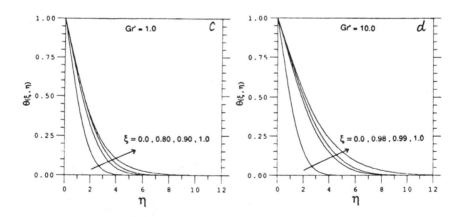

Fig 1.   Transient dimensionless temperature field in semisimilar domain showing an overshoot near the wall for small values of modified Grashof number, Gr', and monotonic behavior far away from the wall for all values of Gr'.

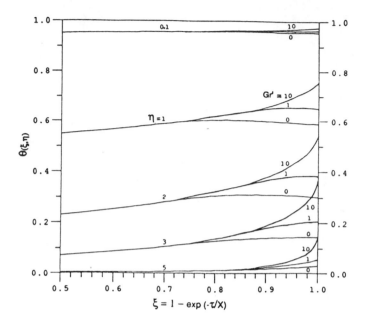

Fig 2.   Transient   dimensionless   temperature   field   obtained   in semisimilar domain for different Gr'.

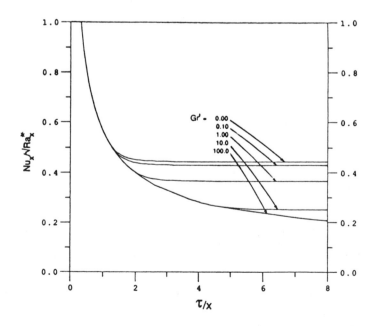

Fig 3.   Transient heat transfer group in physical domain obtained by transformation of solution in (ξ,η), showing a common initial transient period for different values of Gr'.

singular parabolic PDE. The results show that the Nusselt number decreases continuously with time and monotonically approaches the steady-state value for all values of modified Grashof number representing inertia effects. The overshoot in the boundary layer thickness at the end of a transient 1-D conduction, as observed in the classical transient free convection boundary layers, does not exist in the present problem. For a non-Darcian porous medium the inertia losses reduce the flow velocity. This results in a decrease in the heat transfer group and increase in the time required to reach steady-state. The fractional change in the heat transfer group at the end of the initial 1-D transient domain and steady-state also decreases with the increase in modified Grashof number. Thus the Darcian assumption not only over predicts the heat transfer but also under predicts the time required to reach steady state.

REFERENCES

1. Cheng P., Natural Convection in Porous Medium: External Flows, in Kakac, Aung and Viskanta (ed.), Natural Convection, Hemisphere publishing Corporation, New York, 1985. pp 475-513, 1985.

2. Catton I, Natural Convection in Porous Medium, in Kakac, Aung and Viskanta (ed.), Natural Convection, Hemisphere publishing Corporation, New York, 1985. pp 514-543, 1985.

3. Cheng, P. and Minkowycz, W. J., Free Convection about a Vertical Flat Plate Embedded in an Porous Medium with Application to Heat Transfer from a Dike, J. Geophysical Research, Vol. 82, pp 2040-2044, 1977.

4. Merkin J. II. Free Convection Boundary layer on Axi-symmetric and Two-Dimensional bodies of arbitrary shape in a saturated porous medium. Int. J. Heat Mass Transfer, Vol. 22, pp1461-1462, 1979.

5. Ingham D. B. and Brown, S. N., Flow past a suddenly heated vertical plate in porous medium, Proc. R. Soc. Lond. Ser A, Vol. 403, pp 51-80, 1986.

6. Bear J. Dynamics of Fluid in Porous Media, Elsevier, New York, 1972.

7. Fand, R. M., Steinberger, T. E., and P. Cheng, Natural convection heat transfer from a horizontal cylinder embedded in a porous medium, Int. J. Heat Mass Transfer, Vol. 29, No 1, pp 119-133, 1986.

8. Plumb O. A. and Huenefeld, J. C., Non-Darcy Natural Convection from Heated Surfaces in Saturated Porous Media, Int. J. of Heat Mass Transfer, Vol. 24, No 4, pp 765-768, 1981.

9. Bejan A. and Poulikakos, D., The non-Darcy regime for vertical boundary layer natural convection in a porous medium, Int. J. Heat Mass Transfer, Vol. 27, pp 717-722, 1984.

10. Bejan A., The Basic Scales of Natural Convection Heat and Mass Transfer in Fluid Saturated Porous Media, Int. Comm. Heat Mass Transfer, Vol. 14, pp 107-123, 1987.

11. Hong, J. T., Tein C. L., and Kaviany, M., Non-Darcian effects on vertical plate natural convection in porous media with high porosities, Int. J. Heat Mass Transfer, Vol. 28, No 11, pp 2149-2157, 1985.

12. Nanbu, K., Limit of Pure Conduction for Unsteady Free Convection on a Vertical Plate, Int. J. Heat Mass Transfer, Vol. 14, pp 1531-1534, 1971.

13. Ingham, D. B., Singular Parabolic Partial Differential Equations that Arise in Impulsive Motion Problems, Trans. ASME E: J. Appl. Mech., Vol. 44, pp 396-400, 1977.

14. Williams J. C., Mulligan, J. C., and Rhyne, T. B., Semisimilar Solutions for Unsteady Free convection Boundary Layer Flow on a Vertical Flat Plate, J. Fluid Mech., Vol. 175, pp 309-322, 1987.

15. Telionis, D. Unsteady Viscous Flows, Springer-Verlag, New York, 1981.

16. Cheng, P. and Pop, I., Transient Free Convection about Vertical Flat Plate Embedded in a Porous Medium. Int. J. Engng. Sci., Vol. 22, pp 254-264, 1984.

17. Haq, S. and Mulligan, J. C., Transient free convection form an impulsively heated vertical plate embedded in a fluid saturated porous media, Num. Heat Transfer. In press, 1990.

18. Haq, S. Transient Free Convection in a Fluid Saturated Porous Medium, PhD Dissertation, North Carolina State University, North Carolina, USA, 1989.

19. Chen C. K, Hung, C. I., and Horng H. C., Transient Natural Convection on a Vertical Flat Plate Embedded in a High-Porosity Medium, J. of Energy Resources Tech. ASME, Vol. 109, pp 112-118, 1987.

20. Goldstein, R. J. and Briggs, D. G., Transient Free Convection About Vertical Plate and circulate Cylinder, Trans. ASME C: J. Heat Transfer, Vol. 86, pp 490-500, 1964.

21. Wang J. C. T., On the Numerical Methods for the Singular Parabolic Equations in Fluid Dynamics, J. of Computational Physics, Vol. 52 pp 464-479, 1983.

Free Convective Heat Transfer from a Heated Cylinder in an Enclosure with a Cooled Upper Surface

A.R. Elepano, P.H. Oosthuizen
Department of Mechanical Engineering, Queen's University, Kingston, Ontario, Canada

ABSTRACT

Two-dimensional laminar free convective flow in an enclosure containing a heated cylinder and having a cooled upper surface has been numerically studied. The cylinder is heated to a uniform surface temperature. The flow has been assumed to be steady, laminar and two-dimensional and the buoyancy term has been treated using the Buossinesq approximation. The governing equations, written in dimensionless form, have been solved using the finite element method. The solution has as parameters the dimensionless distance between the cylinder and the upper surface, the Rayleigh number and the Prandtl number which was taken as 0.7. Solutions have been obtained for a range of values of parameters. The results indicate that the heat transfer rate increases with Rayleigh number but the distance between the cylinder and the upper surface has a weak effect on the heat transfer rate.

NOMENCLATURE

c_f	specific heat of fluid
D'	diameter of cylinder
g	gravitational acceleration
H'_U	distance between the cylinder and the upper surface
H'_L	distance between the cylinder and the bottom surface
H_U	H'_U/D'
H_L	H'_L/D'
k_f	effective thermal conductivity of fluid
Nu_D	local Nusselt number based on D'
Nu_W	local Nusselt number based on W'
Pr	Prandtl number

p'　　pressure
Q　　dimensionless total heat transfer rate
Ra　　Rayleigh number
T'　　temperature
T'_H　　cylinder surface temperature
T'_c　　upper surface temperature
u'　　velocity component in x' direction
v'　　velocity component in y' direction
x'　　coordinate in horizontal direction
x　　x'/D'
y'　　coordinate in vertical direction
y　　y'/D'
W'　　half-width of the enclosure
W　　W'/D'
α　　thermal diffusivity of fluid
β　　coefficient of thermal expansion of fluid
θ　　dimensionless temperature
ν　　viscosity of fluid
ρ_f　　density of fluid
ϕ　　angular position around cylinder
ψ'　　stream function
ψ　　dimensionless stream function
ω'　　vorticity function
ω　　dimensionless vorticity function

INTRODUCTION

In some crop dryers used in Third-World countries, the combustion gases resulting from the burning of crop residues are passed through a pipe in a trench in the ground and then vented through a chimney. The crop is placed on a support over the trench and the heat transferred from the hot pipe to the crop bed leads to the drying of the crop. In estimating the performance of such a drying system, it is important to know the rate of heat transfer from the pipe to the bed. The heat transfer from the pipe to the bed essentially occurs as a result of the free convective motion in the pit. Since the flow through the bed is very small, the flow situation can be approximately modelled as free convective flow about a heated cylinder in an enclosure with a cooled top. Furthermore, because the trench tends to be relatively long compared to its width and depth, the flow can be assumed to be two-dimensional. Therefore, in order to provide initial information on heat transfer rates that occur in such dryers two-dimensional free convective flow in a rectangular enclosure containing a heated cylinder and with a top at a uniform lower temperature and with all other surfaces adiabatic has been considered. A schematic diagram of the flow situation is shown in Figure 1.

There have been a number of previous studies on natural convection in an enclosure, these being reviewed by Catton [1], Oosthuizen and Paul [2],

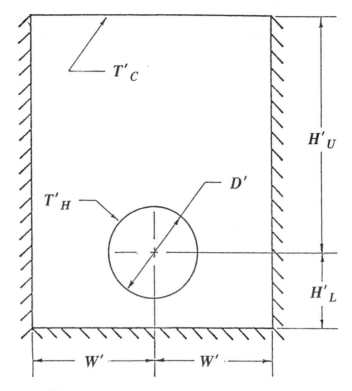

Figure 1. Flow situation considered.

Ostrach [3] and DeVahl Davis and Jones [4].   Heat transfer in enclosures containing a body is discussed by Singh and Chen [5], Ingham [6] and Ozoe et al. [7].   Oosthuizen and Paul [8] described a numerical study of natural convective heat transfer from a prismatic cylinder in a rectangular enclosure while Projahn et al. [9] were concerned with the numerical study of the heat transfer rate from a horizontal circular cylinder, non-symmetrically placed within a horizontal circular cavity.   No previous studies of the situation here considered appear to be available.

GOVERNING EQUATIONS AND SOLUTION PROCEDURE

The study is based on the following assumptions:

- The flow is two-dimensional.
- The flow is steady and laminar.
- Fluid properties are constant except for the density change with temperature which gives rise to the buoyancy forces, this effect being treated using the Boussinesq approximation.
- The side and bottom walls are adiabatic.
- The cylinder and upper surface are at uniform temperatures.

Using these assumptions and assuming symmetry about the vertical centre-line, the governing equations become, in terms of the coordinate system shown in Figure 2:

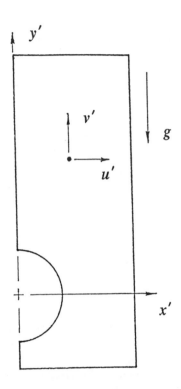

Figure 2. Coordinate system considered.

$$\frac{\partial u'}{\partial x'} + \frac{\partial v'}{\partial y'} = 0 \tag{1}$$

$$u' \frac{\partial u'}{\partial x'} + v' \frac{\partial u'}{\partial y'} = -\frac{1}{\rho_f} \frac{\partial p'}{\partial x'} + \nu \left(\frac{\partial^2 u'}{\partial x'^2} + \frac{\partial^2 u'}{\partial y'^2} \right) \tag{2}$$

$$u' \frac{\partial v'}{\partial x'} + v' \frac{\partial v'}{\partial y'} = -\frac{1}{\rho_f} \frac{\partial p'}{\partial y'} + \nu \left(\frac{\partial^2 v'}{\partial x'^2} + \frac{\partial^2 v'}{\partial y'^2} \right)$$

$$+ \beta g (T' - T'_C) \tag{3}$$

$$u' \frac{\partial T'}{\partial x'} + v' \frac{\partial T'}{\partial y'} = \frac{k_f}{\rho_f c_f} \left(\frac{\partial^2 T'}{\partial x'^2} + \frac{\partial^2 T'}{\partial y'^2} \right) \tag{4}$$

The prime (') is used to denote dimensional quantities and the symbols are as defined in the Nomenclature. The subscript, f, denotes fluid properties. The temperature, T'_c, of the upper surface is used as the reference temperature. The boundary conditions on the solution are:

- on the surface of the cylinder: $u' = 0, \quad v' = 0, \quad T' = T'_H$
- on the line of symmetry: $\frac{\partial u'}{\partial x'} = 0, \quad \frac{\partial T'}{\partial x'} = 0$
- on the upper surface: $u' = 0, \quad v' = 0, \quad T' = T'_c$
- on the vertical adiabatic walls: $u' = 0, \quad v' = 0, \quad \frac{\partial T'}{\partial x'} = 0$
- on the bottom surface: $u' = 0, \quad v' = 0, \quad \frac{\partial T'}{\partial y'} = 0$

The solution has been obtained by introducing the stream and vorticity functions given as usual by:

$$u' = \frac{\partial \psi'}{\partial y'} , \quad v' = - \frac{\partial \psi'}{\partial x'} , \quad \omega' = \frac{\partial v'}{\partial x'} - \frac{\partial u'}{\partial y'}$$

The following dimensionless variables are also defined:

$$\psi = \frac{\psi'}{\alpha} , \quad \omega = \frac{\omega' D'^2}{\alpha}$$

$$x = \frac{x'}{D'} , \quad y = \frac{y'}{D'} , \quad \theta = \frac{T' - T'_c}{T'_H - T'_c}$$

In terms of these variables, the governing equations become:

$$-\omega = \frac{\partial^2 \psi}{\partial x^2} + \frac{\partial^2 \psi}{\partial y^2} \tag{5}$$

$$\frac{1}{Pr} \left(\frac{\partial \psi}{\partial y} \frac{\partial \omega}{\partial x} - \frac{\partial \psi}{\partial x} \frac{\partial \omega}{\partial y} \right) = \frac{\partial^2 \omega}{\partial x^2} + \frac{\partial^2 \omega}{\partial y^2} + Ra \frac{\partial \theta}{\partial x} \tag{6}$$

$$\frac{\partial \psi}{\partial y} \frac{\partial \theta}{\partial x} - \frac{\partial \psi}{\partial x} \frac{\partial \theta}{\partial y} = \frac{\partial^2 \theta}{\partial x^2} + \frac{\partial^2 \theta}{\partial y^2} \tag{7}$$

where:

$$Pr = \frac{\nu}{\alpha} , \quad Ra = \frac{\beta g (T' - T'_c) D'^3}{\nu \alpha}$$

In terms of the dimensionless variables, the boundary conditions on the solution are:

- on the surface of the cylinder: $\psi = 0$, $\theta = 1$
- on the line of symmetry: $\psi = 0$, $\omega = 0$
- on the upper surface: $\psi = 0$, $\theta = 0$
- on the vertical adiabatic wall: $\psi = 0$, $\dfrac{\partial \theta}{\partial x} = 0$
- on the bottom surface: $\psi = 0$, $\dfrac{\partial \theta}{\partial y} = 0$

The governing equations, subject to the boundary conditions discussed above, have been solved using the finite element method with simple linear triangular elements being used. A Gauss-Siedel iteration technique was used to solve the equations. The solution procedure used has been successfully applied before to a number of other problems, e.g. see Oosthuizen and Paul [8].

RESULTS

The solution has, as parameters:

- the dimensionless distance, H_U, between the upper surface and the horizontal axis of the cylinder;
- the Rayleigh number, Ra, based on cylinder diameter;
- the Prandtl number, Pr = 0.7.

Solutions have been obtained for values of Ra between 0 and 100,000 and values of H_U between 1 and 3.

Typical streamline and isotherm patterns are shown in Figure 3. The patterns here have had no smoothing applied to them. Figure 3 illustrates the effect of Ra on the flow at $H_U = 2.57$. Note that at Ra = 0, the heat transfer is purely by conduction, ie. no flow occurs. As Ra is increased the stream function values increase. Hot fluid near the cylinder flows upward to the top surface and flows downwards near the adiabatic wall as it is cooled. There is essentially no flow near the bottom of the cylinder. The fluid motion and temperature change is largely concentrated in the upper portion of the cavity. Isotherms for Ra = 0 are nearly equidistant from the cylinder to the top surface. As flow is induced at higher Ra, the temperature gradients near the constant temperature surfaces increase.

The variation of the total heat transfer rate with Ra will next be discussed. Figure 4 shows the results for H_U of 2.57. The dimensional heat transfer rate, Q, is the product of the mean local Nusselt number and the characteristic length of the surface i.e. $Q = Nu_w \bullet W = Nu_D \bullet \pi\, D/2$. The total heat transfer is nearly constant at low Ra. However, as the Ra is increased, the total heat transfer is increased. It can be seen that Q is approximately proportional to $Ra^{1/4}$ at large Rayleigh numbers. The relationship can

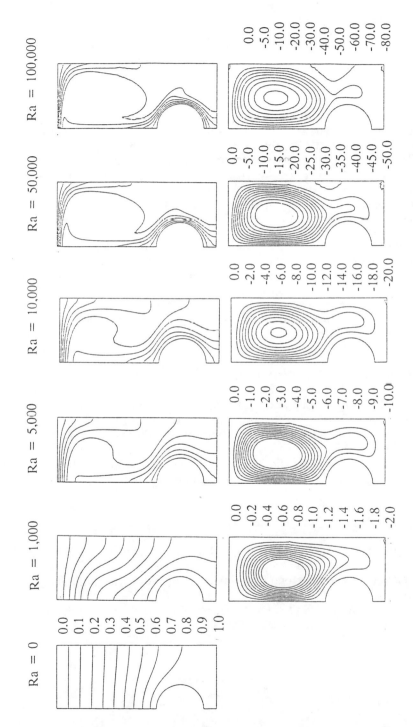

Figure 3. Isotherms and streamlines at various Rayleigh number
for $H_U = 2.57$.

be described as:

$$0 < Ra < 500 \text{ then } Q = 0.6$$

$$500 < Ra < 100{,}000 \text{ then } \quad Q = a + bRa^{1/4}$$
$$\text{where: } a = -0.6743, b = 0.3678$$

The effect of the distance between the upper surface and cylinder is considered next. A typical variation is shown in Figure 5 for Ra = 10,000. It will be seen that there is only a small variation in the total heat transfer rate. The peak in total heat transfer rate occurs when the upper quadrant of the cavity is nearly a square where $H_U = W$. At high H_U the fluid motion is confined to the upper portion of the enclosure while at low H_U the dominant flow occurs at the side of the cylinder. The "blockage" of flow is minimized which then tend to increase the strength of the convective motion.

Typical variations of the local Nusselt number along the upper surface and around the cylinder are shown in Figures 6 and 7, respectively. These results are for $H_U = 2.57$. It will be seen from Figure 6 that the local heat transfer rate on the upper surface tends to remain constant at low Ra value. But that it varies significantly at higher Ra, the higher heat transfer rate being near the line of symmetry at the upper surface. This is of importance in drying. The distribution of heat input over the bed surface must be known in order to ensure that the crop is not overheated during the drying process.

Figure 7 shows the variation of the local Nusselt number along the surface of the cylinder. At low Ra, the gradient of Nu along the surface is not large. Nu is highest at $\phi = 0$, that is on the portion of the cylinder facing the upper surface. At Ra = 10,000 there are two peaks in the Nu_D values. This results because the flow from the bottom surface of the cylinder tends to pass around the upper portion reducing the heat transfer rate near $\phi = 1.0 - 1.2$. At lower Ra the flow is confined to the upper portion of the cavity resulting in a heat transfer profile with a single peak.

CONCLUSIONS

- The distance between the cylinder and the top surface has a weak effect on the total heat transfer rate.
- A relation between the total heat transfer rate and Rayleigh number has been obtained.
- Significant variations in the local heat transfer rate exist along the upper surface.

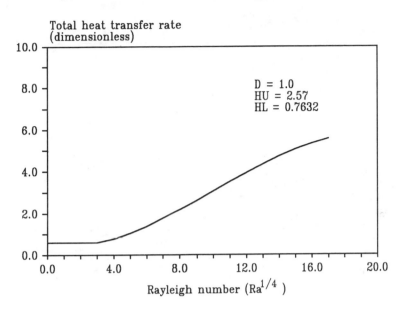

Figure 4. Variation of total heat transfer rate with Rayleigh No.

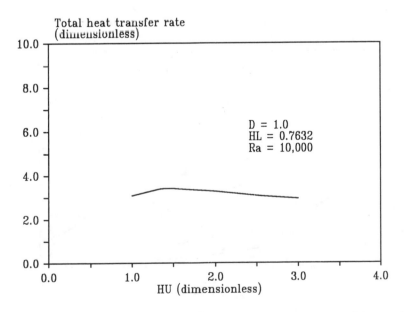

Figure 5. Variation of total heat transfer rate with vertical
distance from the axis of cylinder to top surface.

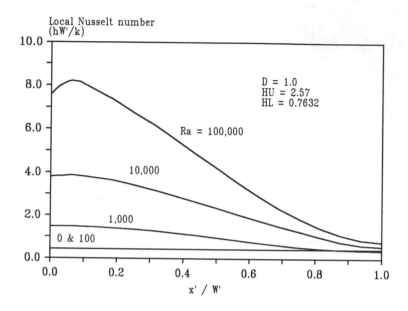

Figure 6. Variations of Nusselt numbers at the top plate at various Rayleigh numbers.

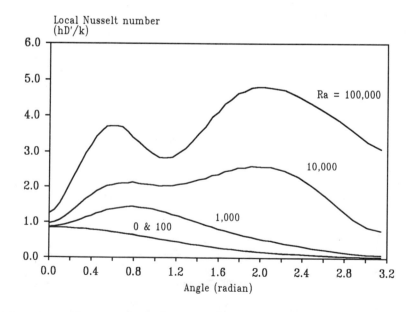

Figure 7. Variations of Nusselt numbers at the surface of the cylinder at various Rayleigh numbers.

REFERENCES

1. Catton, I. Natural Convection Heat Transfer in Enclosures, Proc. 6th Int. Heat Transfer Conference, Vol. 6, pp. 106-120, 1978.

2. Oosthuizen, P. H. and Paul, J. T. Free Convective Flow in an Inclined Square Cavity with a Partially Heated Wall, Heat Transfer in Convective Flows, ASME National Heat Transfer Conference, Vol. 107, pp. 231-238, 1989.

3. Ostrach, S. Natural Convection Heat Transfer in Cavities and Cells, Proc. 7th Int. Heat Transfer Conference, Vol. 1, pp. 365-379, 1982.

4. DeVahl Davis, G. and Jones, I. P. Natural Convection in a Square Cavity: A Comparison Exercise, Int. Journal for Numerical Methods in Fluids, Vol. 3, pp. 227-248, 1983.

5. Singh, S. N. and Chen, J. Numerical Solution for Free Convection between Concentric Spheres at Moderate Grashof Numbers, Numercial Heat Transfer, Vol. 3, pp. 441-459, 1980.

6. Ingham, D. B. Heat Transfer by Natural Convection between Spheres and Cylinders, Numerical Heat Transfer, Vol. 4, pp. 53-67, 1981.

7. Ozoe, H., Fijii, K., Shibata, T. and Kuriyama, H. Three Dimensional Numerical Analysis of Natural Convection in a Spherical Annulus, Numerical Heat Transfer, Vol. 8, pp. 383-406, 1985.

8. Oosthuizen, P. H. and Paul, J. T. Finite Element Study of Natural Convective Heat Transfer for a Prismatic Cylinder in an Enclosure, pp. 13-21, Proceeding of the ASME Winter Meeting, Anaheim, California, 1984.

9. Projahn, V., Reiger, H. and Beer, H. Numerical Analysis of Laminar Natural Convection Between Concentric and Eccentric Cylinders, Numerical Heat Transfer, Vol. 4, pp. 131-146, 1981.

A Numerical Evaluation of the Sensitivity of a New Method of using Heat Transfer Measurements to Determine Wall Shear Stress

P.H. Oosthuizen(*), W.E. Carscallen(**)
(*) Department of Mechanical Engineering, Queen's University, Kingston, Ontario, Canada (**) Gas Dynamics Laboratory, National Research Council, Ottawa, Ontario, Canada K1A 0R6

ABSTRACT

A new method of using heat transfer measurements as the basis for determining local wall shear stresses has recently been proposed. In this method, the wall under consideration is made from a material with a low thermal conductivity. The surface of this wall is covered with a thin metal sheet. A steerable pulsed laser beam is then used to rapidly heat a small portion of the surface of the wall and the temperature-time variation of the heated surface spot following cessation of the irradiation is used as a basis for determining the local wall shear stress. Initial experimental and numerical studies of this proposed method indicated that the method was not sufficiently sensitive when the heat transfer rates at the surface were low such as when there is a laminar boundary layer on the surface. However, it appeared that the sensitivity of the method could be improved by using different surface materials from those initially proposed. The present study was, therefore, undertaken to numerically determine whether this was in fact possible. The laminar, thermal three-dimensional boundary layer flow over a heated spot on a flat plate has been calculated simultaneously with the unsteady conduction in the wall following the essentially instantaneous heating of a small circular portion of the wall surface material. The results allow the variation of the temperature at the centre of the heated spot to be determined as a function of time, this variation being characterized by a suitably defined cooling factor. The sensitivity of the method with various types of surface and wall materials has been examined.

NOMENCLATURE

b_m = thickness of surface material
d = initial diameter of heated spot
h = heat transfer coefficient at surface
k = thermal conductivity of air
k_m = thermal conductivity of surface material
k_s = thermal conductivity of substrate material
r = radial distance

t = time
T = temperature
T_m = temperature of surface material
T_∞ = temperature of air
u = velocity component in x-direction
u_1 = freestream velocity
v = velocity component in y-direction
x = coordinate in flow direction
y = coordinate normal to surface
z = coordinate across surface
α = thermal diffusivity of air
α_m = thermal diffusivity of surface material
α_s = thermal diffusivity of substrate material
β = cooling factor
θ = $T_m - T_\infty$
ρ = density
τ_w = wall shear stress

INTRODUCTION

The use of heat transfer measurements to determine the local wall shear stress acting on the wall has received quite extensive investigation in the past. This method usually involves the mounting of some form of heat transfer gauge on or in the wall. Such gauges tend to be relatively expensive and often only a few gauges are available in any given project with the result that the gauges must be carefully positioned based on some prior knowledge of the flow field involved. Some such heat transfer gauges are also difficult to use in complex flow geometries. A new method of using heat transfer measurements as the basis for determining local wall shear stresses that avoids some of these difficulties has recently been proposed by Carscallen[1]. In this method, the wall under consideration is made from a material with a low thermal conductivity which is covered with a thin metal sheet. A steerable pulsed laser beam is then used to rapidly heat a small portion of the surface of the wall and the temperature-time variation of the heated surface spot following cessation of the irradiation is used as a basis for determining the local wall shear stress. The wall surface temperature can be measured using a thermocouple attached to the underside of the metal surface in the centre of the heated spot or it can be measured using radiation or surface crystal methods. The proposed arrangement is shown schematically in Figure 1. Since the beam can be rapidly steered to any point on the surface, the shear stress at a large number of positions over the entire surface can be rapidly obtained.

Initial experimental and numerical studies of this proposed method indicated that the method had an acceptable sensitivity when the flow over the wall was turbulent and attached. However, these initial studies indicated that the method was not sufficiently sensitive when the heat transfer rates at the surface were low such as when there is a laminar boundary layer on the surface. Since it should be possible to improve the sensitivity of the method by using different surface materials from those initially proposed, the present study was undertaken to numerically determine whether this was in fact possible.

In the present study, the laminar, thermal three-dimensional boundary layer flow over a heated spot on a flat plate has been calculated simultaneously with the unsteady conduction in the wall following the essentially instantaneous heating of a small circular portion of the wall surface material. The velocity field has been assumed to unaffected by the heating and is, therefore, assumed to be known. The results allow the variation of the temperature at the centre of the heated spot to be determined as a function of time, this variation being characterized by a suitably defined cooling factor which should be insensitive to the initial temperature to which the surface is heated. Since the wall shear stress is known from the assumed velocity field, the variation of the cooling factor with wall shear stress can be deduced from the results. This effect of various types of surface and wall materials has been examined to try to determine if any can give an adequate sensitivity.

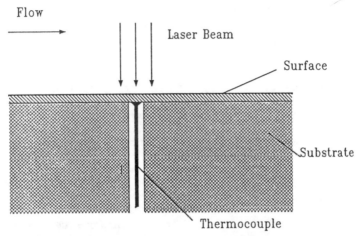

Figure 1   Situation being considered.

Many methods of trying to measure surface shear stress distributions have been used and discussed in the past. Direct methods of measuring the stress using surface elements are described, for example, by Liepmann and Dhawan[2] and by Dhawan[3] . The use of surface pressure probes (i.e. Stanton and Preston probes) is discussed by Preston[4], Trilling and Hakkinen[5] , Abarbanel et al[6], Gadd[7] , Head and Rechenberg[8], Smith et al[9] and Patel[10] for example. Heat transfer based methods of determining the surface shear stress are discussed, for example, by Bellhouse et al[11], Brown[12] , Rubesin et al[13], Freymouth[14] , Menendez and Ramaprian[15] and Khalid[16]. None of the methods described in these studies seems, however, to be as suitable as the method being discussed here for rapidly and fairly simply measuring the wall shear stress at a large number of relatively arbitrarily placed points on a surface of complex shape. The method being considered in the present study has been evaluated by Carscallen et al[17] and Arthur[18] for the case of turbulent boundary layer flow.

BASIC DIFFICULTIES

In the method here being discussed, a small circular section of the wall is essentially instantly heated and the temperature variation at the centre of this heated spot is then measured as it cools. Now the heat transfer from the heated spot, which determines the rate of temperature decrease at the centre point, will be the result of:

- conduction into the substrate material
- conduction in the surface material
- convection to the air flowing over the heated spot

In order for the method to have adequte sensitivity i.e. in order for the the rate of change of the temperature at the centre of the spot to be relatively strongly dependent on the wall shear stress, it is necessary that the convection term be the dominant one. A very rough order of magnitude approach indicates that the relative magnitudes of the three terms in the above list will depend on the relative magnitudes of the following:

- (k_s/k)
- $(k_m/k)\ (b_m/d)$
- $(h\ d/k)$

In the initial implementation of the proposed method, a copper surface sheet was used with a phenolic resin substrate, Using the values of the conductivities of these materials listed in Tables 1 and 2 indicates that since $(h\ d/k_a)$ has been estimated to be roughly 5 with a laminar boundary layer, the convective heat transfer contribution to the total heat transfer rate is likely to only amount to about 5% of the total heat transfer rate, the conduction in the surface material consituting about 85% of the total and the conduction to the substrate being about 10% of the total. Since the convective heat transfer rate itself does not vary very strongly with wall shear stress, these results indicate that with the initially used materials, the cooling rate with laminar flow will be essentially independent of the wall shear stress. To try to improve the sensivity of the method a thinner surface layer with a lower conductivity and a substrate material with a lower conductivity will have to be used. The materials listed in Tables 1 and 2 have therefore been considered. For refererence purposes calculations have, as indicated in these tables, also been undertaken for ideal materials with zero thermal conductivity. The results obtained with these various materials are discussed below. However, it should be noted that according to the above order of magnitude argument, even with the stainless steel surface layer, the convective heat transfer will still constitute less than half of the total heat transfer rate. The sensivity of the method is, therefore, unlikely to be high.

GOVERNING EQUATIONS AND SOLUTION PROCEDURE

In order to try to determine whether the measurement of the surface temperature variation with time would be sensitive enough to variations in the wall shear stress when a laminar boundary flow exists, attention has been given to the case of laminar flow over a flat plate.

Using the coordinate system shown in Figure 2, the equation governing the conduction in the substrate material is:

$$\frac{\partial^2 T}{\partial x^2} + \frac{\partial^2 T}{\partial y^2} + \frac{\partial^2 T}{\partial z^2} = \frac{1}{\alpha_s}\frac{\partial T}{\partial t} \tag{1}$$

Because of the small thickness and relatively high conductivity of the surface material, the variation in temperature across the surface layer has been neglected. The equation governing the temperature in the surface is therefore:

$$b_m \left(\frac{\partial^2 T_m}{\partial x^2} + \frac{\partial^2 T}{\partial z^2} \right) - \left(\frac{k_a}{k_m} \right) \frac{\partial T}{\partial y} \bigg|_{y=0}$$

$$+ \left(\frac{k_s}{k_m} \right) \frac{\partial T}{\partial y} \bigg|_{y=0} = \frac{1}{\alpha_m} \frac{\partial T_m}{\partial t} \qquad (2)$$

The boundary conditions on the above equations are:

$$y \rightarrow -\infty , \qquad T \rightarrow T_\infty \qquad (3)$$

$$r \rightarrow \infty , \qquad T \rightarrow T_\infty \qquad (4)$$

If d is the diameter of the initially heated surface spot then the following initial conditions apply:

$$y = 0 , \quad 0 \le r \le d/2 \quad T = T_0$$

$$y = 0 , \quad r > d/2 \quad T = T_\infty \qquad (5)$$

$$y < 0 , \qquad\qquad T = T_\infty$$

Table 1

Surface Materials

	Copper	Stainless Steel	Ideal
Thickness (mm)	0.076	0.025 B	0.076
Thermal Conductivity (W/mC)	397	15.0	0.0
Specific Heat (kJ/kg C)	380	470	380
Density (kg / m³)	8900	8000	8900
Thermal Diffusivity (m²/s)	$1.17x10^{-4}$	$3.9x10^{-6}$	0.0

Attention is next given to the boundary layer flow. Because the diameter of the heated spot is assumed to be small compared to the distance from the plate leading edge, the variation of the velocity boundary layer thickness over the spot is assumed to be negligible. The equation governing the thermal boundary layer growth over the heated spot is, therefore:

$$u \frac{\partial T}{\partial x} + v \frac{\partial T}{\partial y} = \alpha \left[\frac{\partial^2 T}{\partial y^2} + \frac{\partial^2 T}{\partial z^2} \right] \qquad (6)$$

The coordinate system used is indicated in Figure 2. The velocity field is assumed to be two-dimensional. In this equation u and v are assumed to be known functions of y. However, because the thermal boundary layer thickness is small compared to the velocity layer thickness, the term $v \partial T / \partial x$ has been neglected and the equation governing the temperature boundary layer has, therefore, been assumed to be

$$u \frac{\partial T}{\partial x} = \alpha \left[\frac{\partial^2 T}{\partial y^2} + \frac{\partial^2 T}{\partial z^2} \right] \qquad (7)$$

Table 2

Substrate Materials

	Phenolic Resin	Bakelite	Ideal
Thermal Conductivity (W/mC)	0.36	0.062	0.0
Thermal Diffusivity (m²/s)	$1.86x10^{-7}$	$3.2x10^{-9}$	0.0

The above equations for the temperatures in the wall and in the boundary layer have been simultaneously solved numerically to give the varaiation of the temperture of the surface material at the centre of the heated spot with time. A finite-difference procedure has been used, the procedure being fully implicit in the y-direction. In order to carry out this solution, the variation of u with y in the boundary layer must be specified. For the flat plate flow here being considered, the following approximate equation has been used:

$$(u / u_1) = 2 (y / \delta) + (y / \delta)^2 \qquad (8)$$

where δ is the boundary layer thickness, assumed to be given by:

$$\delta / x = 5 / Re_x^{0.5} \qquad (9)$$

While these equations are relativelu crude, they should be quite adequate for assessing the effects of changes in wall materials on the sensitivity of the method here being considered. Equations (8) and (9) together give the wall shear stress as:

$$\tau_w / \rho \, u_1^2 = 0.4 / Re_x^{0.5} \qquad (9)$$

The above equations were used to solve for the variation of surface temperature with time for the materials indicated in Tables 1 and 2. for an initial heated spot diameter of 12.5 mm. A typical such variation is shown in Figure

3.

Now if the method of determining wall shear stress here being discussed is to be conveniently applied it is necessary to derive a parameter that characterizes the cooling rate and which is insensitive to the initial temperature of the heated spot, this temperature being dependent on the laser power which can vary somewhat between shots. It was decided to try to use a cooling factor, β , defined by:

$$\beta = \frac{1}{(T_m - T_\infty)} \frac{\partial T_m}{\partial t} = \frac{1}{\theta} \frac{\partial \theta}{\partial t} \qquad (11)$$

for this purpose. Using the results of the numerical conduction calculations, the values of β at various times after the initial heating were calculated. A typical variation of β, with time is shown in Figure 4. It was felt that at very short times the actual values of β obtained in a real implementation of the method would be strongly influenced by nonuniformities in the initial heated area At long times after the initial heating β becomes relatively small and difficult to measure accurately. It was decided, therefore, that the values of β at 0.50 seconds after the initial heating would be used to characterize the rate of cooling.

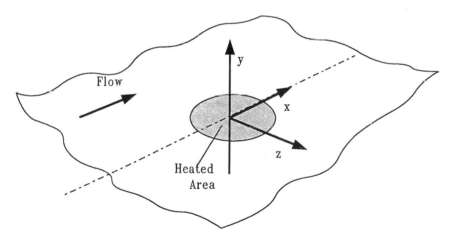

Figure 2   Coordinate system used.

RESULTS

As mentioned above, attention has been restricted here to flow over a flat plate. To fit in approximately with the experimental study described by Arthur[18] , a plate with an overall length of one meter has been considered. For relatively arbitarily selected velocites of 1 and 5 m/s the values of the cooling constant at various distances have been calculated for various combinations of surface materials, the assumed properties of these surface materials being as given in Tables 1 and 2. Typical variations of β with x are given in Figures 5 to 10. It will be seen from the results given in Figure 5, which are for "ideal" surface and substrate materials, that because the conduction heat transfer is zero with these materials, the cooling factor varies sufficiently to allow the cooling factor to be used for the measurement of surface shear stress. However, it will be seen from Figures 6 to 9 that when either a real surface material or a real substrate material is used, the cooling factor is, due to conduction heat transfer, much larger than with the "ideal" materials and much less sensitive to changes in wall shear stress. In particular, it will be noted that when a copper surface material is used with a phenolic resin substrate material, the combination that was used in the initial turbulent flow tests, the conduction heat transfer completely swamps the convective heat transfer and the cooling factor becomes almost independent of the wall shear stress. Even when the thinner and lower conductivity stainless steel surface material is used with the two real substrate materials here being considered, the sensitivity is still to small for practical purposes. However, the results given in Figure 10, which are for a stainless steel surface material with an "ideal" substrate material, indicate that with this combination the sensitivity of the method is acceptable. This indicates that if a substrate material with a low enough conductivity is used, an adequate sensitivity can be achieved. The actual substrate material used will depend on the application.

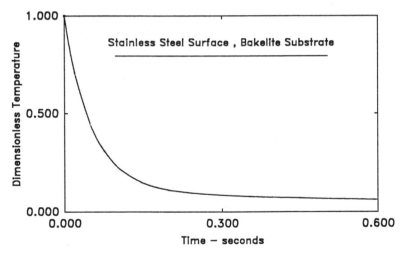

Figure 3   Typical variation of temperature difference ratio with time.

CONCLUSIONS

The present results indicate that the proposed method of measuring local shear stress using a laser heated spot is not sufficiently sensitive for laminar flows with the surface and substrate materials adopted in past work. The results do, however, indicate that the method could potentially be made to have an adequate sensitivity by the use of other materials.

ACKNOWLEDGEMENTS

This work was supported by the National Research Council of Canada.

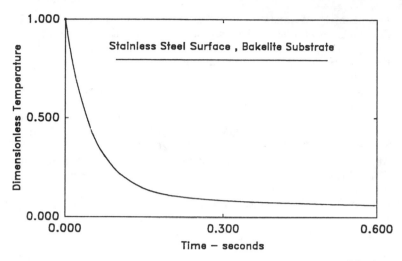

Figure 4   Typical variation of cooling rate parameter with time.

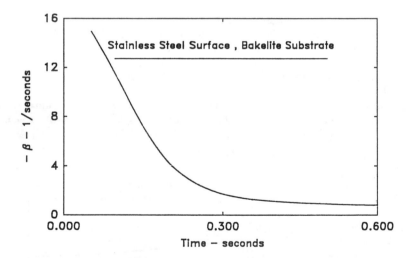

Figure 5   Variation of cooling rate factor at 0.5 secs. after heating with distance from the leading edge for different free stream velocities for the case of an "ideal" surface material and an "ideal" substrate material.

REFERENCES

1.   Carscallen, W.E., A Novel Method of Wall Shear Stress Measurement Using a High Energy Laser Beam, *Personal Communication*, National Research Council, Ottawa, Canada, 1987.

F.

120   Natural and Forced Convection

2.   Liepmann, H. W. and Dhawan, S., Direct Measurements of Local Skin Friction in Low Speed and High Speed Flow, *Proceedings, 1st U.S. National Congress of Applied Mechanics*, Chicago, 869-874, 1951.
3.   Dhawan, S., Direct Measurements of Skin Friction, National Advisory Committee for Aeronautics, *NACA Report 1121*, 1953.
4.   Preston, J.H., The Determination of Turbulent Skin Friction by Means of Pitoc Tubes, *Journal of the Royal Aeronautical Society*, 58, 109, 1954.
5.   Trilling, L. and Hakkinen, R.J., The Calibration of the Stanton Tube as a Skin Friction Meter, *50 Jahre Grenzschichtforshung*, H. Gortler, W. Tollimen, eds., Friedr, Vieweg and Sohn, 201-209, 1955.
6.   Abarbanel, S.S., Hakkinen, R.J. and Trilling, L., Use of a Stanton Tube for Skin Friction Measurements, National Aeronautical and Space Administration, NASA MEMO 2-17-59-W, 1959.
7.   Gadd, G.E., A Note on the Theory of the Stanton Tube, Aeronautical Research Council, *ARC 20*, 471, FM 2740, TP 605, 1958.
8.   Head, M.R. and Rechenberg, I., The Preston Tube as a Means of Measuring Skin Friction, *Journal of Fluid Mechanics*, 14, 1, 1962.
9.   Smith, K.G., Gaudet, L. and Winter, K.G., The Use of Surface Pitot Tubes as Skin Friction Meters at Subsonic Speeds, Royal Aeronautical Establishment, RAE Aero. 2665, 1962.
10.   Patel, V.C., Calibration of the Preston Tube and Limitations on its us in Pressure Gradients, *Journal of Fluid Mechanics*, 23, 185, 1965.
11.   Bellhouse, B.J. and Schultz, D.L., Determination of Mean and Dynamic Skin Friction, Separation and Transition in Low-Speed Flow with a Thin-Film Heated Element, *Journal of Fluid Mechanics*, 24, 2: 379-400, 1966.
12.   Brown, G.L., Theory and Application of Heated Films for Skin Friction Measurement, *Proceedings, 1967 Heat Transfer and Fluid Mechanics Institute*, Stanford University Press, 361-381, 1967.

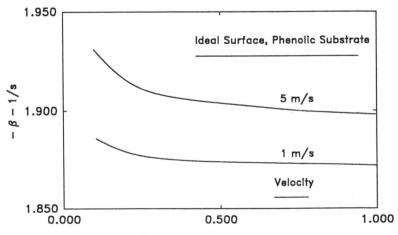

Figure 6   Variation of cooling rate factor at 0.5 secs. after heating with distance from the leading edge for different free stream velocities for the case of an "ideal" surface material and a phenolic resin substrate material.

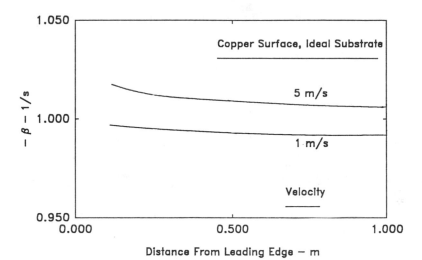

Figure 7   Variation of cooling rate factor at 0.5 secs. after heating with distance from the leading edge for different free stream velocities for the case of a copper surface and an "ideal" substrate material.

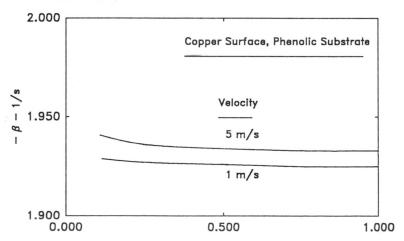

Figure 8   Variation of cooling rate factor at 0.5 secs. after heating with distance from the leading edge for different free stream velocities for the case of a copper surface and a phenolic resin substrate material.

13. Rubesin, M.W., Okuno, A.F. and Mateer, G.G., A Hot-Wire Surface Gage for Skin Friction and Seeration Detection Measurements, National Aeronautics and Space Administration, NASA TM X-62465, California, 1975.

14. Freymouth, P., Modelling of Hot-Films by an Extended Bellhouse- Schultz Model, *Proceedings, 2nd Symposium; Flow, Its Measurement and Control in Science and Industry*, 69-75, 1981.

16. Inaba, H. and Fukuda, T., Natural Convection in an Inclined Square Cavity in Regions of Density Inversion of Water, *Journal of Fluid Mechanics*, 142, 363-381, 1984.

17. Ivey, G.N. and Hamblin, P.F., Convection Near the Temperature of Maximum Density for High Rayleigh Number, Low Aspect Ratio, Rectangular Cavities, *Journal of Heat Transfer*, 111, 100-104, 1989.

18. Lankford, K.E. and Bejan, A., Natural Convection in a Vertical Enclosure Filled with Water Near 4° C, *Journal of Heat Transfer*, 108, 755-763, 1986.

19. Watson, A., The Effect of the Inversion Temperature on the Convection of Water in an Enclosed Rectangular Cavity, *Quarterly Journal of Mechanics and Applied Mathematics*, 25, 423-446, 1972.

20. Hung Nguyen, T., Vasseur, P. and Robillard, L., Natural Convection of Cold Water in Concentric Annular Spaces: Effect of the Maximum Density, *Proceedings, 7th International Heat Transfer Conference*, 2, 251-256, 1982.

21. Robillard, L. and Vasseur, P., Transient Natural Convection Heat Transfer of Water with Maximum Density and Supercooling, *Journal of Heat Transfer*, 103, 528-534, 1981.

22. Vasseur, P. and Robillard, L., Transient Natural Convection Heat Transfer in a Mass of Water Cooled through 4° C, *International Journal of Heat and Mass Transfer*, 23, 1195 -1205, 1980.

23. Oosthuizen, P. H. and Paul, J. T., Unsteady Free Convective Flow in an Enclosure Containing Water Near its Density Maximum, To be Presented at the 1990 AIAA/ASME Thermophysics and Heat Transfer Conference.   Heat Transfer Conference, San Francisco, California, 1527-1532, 1986.

24. Blake, K.R., Bejan, A. and Poulikakos, D., Natural Convection Near 4° C in Water Saturated Porous Layer Heated From Below, *International Journal of Heat and Mass Transfer*, 27, 2355-2364, 1984.

25. Philip, J. R., Free Convection in Porous Cavities Near the Temperature of Maximum Density, *PhysicoChemical Hydrodynamics*, 10, 283-294, 1988.

26. Oosthuizen, P.H. and Paul, J.T., Heat Transfer through a Square Container Filled with a Liquid and a Gas," *ASME Paper 83-WA/HT-101*, 1982.

27. Ivey, G.N., Experiments on Convection Near the Temperature of Maximum Density and Their Application, *Proceedings, 8th Australasian Fluid Mechanics Conference*, 1, 2A.4-2A.6., 1983.

Table 1

Surface Materials

θ_i	θ_{ct}	ψ_{ct}	Nu
-0.5	-0.17	5.4	2.08
0	0	0	2.15
0.5	0.17	-5.4	2.08

Natural Convective Flow in an Enclosure Containing a Liquid near its Density Maximum with a Free Surface

P.H. Oosthuizen, J.T. Paul

Department of Mechanical Engineering, Queen's University, Kingston, Ontario, Canada K7L 3N6

ABSTRACT

Unsteady natural convective flow in an enclosure with one wall heated to a uniform temperature and another parallel wall cooled to a uniform lower temperature has been numerically considered. The bottom and free surface of the enclosure are assumed to be adiabatic. The enclosure is filled with a liquid which has a free surface and which is at a temperature near that at which a a density maximum exists, the relationship between density and temperature being non-linear. The hot wall is heated to a temperature above that at which the density maximum occurs and the cold is at a temperature that is lower than this value. The flow has been assumed to be laminar and two-dimensional. Fluid properties have been assumed constant except for the density change with temperature that gives rise to the buoyancy forces, this being treated by assuming a parabolic type relationship. The governing equations, expressed in terms of stream function and vorticity, have been written in dimensionless form. The resultant equations have been solved using a finite-difference method. Results have been obtained for modified Rayleigh numbers between 10,000 and 1,000,000 for a Prandtl number of 12 for aspect ratios of between 0.5 and 1 for two different initial dimensionless fluid temperatures.

NOMENCLATURE

A = aspect ratio, h'/w'
a = constant in density temperature relationship
h' = height of "vertical" walls of cavity
k = thermal conductivity
Nu = mean Nusselt number based on w'
Pr = Prandtl number
q = local dimensionless heat transfer rate
$\bar{q}$ = mean heat transfer rate
Ra^* = modified Rayleigh number based on w'
t' = time
t = dimensionless time
T = dimensionless temperature
T' = temperature
T_H' = temperature of hot wall
T_C' = temperature of cold wall
T_m' = temperature at which maximum density occurs
u' = velocity component in x' direction
v' = velocity component in y' direction
w' = width of cavity
x = dimensionless x' coordinate
x' = horizontal coordinate position
y = dimensionless y' coordinate
y' = vertical coordinate position
α = thermal diffusivity
θ = dimensionless temperature

θ_H = dimensionless hot wall temperature
θ_i = dimensionless initial temperature
θ_C = dimensionless cold wall temperature
θ_{ct} = dimensionless temperature at centre point
ν = kinematic viscosity
ρ = density
ρ_m = maximum density
ψ = dimensionless stream function
ψ_{ct} = dimensionless stream function at centre point
ψ' = stream function
ω = dimensionless vorticity
ω' = vorticity

INTRODUCTION

Free convective flow in a rectangular cavity which has one vertical wall heated to a uniform temperature, T'_H, and another parallel wall cooled to a uniform lower temperature, T'_C, and with the top and bottom walls adiabatic has been the subject of many studies. In most such studies it is assumed that the that the cavity contains a fluid with a linear relationship between density and temperature, i.e. the Boussinesq approximation is used. For example, steady flow in such a case is discussed by Catton[1], de Vahl Davis and Jones[2], de Vahl Davis[3], Ostrach[4] and Wong and Wraithby[5] while unsteady flow is discussed, for example, by Kuhn and Oosthuizen[6,7,8]. The related problem of free convective heat transfer across an enclosure with a free surface has also received attention, e.g see Oosthuizen and Paul[9]. As is well known, however, such a linear density-temperature relationship does not describe the behaviour of water and certain other liquids near the freezing point, a density maximum existing in such cases and the relationship between density and temperature being non-linear. In the present study, unsteady flow in a such a case in an enclosure with one wall heated to a uniform temperature, T'_H, and another parallel wall cooled to a uniform lower temperature, T'_C, has been has been numerically studied. The enclosure is filled with a liquid which has a free surface. The bottom of the enclosure is assumed to be adiabatic. The flow has been assumed to be laminar and two-dimensional.

There have been many previous studies of free convective heat transfer to water which is at a temperature near that at which the maximum density occurs. External natural convective flows have been considered, for example, by Carey and Gebhart[10], El-Henawa et al.[11], Gebhart and Mollendorf[12], Vanier and Tien[13] and Wilson and Lee[14]. Steady natural convective heat transfer across cavities has been studied by Fukumari and Wake[15], Inaba and Fukuda[16], Ivey and Hamblin[17], Lankford and Bejan[18] and Watson[19]. Natural convection in an annular cavity was considered by Hung Nguyen et al.[20]. Most of the above studies have basically dealt with steady flow situations. Transient natural convection in cavities has been considered by Robillard and Vasseur[21] and Vasseur and Robillard[22]. In both of these papers, a container of water whose surface is suddenly cooled at time zero is considered i.e. the flow situation differs from that dealt with in the present study. In a recent paper Oosthuizen and Paul[23] considered unsteady flow in a closed cavity and concluded that the symmetrical flow that is usually assumed to exist when the difference between the hot wall temperature and the maximum density temperature is equal to the difference between the maximum density temperature and the cold wall temperature is unlikely to exist in real situations. They also concluded that the final steady state solution could depend on the initial conditions. The purpose of this study is to further investigate these possibilities and to determine if the presence of a free surface and changes in aspect ratio have an effect on the conclusions.

GOVERNING EQUATIONS AND SOLUTION PROCEDURE

The flow has been assumed to be laminar and two-dimensional. The solution is based on the use of the two-dimensional Navier-Stokes, continuity and energy equations. Fluid properties have been assumed constant except for the density change with temperature that gives rise to the buoyancy force, this being treated by assuming a quadratic type relationship. It has also been assumed that the because the velocities are low, the free surface remains effectively flat. The governing equations are then:

$$\frac{\partial u'}{\partial x'} + \frac{\partial v'}{\partial y'} = 0 \tag{1}$$

$$\frac{\partial v'}{\partial t'} + u' \frac{\partial v'}{\partial x} + v' \frac{\partial v'}{\partial y} = g(\rho_m - \rho)/\rho$$

$$- \frac{1}{\rho} \frac{\partial p'}{\partial y'} + \nu \left(\frac{\partial^2 v'}{\partial x'^2} + \frac{\partial^2 v'}{\partial y'^2} \right) \tag{2}$$

$$\frac{\partial u'}{\partial t'} + u' \frac{\partial u'}{\partial x} + v' \frac{\partial u'}{\partial y} =$$

$$- \frac{1}{\rho} \frac{\partial p'}{\partial x'} + \nu \left(\frac{\partial^2 v'}{\partial x'^2} + \frac{\partial^2 v'}{\partial y'^2} \right) \tag{3}$$

$$\frac{\partial T'}{\partial t'} + u' \frac{\partial T'}{\partial x'} + v' \frac{\partial T'}{\partial y'}$$

$$= \left(\frac{k}{\rho c} \right) \left(\frac{\partial^2 T'}{\partial x'^2} + \frac{\partial^2 T'}{\partial y'^2} \right) \tag{4}$$

The prime (′) denotes a dimensional quantity and the coordinates are as defined in Figure 1. It will be noted that only the density change with temperature has been considered, the other fluid properties being assumed constant. The relation between density and temperature is assumed to of the form:

$$(\rho_m - \rho)/\rho = a(T' - T_m')^2 \tag{5}$$

The adequacy of this form of relation for describing the behaviour of water near the temperature of maximum density is discussed, for example, by Blake et al.[24] and Philip[25] .

The solution has been obtained in terms of the stream function and vorticity defined, as usual, by:

$$u' = \frac{\partial \psi'}{\partial y'} , \quad v' = - \frac{\partial \psi'}{\partial x'}$$

$$\omega' = \frac{\partial v'}{\partial x'} - \frac{\partial u'}{\partial y'} \tag{6}$$

In terms of these variables, the governing equations become:

$$\frac{\partial^2 \psi'}{\partial x'^2} + \frac{\partial^2 \psi'}{\partial y'^2} = - \omega' \tag{7}$$

$$\frac{\partial \omega'}{\partial t'} + \frac{\partial \psi'}{\partial y'} \frac{\partial \omega'}{\partial x'} - \frac{\partial \psi'}{\partial x'} \frac{\partial \omega'}{\partial y'}$$

$$= 2 a g(T' - T_m') \frac{\partial T'}{\partial x'} + \nu \left(\frac{\partial^2 \omega'}{\partial x'^2} + \frac{\partial^2 \omega'}{\partial y'^2} \right) \tag{8}$$

$$\frac{\partial T'}{\partial t'} + \frac{\partial \psi'}{\partial y'} \frac{\partial T'}{\partial x'} - \frac{\partial \psi'}{\partial x'} \frac{\partial T'}{\partial y'}$$

$$= \left(\frac{k}{\rho c} \right) \left(\frac{\partial^2 T'}{\partial x'^2} + \frac{\partial^2 T'}{\partial y'^2} \right) \tag{9}$$

The following dimensionless variables have been introduced:

$$x = x'/w' \qquad y = y'/w'$$

$$\psi = \psi' Pr/\nu \qquad \omega = \omega' w'^2 Pr/\nu \tag{10}$$

$$\theta = (T' - T_m')/(T_H' - T_C') \qquad t = t'\,\nu \,/\,Pr\,w'^2$$

the cavity width, w', thus being used as the characteristic length scale.

In terms of these variables, the governing equations become:

$$\frac{\partial^2 \psi}{\partial x^2} + \frac{\partial^2 \psi}{\partial y^2} = -\omega \tag{11}$$

$$\frac{\partial \omega}{\partial t} + \frac{1}{Pr}\left(\frac{\partial \psi}{\partial y}\frac{\partial \omega}{\partial x} - \frac{\partial \psi}{\partial x}\frac{\partial \omega}{\partial y}\right)$$

$$= \frac{\partial^2 \omega}{\partial x^2} + \frac{\partial^2 \omega}{\partial y^2} + 2\,Ra^*\,\theta\,\frac{\partial \theta}{\partial x} \tag{12}$$

$$\frac{\partial T}{\partial t} + \frac{\partial \psi}{\partial y}\frac{\partial \theta}{\partial x} - \frac{\partial \psi}{\partial x}\frac{\partial \theta}{\partial y} = \frac{\partial^2 \theta}{\partial x^2} + \frac{\partial^2 \theta}{\partial y^2} \tag{13}$$

where Ra^* is a modified Rayleigh number based on the cavity width, w', i.e.:

$$Ra^* = a\,g\,(T_r' - T_C')^2\,w'^3\,/\,\nu\,\alpha \tag{14}$$

The boundary conditions on the solution are as follows, the lettered wall segments being as defined in Figure 1:

on all walls and on free surface:

$$\psi = 0$$

on BC:

$$\omega = 0$$

on A B:

$$\theta = \theta_H \tag{15}$$

on C D:

$$\theta = \theta_C$$

on end wall and free surface, DA and BC, (which are assumed adiabatic):

$$\partial \theta \,/\,\partial y = 0$$

The initial conditions are:

$$\theta = \theta_i \tag{16}$$

and the liquid being at rest at this initial time.

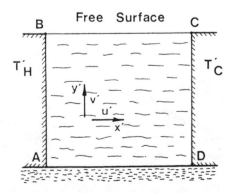

Figure 1   Flow situation considered

The above dimensionless governing equations, subject to these boundary conditions, have been solved using a finite difference method similar to that described, for example, by Oosthuizen and Paul[26] and Kuhn and Oosthuizen[7]. A uniform grid was used in all calculations and calculations were carried out for several different grid sizes to ensure that the results obtained were grid independent. It was found that a 57 x 57 grid was adequate. With this grid size the results given here are grid independent in all cases to significantly better than 1%. The time step used depended on the value of the modified Rayleigh number. Calculations with various time steps were undertaken to ensure that the results presented below are essentially independent of the time step used.

The solution directly gives the local dimensionless heat transfer rate distributions on the walls of the cavity, i.e. the local Nusselt number distributions on the hot and cold walls. This Nusselt number is given by:

$$Nu_\ell = \frac{qw'}{k(T_H' - T_C')} \tag{17}$$

These distributions can then be integrated to give the mean dimensionless heat transfer rates on the hot and the cold walls. This mean heat transfer rate has been expressed in the form of a mean Nusselt number, i.e. by:

$$Nu = \frac{\bar{q}w'}{k(T_H' - T_C')} \tag{18}$$

RESULTS

The solution to the equations given in the previous section has, in general, the following parameters:

- the modified Rayleigh number based on the cavity width, Ra^*
- the Prandtl number, Pr
- the cavity aspect ratio, $A = h'/w'$
- the dimensionless temperature to which the hot wall is heated, θ_H
- the dimensionless initial temperature of the liquid θ_i

Results have been obtained for modified Rayleigh numbers between 10,000 and 1,000,000 for aspect ratios between 0.5 and 1 for a Prandtl number of 12. This is approximately the Prandtl number of water at a temperature near that at which the maximum density occurs. Results have been obtained for dimensionless hot wall temperatures of from 0.2 to 0.8. Because of the way in which the dimensionless temperature is defined, these values correspond to dimensionless cold wall temperatures of -0.8 to -0.2 because $\theta_H = 1.0 + \theta_C$. Since a dimensionless hot wall temperature of 0 corresponds to to the case where the hot wall is at the temperature at which the density maximum occurs, it will be noted that, in the present study, attention has been restricted to the situation where the hot wall temperature is above that at which the density maximum occurs while the cold wall is at a temperature below this value. Most calculations were undertaken for an initial temperature that that was the average of the hot and cold wall temperatures.

Although the unsteady flow following a step change in the wall temperatures has been calculated, attention will here be given to the values the following variables in the final steady state:

- the dimensionless temperature, θ_{ct}, at the centre point of the cavity
- the stream function at the centre point of the cavity, ψ_{ct}
- the mean Nusselt number, Nu

Attention will first be given to the case where the initial temperature in the enclosure is the average of the hot and cold wall temperatures. Consideration will later be given to the possibility that this initial temperature effects the final steady state results. The variation of the final steady-state centre point temperature with hot wall temperature will first be considered. Typical variations for different modified Rayleigh numbers are shown in Figures 2, 3 and 4 for an aspect ratio of 1. Now it will be noted that when the dimensionless hot wall temperature is 0.5, the dimensionless cold wall temperature will be -0.5 and in the steady state condition the dimensionless centre point temperature will in this case be 0, the flow in the cavity then consisting of two identical but counter-rotating vortices. However, it is apparent from the results given that the centre point temperature does not vary monotonically with hot wall temperature for dimensionless hot wall temperatures near 0.5. Instead, the centre point temperature varies rapidly with change in wall temperature for dimensionless wall temperatures near 0.5. As a result, a flow with two nearly identical counter-rotating vortices is relatively unstable and is not likely to exist in most practical situations.

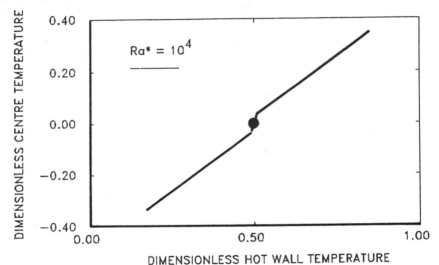

Figure 2 Variation of dimensionless centre point temperature with dimensionless hot wall temperature for a modified Rayleigh number of 10,000 and for an aspect ratio of 1

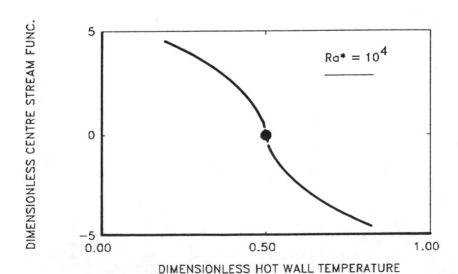

Figure 3 Variation of dimensionless centre point temperature with dimensionless hot wall temperature for a modified Rayleigh number of 100,000 and for an aspect ratio of 1

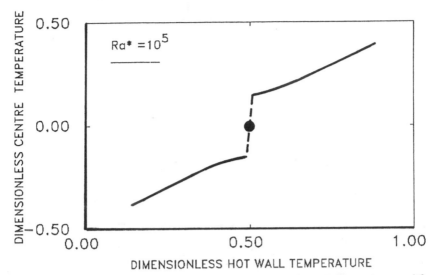

Figure 4 **Variation of dimensionless centre point temperature with dimensionless hot wall temperature for a modified Rayleigh number of 1,000,000 and for an aspect ratio of 1**

The dimensionless centre point stream functions will next be considered. Typical variations with hot wall temperature for various values of the modified Rayleigh number are shown in Figures 5, 6 and 7. In interpreting these results, it should be noted that positive values of the dimensionless stream function are associated with an upward flow along the cold wall while negative values of the stream function are associated with an upward flow along the hot wall for the coordinate system here being used. The results given in Figures 5, 6 and 7, therefore, indicate as expected that when the hot wall temperature is below 0.5, the flow near the centre is dominated by a vortex associated with the upward flow along the cold wall while when the hot wall temperature is below 0.5, the upward flow along the hot wall becomes the dominant motion in the centre region. Now, as mentioned before, for a dimensionless hot wall temperature of 0.5, two vortices equal strength will exist in the steady state and, therefore, the centre point stream function will be 0 in the steady state. The results given in Figures 7, 8 and 9 again show that very small changes in wall temperature for wall temperatures near 0.5 produce relatively large changes in centre point stream function. This again indicates that a flow with two equal but oppositely rotating vortices is essentially unstable. The effect of the cavity aspect ratio on the form of the results is illustrated by the results given in Figures 8 and 9. While the values of the centre point temperature and stream function are effected by aspect ratio, the basic form of the variations with hot wall temperature are the same as with an aspect ratio of 1.

Typical variations of the Nusselt number are shown in Figures 10 and 11. Figure 8 shows the variation of Nusselt number with dimensionless hot wall temperature for various values of the Rayleigh number. As has been noted by many others, the Nusselt number is lowest for for wall temperatures near 0.5. It will also be noted that for dimensionless temperatures near 0.5, the Nusselt number varies little with wall temperature, rising in fact slightly to a weak maximum at a wall temperature of 0.5. The effect of aspect ratio on the Nusselt number is illustrated by the results given in Figure 11. The effect is relatively small but the extent of the near constant Nusselt number region near a wall temperature of 0.5 tends to increase with decreasing aspect ratio.

The results presented above were all for the case where the initial temperature in the enclosure is the average of the hot and cold wall temperatures. In order to determine whether the initial temperature effects the final steady state results calculations were also undertaken for the cases where the initial temperature is equal to hot wall temperature and where it is equal to the cold wall temperature. Typical results for the

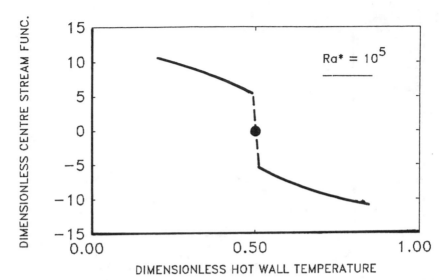

Figure 5   Variation of dimensionless centre point stream function with dimensionless hot wall temperature for a modified Rayleigh number of 10,000 and for an aspect ratio of 1

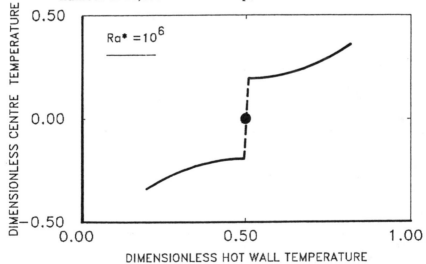

Figure 6   Variation of dimensionless centre point stream function with dimensionless hot wall temperature for a modified Rayleigh number of 100,000 and for an aspect ratio of 1

case where the hot wall temperature is 0.5 are given in Table 1. It will be seen that the initial temperature does indeed effect the form of the steady state solution.

CONCLUSIONS

The results of the present study indicate that:

- The near symmetrical flow that exists when the dimensionless hot wall temperature is near 0.5 only exists over a very narrow range of wall temperatures,

this being true at all aspect ratios considered.
- The steady state solution can be influenced by the initial temperature.
- In broad outline, the form of the results for a cavity with a free surface are similar to those for a full enclosure.

ACKNOWLEDGEMENTS

This work was supported by the Natural Sciences and Engineering Research Council of Can   ι.

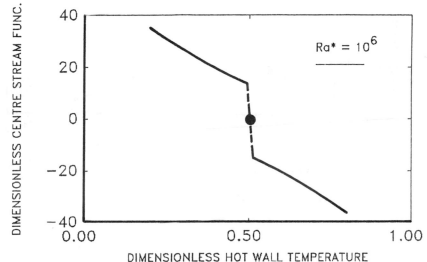

Figure 7   Variation of dimensionless centre point stream function with dimensionless hot wall temperature for a modified Rayleigh number of 1,000,000 and for an aspect ratio of 1

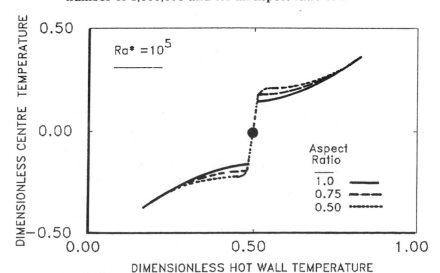

Figure 8   Variation of dimensionless centre point temperature with dimensionless hot wall temperature for a modified Rayleigh number of 100,000 for various aspect ratios

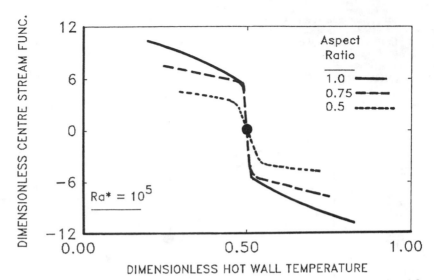

Figure 9 **Variation of dimensionless centre point stream function with dimensionless hot wall temperature for a modified Rayleigh number of 100,000 for various aspect ratios**

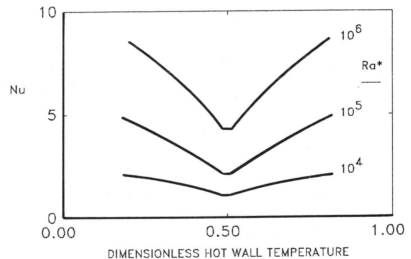

Figure 10 **Variation of Nusselt number with dimensionless hot wall temperature for an aspect ratio of 1 and for various modified Rayleigh numbers**

REFERENCES

1. Catton, I., Natural Convection in Enclosures,*Proceedings, 6th International Heat Transfer Conference*, 6, 13-43, 1978.
2. de Vahl Davis, G. and Jones, I. P., Natural Convection in a Square Cavity: A Comparison Exercise, *International Journal of Numerical Methods in Fluids*, 3, 227-248, 1983.

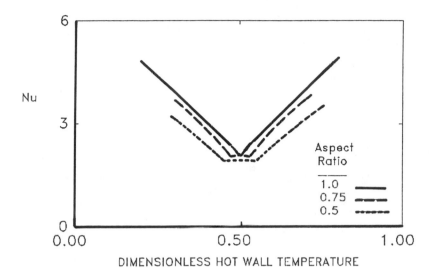

Figure 11 Variation of Nusselt number with dimensionless hot wall temperature for a modified Rayleigh number of 100,000 and for various aspect ratios

3. de Vahl Davis, G., Finite Difference Methods for Natural and Mixed Convection in Enclosures, *Proceedings, 8th International Heat Transfer Conference.*, 1, 101-109, 1986.

4. Ostrach, S., Natural Convection Heat Transfer in Cavities and Cells, *Proceedings, 7th International Heat Transfer Conference*, 1, 365-379, 1982.

5. Wong, H. H. and Raithby, G. D., Improved Finite-Difference Methods Based on a Critical Evaluation of the Approximation Errors, *Numerical Heat Transfer.*, 2, 139-163, 1979.

6. Kuhn, D. C. S. and Oosthuizen, P. H., Transient Three-Dimensional Natural Convective Flow in a Rectangular Enclosure with Two Heated Elements on a Vertical Wall, *Proceedings, 5th International Conference on Numerical Methods in Thermal Problems*, V, 1: 524-535, 1987.

7. Kuhn, D. C. S. and Oosthuizen, P. H., Unsteady Natural Convection in a Partially Heated Rectangular Cavity, *Journal of Heat Transfer*, 109, 3: 798-801, 1987b.

8. Kuhn, D. C. S. and Oosthuizen, P. H., Transient Three-Dimensional Flow in an Enclosure with a Hot Spot on a Vertical Wall, *International Journal of Numerical Methods in Fluids*, 8, 369-385, 1988.

9. Oosthuizen, P. H. and Paul, J. T., Natural Convective Heat Transfer Across a Liquid-Filled Cavity, *Proceedings, 8th International Heat Transfer Conference*, San Francisco, California, 1527-1532, 1986.

10. Carey, V.P. and Gebhart, B., Visualization of the Flow Adjacent to a Vertical Ice Surface Melting in Cold Pure Water, *Journal of Fluid Mechanics*, 107, 37-55, 1981.

11. El-Henawy, I.M., Hassard, B.D. and Kazarinoff, N.D., A Stability Analysis of Non-Time-Periodic Perturbations of Buoyancy-Induced Flows in Pure Water Near 4° C, *Journal of Fluid Mechanics*, 163, 1-20, 1986.

12. Gebhart, B. and Mollendorf, J.C., Buoyancy-Induced Flow in Water under Conditions in which Density Extrema May Arise, *Journal of Fluid Mechanics*, 89, 4: 673-707, 1978.

13. Vanier, C.R. and Tien, C., Effect of Maximum Density and Melting on Natural Convection Heat Transfer from a Vertical Plate, *Chemical Engineering Progress Symposium*, Series 64, 240-254, 1968.

14. Wilson, N. W. and Lee, J. J., Melting of a Vertical Ice Wall by Free Convection into Fresh Water, *Journal of Heat Transfer*, 103, 13-17, 1981.

15. Fukumori, E. and Wake, A., The Natural Convection in Pure Water Near the Density Maximum, *AIChE Symposium Series*, Heat Transfer- Philadelphia, 85, 269: 448-453, 1989.

16. Inaba, H. and Fukuda, T., Natural Convection in an Inclined Square Cavity in Regions of Density Inversion of Water, *Journal of Fluid Mechanics*, 142, 363-381, 1984.

17. Ivey, G.N. and Hamblin, P.F., Convection Near the Temperature of Maximum Density for High Rayleigh Number, Low Aspect Ratio, Rectangular Cavities, *Journal of Heat Transfer*, 111, 100-104, 1989.

18. Lankford, K.E. and Bejan, A., Natural Convection in a Vertical Enclosure Filled with Water Near 4° C, *Journal of Heat Transfer*, 108, 755-763, 1986.

19. Watson, A., The Effect of the Inversion Temperature on the Convection of Water in an Enclosed Rectangular Cavity, *Quarterly Journal of Mechanics and Applied Mathematics*, 25, 423-446, 1972.

20. Hung Nguyen, T., Vasseur, P. and Robillard, L., Natural Convection of Cold Water in Concentric Annular Spaces: Effect of the Maximum Density, *Proceedings, 7th International Heat Transfer Conference*, 2, 251-256, 1982.

21. Robillard, L. and Vasseur, P., Transient Natural Convection Heat Transfer of Water with Maximum Density and Supercooling, *Journal of Heat Transfer*, 103, 528-534, 1981.

22. Vasseur, P. and Robillard, L., Transient Natural Convection Heat Transfer in a Mass of Water Cooled through 4° C, *International Journal of Heat and Mass Transfer*, 23, 1195 -1205, 1980.

23. Oosthuizen, P. H. and Paul, J. T., Unsteady Free Convective Flow in an Enclosure Containing Water Near its Density Maximum, To be Presented at the 1990 AIAA/ASME Thermophysics and Heat Transfer Conference. Heat Transfer Conference, San Francisco, California, 1527-1532, 1986.

24. Blake, K.R., Bejan, A. and Poulikakos, D., Natural Convection Near 4° C in Water Saturated Porous Layer Heated From Below, *International Journal of Heat and Mass Transfer*, 27, 2355-2364, 1984.

25. Philip, J. R., Free Convection in Porous Cavities Near the Temperature of Maximum Density, *PhysicoChemical Hydrodynamics*, 10, 283-294, 1988.

26. Oosthuizen, P.H. and Paul, J.T., Heat Transfer through a Square Container Filled with a Liquid and a Gas," *ASME Paper 83-WA/HT-101*, 1982.

27. Ivey, G.N., Experiments on Convection Near the Temperature of Maximum Density and Their Application, *Proceedings, 8th Australasian Fluid Mechanics Conference*, 1, 2A.4-2A.6., 1983.

Transient Natural Convection Boundary Layer Induced by Radiant Heating

X. Li(*), P. Durbetaki(**)

The George W. Woodruff School of Mechanical Engineering, Georgia Institute of Technology, Atlanta, Georgia, 30332-04540, USA

ABSTRACT

Transient temperature and velocity distributions for the solid and the gaseous boundary layers were obtained for the vertical surface of a semi-infinite solid suddenly exposed to a uniform radiant flux. The numerical scheme described in this paper treated the solid and the gas phases as a single computational domain, thus eliminating the need for matching the interface. This single-domain approach was made possible by applying different material properties for the two physical domains, and by properly treating the discontinuity in properties at the interface. The governing equations for the two phases were discretized by a finite control volume method, which assured that appropriate conservation principles were satisfied regardless of the grid size. The computational domain in the gas phase had an adaptive grid to ensure numerical accuracy with minimal nodes, and to take advantages of the parabolic nature of boundary layer equations. The results showed that such a conjugate scheme was capable of handling problems involving more than one medium, and it could also be employed to deal with more complex situations including chemical reactions and gas phase absorption of the incoming radiation.

NOMENCLATURE

a	normalized thermal diffusivity; discretization coefficient
b	source term in discretization equations
c_p, c_s	specific heats of gas and solid
D	binary diffusion coefficient
S	source term
t	time
u, v	x- and y-components of velocity
x, ξ	coordinate variable along the surface
y, η	coordinate variable normal to the surface
ϕ	generic dependent variable
ψ	stream function
θ	dimensionless temperature
ρ, μ, λ	density, dynamic viscosity, thermal conductivity

(*) Graduate Research Assistant
(**) Professor

$[a, b]$ to take the larger member of a and b

Subscripts

E, I external and internal boundary layer edges
S, N, P, U south, north, present, upstream nodes
s solid phase

Superscripts

$*$ normalized variables
$\circ$ value at previous time step
∞ free stream

1. INTRODUCTION

The vertical surface of a semi-infinite solid is suddenly exposed to a uniform radiant flux. A thermal wave propagates into the solid, a hydrodynamic and a thermal boundary layer develop adjacent to the surface. Such natural convection problems involving thermal radiation must deal with the nonlinear behavior of radiation as well as the inherent hydrodynamic and thermal coupling. In the absence of thermal radiation, analyses of the solid and the gaseous boundary layer are separable, and the results are matched at the interface to form the complete solution of the entire solid-gas natural convection problem [1, 2]. However, in natural convection problems typified by the aforementioned situation, radiant exchange rate between the solid surface and the environmental enclosure is as important as other modes of energy exchange. The nonlinear behavior, together with additional complications introduced by the use of variable properties requires for a more accurate prediction and often a numerical solution is the only choice.

2. THEORETICAL MODEL

Let the origin of the cartesian coordinate system be at the lower border of the irradiated vertical plane. Then, the positive x-direction is measured along the vertical surface. The positive y-direction coincides with the surface normal. Assume that air is an ideal and calorific gas, transparent to the incoming radiation, and that pressure distribution is hydrostatic. Assume that heat flow in the solid occurs only in the direction parallel to the surface normal. Then, the conservations of mass, momentum and energy in the boundary layer result in the following equations

$$\frac{\partial \rho}{\partial t} + \frac{\partial (\rho u)}{\partial x} + \frac{\partial (\rho v)}{\partial y} = 0 \tag{1}$$

$$\rho \frac{\partial u}{\partial t} + \rho u \frac{\partial u}{\partial x} + \rho v \frac{\partial u}{\partial y} = \frac{\partial}{\partial y} \left(\mu \frac{\partial u}{\partial y} \right) + \rho g (T/T_\infty - 1) \tag{2}$$

$$\rho \frac{\partial T}{\partial t} + \rho u \frac{\partial T}{\partial x} + \rho v \frac{\partial T}{\partial y} = \frac{\partial}{\partial y} \left(\frac{\lambda}{c_p} \frac{\partial T}{\partial y} \right) \tag{3}$$

Conservation of energy inside the solid would result in

$$\frac{\partial T_s}{\partial t} = \frac{\partial}{\partial y} \left(\frac{\lambda_s}{\rho_s c_s} \frac{\partial T_s}{\partial y} \right) \tag{4}$$

Initially, the air adjacent to the solid is quiescent at 300 K and in thermal equilibrium with the environment

$$t = 0^+ : \qquad u = 0, \quad v = 0, \quad T = T_\infty, \quad T_s = T_\infty \tag{5}$$

These conditions must also be satisfied at $x = 0$ all the time. At the interface, a no-slip condition and thermal continuity are preserved

$$t \geq 0 \text{ and } y = 0 : \qquad u = 0, \quad v = 0, \quad T(0) = T_s(0) \equiv T_{sw}$$

$$-\lambda_s \frac{\partial T_s}{\partial y}\Big)_0 = -\lambda \frac{\partial T}{\partial y}\Big)_0 + \epsilon \sigma T_{sw}^4 - (1 - r)(I_o + \sigma T_{sw}^4) \tag{6}$$

At the free stream, velocity and temperature values approach those of the undisturbed environment

$$t \geq 0 \text{ and } y \longrightarrow \infty : \qquad u = 0, \quad T = T_\infty \tag{7}$$

Similarly, deep inside the solid, the temperature remains unchanged

$$t \geq 0 \text{ and } y \longrightarrow -\infty : \qquad T_s = T_\infty \tag{8}$$

The height of the irradiated surface L is selected as the length scale, and normalized variables are defined

$$\xi = x/L, \qquad \eta = y/L, \qquad t^* = \alpha_s t/L^2 \tag{9a}$$

$$u^* = uL/\nu_\infty, \qquad v^* = vL/\nu_\infty, \qquad \theta = T/T_\infty, \qquad \theta_s = T_s/T_\infty \tag{9b}$$

The following dimensionless parameters are also introduced

$$I_o^* = I_o/\sigma T_\infty^4, \quad Gr = gL^3/\nu_\infty^2, \quad Pr_s = \nu_\infty/\alpha_s, \quad \Lambda = \frac{\sigma T_\infty^3 L}{\mu_\infty c_p} \tag{10a}$$

$$\rho^* = \rho/\rho_\infty = 1/\theta, \quad \mu^* = \mu/\mu_\infty, \quad a^* = \lambda/c_p\mu_\infty, \quad a_s^* = \lambda_s/\mu_\infty c_p \tag{10b}$$

With these definitions, the governing equations are

$$\frac{1}{Pr_s} \frac{\partial \rho^*}{\partial t^*} + \frac{\partial(\rho^* u^*)}{\partial \xi} + \frac{\partial(\rho^* v^*)}{\partial \eta} = 0 \tag{11}$$

$$\frac{1}{Pr_s} \rho^* \frac{\partial u^*}{\partial t^*} + \rho^* u^* \frac{\partial u^*}{\partial \xi} + \rho^* v^* \frac{\partial u^*}{\partial \eta} = \frac{\partial}{\partial \eta}\left(\mu^* \frac{\partial u^*}{\partial \eta}\right) + Gr(\theta - 1)\rho^* \tag{12}$$

$$\frac{1}{Pr_s} \rho^* \frac{\partial \theta}{\partial t^*} + \rho^* u^* \frac{\partial \theta}{\partial \xi} + \rho^* v^* \frac{\partial \theta}{\partial \eta} = \frac{\partial}{\partial \eta}\left(a^* \frac{\partial \theta}{\partial \eta}\right) \tag{13}$$

$$\frac{\partial \theta_s}{\partial t^*} = \frac{\partial}{\partial \eta}\left(\frac{\partial \theta_s}{\partial \eta}\right) \tag{14}$$

The normalized auxiliary conditions are

$$t = 0^- : \quad u^* = 0, \quad v^* = 0, \quad \theta = 1, \quad \theta_s = 1$$

$$t \geq 0 \text{ and } \xi = 0 : \quad u^* = 0, \quad v^* = 0, \quad \theta = 1$$

$$t \geq 0 \text{ and } \eta = 0 : \quad u^* = 0, \quad v^* = 0$$

$$\theta(0) = \theta_s(0) \equiv \theta_{sw}$$

$$-a_s^* \left.\frac{\partial \theta_s}{\partial \eta}\right)_0 = -a^* \left.\frac{\partial \theta}{\partial \eta}\right)_0 + \Lambda \epsilon \theta_{sw}^4 - (1-r)(I_o^* + 1)\Lambda$$

$$t \geq 0 \text{ and } \eta \longrightarrow \infty : \quad u^* = 0, \quad \theta = 1$$

$$t \geq 0 \text{ and } \eta \longrightarrow -\infty : \quad \theta_s = 1$$

$$(15)$$

The parameter Pr_s is actually the ratio of the solid response time constant to that of the gas phase. For the system under consideration, air in the boundary layer responds much faster than the solid, leaving the solid response as the controling factor. In other words, any disturbance caused by changes in the solid would be equilibrated in the gas phase so quickly that changes in the gas phase are essentially instantaneous. Consequently, the time derivative term in all gas phase equations could be dropped. However, numerical experience shows that retaining the time derivative in all equations except the continuity improves the rate of convergence. This topic shall be discussed later in more detail.

For the numerical computation, polymethyl methacrylate (PMMA) was selected as the solid. PMMA is widely used in ignition studies due to its unique monomer-forming decomposition behavior. In the present study however, PMMA is assumed to be inert, avoiding difficulties associated with composition changes and chemical reactions. Thus the properties of interest are [3, 4]: density (ρ_s) 1200 kg/m^3, thermal conductivity (λ_s) 0.1875 W/m·K, specific heat (c_s) 1464 J/kg·K, surface reflectivity (r) 0.05, and surface emissivity (ϵ) 0.94. For air, the Lewis number is assumed to be unity, thus $a^* = \rho^* D^* = D^*/\theta$. Relevant air property data were found in reference [1] and polynomial expressions were fitted to facilitate computer implementation. For the dimensionless viscosity (μ^*) and binary diffusivity (D^*) the following were used in the computations

$$\mu^* = 1.00 + 2.91\beta - 1.71\beta^2 + 0.73\beta^3 \tag{16}$$

$$D^* = 1.39 + 11.07\beta + 8.30\beta^2 \tag{17}$$

where $\beta = (\theta - 1)/4$.

3. NUMERICAL PROCEDURE

Equations (11)-(14) are coupled and nonlinear, and together with auxiliary conditions (15) form a two-point boundary value problem. In addition, this set is transient and two-phased. The similarity based shooting method with a continuation technique [5, 6], used for related problems, proved efficient only if the number of equations was small, in addition to having a similarly transformable set of equations and auxiliary conditions. The present problem is not a similar set. A new approach had to be developed.

3.1 General discretization procedures

Patankar and Spalding [7, 8, 9] developed a direct solution technique which applies the conservation principles at every node point. This finite control volume approach is different from the usual finite difference method. It is more like a finite element method. Application of this method to specific problems require a detailed analysis with clear physical implications, and the understanding that the resulting discretization equations must produce physically meaningful solutions regardless of the size of the control volume. The underlying structure of all equations describing thermal science phenomena is recognized to be

$$\frac{\partial}{\partial t}(\rho\phi) + \frac{\partial}{\partial x_j}(\rho u_j\phi) = \frac{\partial}{\partial x_j}\left(\Gamma\frac{\partial\phi}{\partial x_j}\right) + S \qquad (18)$$

where ϕ is a generic dependent variable, Γ is a generic diffusion coefficient, S is the source term, which can also include anything unaccountable by the other terms. ϕ, Γ and S together uniquely specify one equation.

A solution technique for steady state two-dimensional parabolic problems is described in references [7, 8]. The gas phase of the present problem is in fact parabolic in the spatial coordinate. In the discussions to follow, the time derivative in Eq. (11)-(13) is dropped unless specified otherwise. A new coordinate variable ω, in terms of the stream function ψ, is introduced to replace y so that constant ω lines fit the shape of the boundary layer

$$\omega \equiv \frac{\psi - \psi_I}{\psi_E - \psi_I} \qquad (19)$$

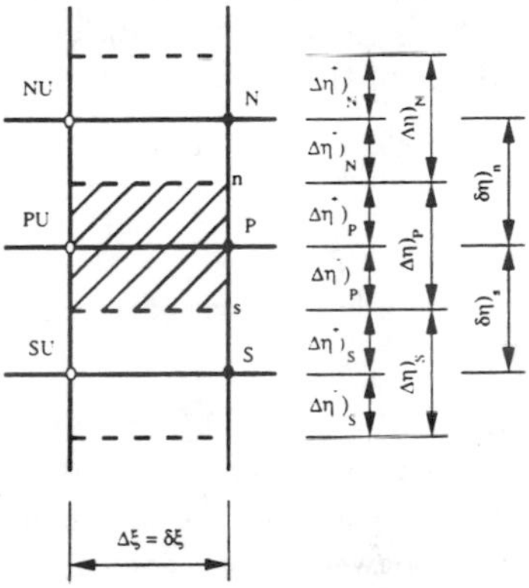

Figure 1. Control Volumes and Nodes

Various distances and the relationship between a node point and a control volume are presented in Figure 1. The general form of the discretization equations is

$$a_P\phi_P = a_N\phi_N + a_S\phi_S + b \qquad (20)$$

The coefficients are derived as

$$D_n = \frac{\Gamma_n\Delta\xi}{(\delta\eta)_n} = \Delta\xi\bigg/\left[\frac{(\Delta\eta^+)_P}{\Gamma_P} + \frac{(\Delta\eta^-)_N}{\Gamma_N}\right]$$

$$D_s = \frac{\Gamma_s\Delta\xi}{(\delta\eta)_s} = \Delta\xi\bigg/\left[\frac{(\Delta\eta^+)_S}{\Gamma_S} + \frac{(\Delta\eta^-)_P}{\Gamma_P}\right]$$

$$F_n = \dot{m}''_n\Delta\xi, \qquad P_n = F_n/D_n$$

$$F_s = \dot{m}''_s\Delta\xi, \qquad P_s = F_s/D_s$$

$$A(|P|) = \left[0, (1 - 0.1|P|)^5\right] \qquad (21)$$

$$a_N = D_n A(|P_n|) + \left[-F_n, 0\right]$$

$$a_S = D_s A(|P_s|) + \left[F_s, 0\right]$$

$$b = S_C(\Delta\eta)_P\Delta\xi + a_{PU}\phi_{PU}$$

$$a_P = a_N + a_S + a_{PU} - S_P(\Delta\eta)_P\Delta\xi$$

$$a_{PU} = \psi_{EI,U}(\Delta\omega)_P$$

where D is the diffusion conductance obtained by taking a harmonic mean of the diffusivities at the neighboring node points, F is the mass flow rate through the control volume face or the convection strength, P is the control volume Peclet number, S_C and S_P are the coefficients of the source term linearization ($S = S_C + S_P \phi$), and $\dot{m}_n''$ and $\dot{m}_s''$ are the mass flux through the north and the south faces of the control volume, respectively

$$\dot{m}_n'' = (1 - \omega_n)\dot{m}_I'' + \omega_n \dot{m}_E''$$
$$\dot{m}_s'' = (1 - \omega_s)\dot{m}_I'' + \omega_s \dot{m}_E'' \tag{22}$$

$\dot{m}_I''$ and $\dot{m}_E''$ are the mass flux values at the internal and the external boundaries respectively. The relationship between ω and η is determined by combining the definitions of the stream function ψ and ω itself

$$\eta = \psi_{EI} \int_0^\omega \frac{1}{\rho^* u^*} d\omega \tag{23}$$

Further details of the general discretization equations may be found in references [7, 8].

3.2 Special treatment of the energy equation

There are two energy equations, one for the gas phase, the other for the solid. Since both equations are statements of the energy conservation principle, they could be viewed as one equation with two solution domains. In other words, the two energy equations form a conjugate problem. To avoid the difficulties of matching solutions at the interface, the two energy equations must be solved simultaneously. The discretization equations for the interior node points in the gas phase for the energy equation are those derived in section 3.1. In this section, the discretization equations for the interior nodes in the solid are derived.

The solid phase energy equation could be rewritten as

$$\rho^* u^* \frac{\partial \theta_s}{\partial \xi} + \rho^* v^* \frac{\partial \theta_s}{\partial \eta} = \frac{\partial}{\partial \eta}\left(\frac{\partial \theta_s}{\partial \eta}\right) - \frac{\partial \theta_s}{\partial t^*} \tag{24}$$

where u^* and v^* are zero inside the solid. Thus the problem of unsteady conduction without a source term is treated as a steady state conduction problem with a source term

$$S = -\frac{\partial \theta_s}{\partial t^*} \approx -\frac{\theta_P - \theta_{PU}}{\Delta t^*} \tag{25}$$

Therefore,

$$S_C = \frac{\theta_{PU}}{\Delta t^*}, \qquad S_P = -\frac{1}{\Delta t^*} \tag{26}$$

This is in fact a central difference scheme in time. If S_C is set to zero, the scheme becomes fully implicit. Finally with $\Gamma = 1$ in the solid, the discretization equations are

$$a_P \theta_P = a_N \theta_N + a_S \theta_S + b \tag{27}$$

where

$$a_N = \frac{1}{(\delta\eta)_n} = 1/[(\Delta\eta^-)_N + (\Delta\eta^+)_P]$$

$$a_S = \frac{1}{(\delta\eta)_s} = 1/[(\Delta\eta^-)_P + (\Delta\eta^+)_s]$$

$$a_P = a_N + a_S + \frac{1}{\Delta t^*}(\Delta\eta)_P \tag{28}$$

$$b = \frac{\theta_{PU}}{\Delta t^*}(\Delta\eta)_P$$

At this point, the questions that remain unanswered are related to the treatment of the interface node points. The nodes and control volumes are depicted in Figure 2. The vertical interface is shown as the shaded area. By combining the two phases, the interface becomes a control volume surface. However, there are radiation exchanges at the interface, making it somewhat different from the other nodes.

Point 2 in Figure 2 is the first node in the gas side of the interface. The heat flux through the north surface of the control volume for this node, expressed in terms of the nodal temperatures, is

$$q_n = -a_n^* \frac{\theta_3 - \theta_2}{(\delta\eta)_n} \tag{29}$$

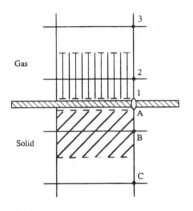

Figure 2. The Interface

The heat flux through the south surface is established from the energy balance on the interface and included in the auxiliary conditions, Equation (15). Then

$$-a_s^* \frac{\partial\theta_s}{\partial\eta}\Big)_0 = -a^* \frac{\partial\theta}{\partial\eta}\Big)_0 + \Lambda\epsilon\theta_{sw}^4 - (1 - r)(I_o^* + 1)\Lambda$$

$$\equiv -a^* \frac{\partial\theta}{\partial\eta}\Big)_0 + f(\theta) \tag{30}$$

where f_0 is the lumped source term at the interface. Explicitly

$$\begin{bmatrix} \text{solid} \\ \text{conduction} \end{bmatrix} = \begin{bmatrix} \text{gas} \\ \text{convection} \end{bmatrix} + \begin{bmatrix} \text{other} \\ \text{sources} \end{bmatrix} \tag{31}$$

The left-hand-side of Equation (30), expressed in terms of the nodal temperatures, is

$$q_0 = -a_0^* \frac{\theta_2 - \theta_B}{(\delta\eta)_0} \tag{32}$$

Let $\theta_{sw} = (\theta_2 + \theta_B)/2$ and evaluate θ_{sw} at the most recently available values, then f_0 could be determined and treated as constant in the current time step. Thus, the energy balance for the control volume surrounding node 2 becomes

$$q_0 - q_n + S(\Delta\eta)_2 - \dot{m}_n''\theta_n + \dot{m}_0''\theta_0 = 0 \tag{33}$$

Let $J_0 = q_{0,cv} + \dot{m}_0'' \theta_0$, and $J_n = q_n + \dot{m}_n'' \theta_n$, where $q_{0,cv}$ is the convection part of q_0, and the energy balance could be rewritten as

$$J_n - J_0 = \left(S_C + S_P \theta_2\right)(\triangle \eta)_2 \tag{34}$$

Equation (34) is in an identical form to energy balance equations for all the general nodes inside the boundary layer.

The first node point in the solid side of the interface is node B. The energy balance for the corresponding control volume is

$$q_s - q_0' + S(\triangle \eta)_B = 0 \tag{35}$$

Because of the normalization, $\Gamma = 1$ in solid, thus

$$
\begin{aligned}
q_s &= -\frac{1}{(\delta \eta)_s}(\theta_B - \theta_C) \\
q_0' &= -\frac{1}{(\delta \eta)_0}(\theta_2 - \theta_B) \equiv \frac{q_0}{a_s^*} = \frac{1}{a_s^*}(q_{0,cv} + f_0)
\end{aligned}
\tag{36}
$$

therefore,

$$q_s - \frac{1}{a_s^*} q_{0,cv} - \frac{f_0}{a_s^*} + (S_C + S_P \theta_B)(\triangle \eta)_B = 0 \tag{37}$$

Equation (37) is rearranged, then

$$a_B \theta_B = a_2 \theta_2 + a_C \theta_C + d \tag{38}$$

where

$$
\begin{aligned}
a_2 &= \frac{1}{a_s^*} / \left[\frac{(\triangle \eta^-)_2}{a^*} + \frac{(\triangle \eta^+)_B}{1}\right] \\
a_C &= \frac{1}{(\delta \eta)_s} = 1/\left[(\triangle \eta^-)_B + (\triangle \eta^+)_C\right] \\
a_B &= a_2 + a_C - S_P(\triangle \eta)_B \\
d &= -\frac{f_0}{a_s^*} + S_C(\triangle \eta)_B
\end{aligned}
\tag{39}
$$

S_C and S_P for the solid are given in Eq. (26), and a_s^* in Eq. (10b).

3.3 Treatment of the transient behavior

The derived discretization equations are in the steady state form both in the boundary layer and in the solid. However, for each established interface temperature the one-dimensional transient conduction problem must be solved at every ξ location. Because of the assumed one-dimensional nature, the value of any node point inside the solid is affected only by its past value and its immediate neighbors along the direction of the surface normal. On the other hand, the node points in the gas phase have only spatial influence. As a result, numerical instability develops at

Table 1. Specific Values for ϕ, Γ, S

ϕ	Γ	S
u^*	μ^*	$Gr(\theta - 1)\rho^* - \frac{\rho^*}{Pr_s}\frac{\partial u^*}{\partial t^*}$
θ	$a^* = \rho^* D^*$	$-\frac{\rho^*}{Pr_s}\frac{\partial \theta}{\partial t^*}$
θ_s	1	$-\partial \theta_s / \partial t^*$

the interface. To account for the transient behavior in the gas phase, the time derivatives are re-introduced, although their physical importance has proven to be secondary. Similar to the solid phase energy equation, the time derivatives are treated as additional source terms in the gas phase equations. The identifications of Φ, Γ and S are described in Table 1.

As a result of this modification, the expressions of b and a_P in the gas phase discretization equations became

$$
\begin{aligned}
b &= S_C (\triangle\eta)_P \triangle\xi + a_{PU}\phi_{PU} + a_P^\circ \phi_P^\circ \\
a_P &= a_N + a_S + a_{PU} - S_P (\triangle\eta)_P \triangle\xi + a_P^\circ \\
a_P^\circ &= \frac{\rho_P^\circ (\triangle\eta)_P \triangle\xi}{Pr_s \triangle t^*}
\end{aligned}
\tag{40}
$$

Numerically, the modified discretization equations are very stable, but extremely slow to converge. Timewise acceleration or overrelaxation must be applied. The inertia relaxation scheme [7] changes the nodal equation to

$$
(a_P + i)\phi_p = a_S \phi_S + a_N \phi_N + b + i\phi_P^\circ
\tag{41}
$$

Let the inertia $i = \beta a_P^\circ$, then

$$
\begin{aligned}
b &= S_C (\triangle\eta)_P \triangle\xi + a_{PU}\phi_{PU} + (1+\beta)a_P^\circ \phi_P^\circ \\
a_P &= a_N + a_S + a_{PU} - S_P (\triangle\eta)_P \triangle\xi + (1+\beta)a_P^\circ
\end{aligned}
\tag{42}
$$

where β is the relaxation factor implemented in the computer. When $\beta > 0$, equations are underrelaxed; when $\beta < 0$, overrelaxation results. $\beta = -0.9$ is a typical situation in the present study.

4. RESULTS AND DISCUSSION

The classical Blasius problem was solved by the present technique. Preliminary computations were used to establish the parameters needed for error control, entrainment evaluation and grid allocation. For 20 nodes in the gas phase, calculated velocity distribution agreed with benchmark data [1] to within 0.8%. Further computations were made for the steady state natural convection boundary layer adjacent to a vertical hot surface. The calculated heat transfer results were compared with Ede's formula [1]. For 20 nodes, the results agreed within 1.0%.

While comparisons of computed results for the simpler problems with available data established the validity of the numerical technique, further exploratory computations of the transient conjugate problem were required to determine the number of node points necessary to control the error due to control volume sizes, and values of the relaxation factor. It was found that indeed the value of the relaxation factor affected the rate of convergence only, not the value of the converged solutions. For the present computations, $\beta = -0.98$ was found to be satisfactory. For the results presented in this paper, 40 nodes in the gas phase and 20 nodes in the solid were found to be adequate. When the number of nodes in the gas phase was increased from 30 to 40, the maximum change in any of the calculated parameters was of the order of 10^{-4}. A time step size of 0.1 second, which corresponded to a dimensionless size 4.74×10^{-7}, was used. It was found that below this level, roundoff error offsets the benefits of a smaller step size.

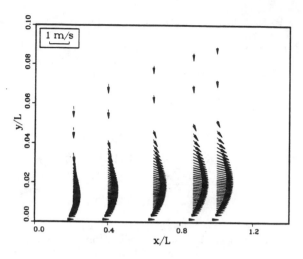

Figure 3. Velocity Vectors at t=5 sec.

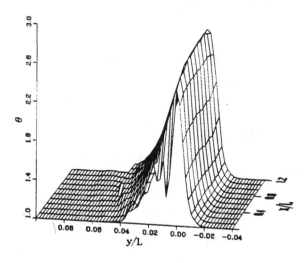

Figure 4. x−y−θ Surface at t=5 sec.

Shown in Figure 3 is the boundary layer velocity vector at 5 seconds after the onset of exposure. The horizontal axis is actually along the vertical surface, and the vertical axis shows the boundary layer thickness. The boundary layer grows due to entrainment of free stream air. Near the surface, air movement is hindered by the dominant viscous force. Somewhere in the middle of the boundary layer, a maximum velocity occurs, which is accelerated due to the buoyancy force. In Figure 4, the dimensionless temperature $(T/300)$ at $t = 5$ sec. was presented. The solid surface is located at $y/L = 0$ where the temperature achieved its maximum. At the right hand side, temperature drops sharply in the solid due to the large thermal inertia in the solid. Temperature decreases to the free stream value more gradually in the gas phase. Note that the sharp spike downward in the middle is merely a result of the interpolation techniques used to produce this plot. The surface temperature varied only slightly along the x-direction, as could be seen in Figure 5. While a thicker boundary layer might be expected to yield a lower surface temperature, the radiation transparency of air supports the present results.

Figure 6 compares the surface temperatures at $\xi = 1.0$ for the models with and without surface reradiation. Initially the differences are small. But as time progresses, the heat loss due to surface reradiation becomes more and more important because the surface temperature increases. This figure indicates that while neglecting the reradiation could linearize the problem, the results of such a linear analysis should be interpreted with caution.

5. CONCLUSION

A finite control volume technique for solving transient conjugate problems has been described. This technique has its origin in the two-dimensional steady state boundary layer technique known as the Patankar-Spalding method [7–9]. The present scheme takes a direct solution approach, produces physically meaningful results regardless of the control volume size. It is capable of handling variable properties, multiple solution domains, and transient behavior. The physical problem studied in the present paper involved natural convection induced by radiant heating of an inert solid. The transition of this solid-gas assembly from the quiescent initial state at the onset of exposure to a state of a well-defined yet still developing boundary layer was described. This finite control volume technique is currently being used to study the radiant ignition process of solid fuels. Included in the analysis are the interface and gas phase chemical activities, and gas radiation.

REFERENCES

1. Kays, W.M. and Crawford, M.E. *Convective Heat and Mass Transfer,* (2nd ed.) McGraw-Hill, New York, 1980.

2. Bejan, A. *Convection Heat Transfer,* John Wiley and Sons, New York, 1984.

3. Rohm & Hass Co. *Plexiglas Design and Fabrication Data,* PL-1p, 1983.

4. Kashiwagi, T. *Fire Safety Journal,* Vol. 3, pp. 185-200, 1981.

5. Phuoc, T.X. Ignition of Polymeric Materials under Radiative and Convective Exposure, *Ph.D. Thesis,* School of Mechanical Engineering, Georgia Institute of Technology, Atlanta, 1985.

6. Li, X. and Durbetaki, P. *Numerical Methods in Thermal Problems,* Vol. VI, Ed: Lewis, R.W. and Morgan, K., Pineridge Press Limited, Swansea, U.K., pp. 1247-57, 1989.

7. Patankar, S.V. *Numerical Heat Transfer and Fluid Flow,* Hemisphere Publishing Corporation, Washington, D.C., 1980.

8. Patankar, S.V. *Handbook of Numerical Heat Transfer,* Ed:Minkowycz, W.J., Sparrow, E.M., Schneider, G.E. and Pletcher, R.H., John Wiley and Sons, New York, pp. 89–115, 1988.

9. Spalding, D.B. *GENMIX: A General Computer Program for Two-Dimensional Parabolic Phenomena,* Pergamon Press, Oxford, 1977.

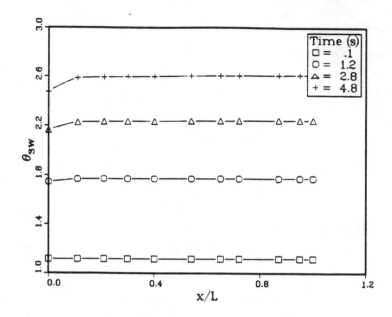

Figure 5. Surface Temperature

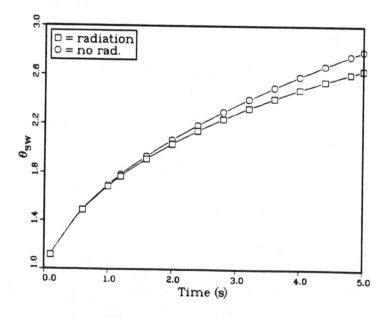

Figure 6. Radiation Effect (θ_{sw} @ x=L)

The Development of a Vectorized Computer Code for Solving Three-Dimensional, Transient Heat Convection Problems

C.H. Chuan, W.C. Schreiber
Mechanical Engineering Department, The University of Alabama at Tuscaloosa, USA

ABSTRACT

The present article describes an ongoing effort to develop an efficient computational procedure for solving transient, three-dimensional, coupled fluid flow and heat transfer with a supercomputer. Two different algorithms, SIMPLER and PISO, have been vectorized for optimal performance on a supercomputer's pipelined architecture. The methods used for the purpose of vectorizing the algorithms are described, and the two vectorized versions are considered for their efficacy in solving both recirculating forced convection and internal natural convection.

SUMMARY

A computer program has been developed to solve the transient, three-dimensional Navier-Stokes and energy equations in primitive variables for simulating either forced or natural convection heat transfer. The finite volume method has been used to derive, on a staggered grid, the finite difference equations which relate the velocity, pressure, and temperature unknowns in these equations.

In an effort to obtain the most efficient code for optimum performance on a supercompter, the authors have investigated two different strategies to solve the coupled pressure, velocity, and temperature fields. In consideration of the supercomputer's pipelined architecture, these two techniques were chosen partly for their ability to be vectorized. The first of the techniques uses the Semi-Implicit Method for Pressure-Linked Equations (SIMPLER) method to generate systems of algebraic equations in which the pressure, velocity, or temperature are implicitly defined in finite difference forms of the Poisson equation. An iterative procedure known as the point-Jacobi method is engaged to solve the Poisson equations. The point-Jacobi technique is combined with dynamic storage allocation to render the field variables of the computational domain into one long vector whose scalar components can be solved iteratively and non-recursively. This vectorized code has been programmed to provide efficient performance on the CRAY XMP/2.

More recently, a second solution algorithm, Pressure-Implicit with Splitting of Operators (PISO), has been vectorized to calculate the solution to transient, convective heat transfer problems much more rapidly than the first method. While the point-Jacobi method combined with dynamic storage allocation can be optimally vectorized, its convergence is slowed by the numerous iterations the SIMPLER method requires to calculate the pressure field at each time step. The PISO method, on the other hand, is a non-iterative technique for uncoupling the pressure and velocity found in the implicitly discretized fluid flow equations. PISO updates the pressure, velocity, and temperature fields over a time step using the sequence of an initial prediction to estimate an approximate velocity field, two correction steps to improve the prediction of the pressure and velocity fields, and solution of the energy equation. The Alternating Direction Implicit (ADI) method has been implemented to solve the velocity and temperature fields while a Gauss-Seidel Semi-Iterative method (GS-SI) has been used with the pressure correction steps. Both of these methods are fully vectorizable for maximum exploitation of supercomputer parallel architecture for gains in speed-up; at the same time, these methods minimize the need for iteration between time steps. The potential for vectorization combined with the minimization of iteration results in an algorithm which is optimized for swift and accurate solution of transient, three-dimensional convective heat transfer.

The present study examines the efficiencies of vectorized versus scalar routines for both the SIMPLER and PISO routines. Vectorization is found to enhance greatly performance on the supercomputer architecture for either of the routines. With the SIMPLER method, the use of the point-Jacobi method allows more than 90% of the code to be vectorized to yield a solution rate that is faster by a factor of ten over a scalar Gauss-Seidel iteration; however, the PISO routine, incorporating vectorized versions of the ADI method and the GS-SI method, far surpasses SIMPLER for expedient simulation of transient, three-dimensional problems in either natural or forced convection.

NUMERICAL METHOD

The notation used here will be the same as that found in reference 1 which provides a working description of the control volume method as applied to the SIMPLER algorithm. References 2 and 3 describe in detail the theory and application of the PISO method. In reference 4, both of these methods are clearly summarized and compared for their respective abilities to solve steady state heat convection problems using a non-vectorized algorithm. The SIMPLER and PISO methods are outlined here for completeness and in order to specify how vectorization has been effected. In an effort to simplify the discussion of vectorization, only the two-dimensional problem is considered; extension to the three-dimensional solvers, which were coded for this paper's results, is straightforward.

The convective transport equation for momentum can be expressed as:

$$\partial (\rho \phi)/\partial t + \nabla \cdot (\rho U \phi) = -\nabla p + \nabla \cdot (\Gamma \nabla \phi) + F_B$$

(1)

where the variable, ϕ, is replaced by one of the components of velocity, u, v, or w; Γ by the dynamic viscosity, μ, and the body force term, F_B, by the Boussinesq expression for buoyancy. Using the finite volume method, the momentum equation is discretized with respect to a staggered grid as pictured in figure 1. Integrating about the control volume for u_e results in the discretization equation which describes momentum in the x direction; the equation for v_n can be similarly derived:

$$(a_e + \rho\Delta V/\Delta t)\, u_e = \Sigma\, a_{nb}\, u_{nb} + A_e\,(p_P - p_E) + q\Delta V/\Delta t\, u_e{}^o \tag{2}$$

$$(a_n + \rho\Delta V/\Delta t)\, v_n = \Sigma\, a_{nb}\, v_{nb} + A_n\,(p_P - p_N) + q\Delta V/\Delta t\, v_n{}^o + S_c\Delta V \tag{3}$$

where a_e and a_n are the coefficient for the velocities, u_e and v_n, respectively, at the control volume's center, ΔV is the volume of the control volume, S_c is the expression for body force per volume, and a_{nb} stands for a coefficient of any one of the x-direction (or y-direction) velocities, u_{nb} (or v_{nb}), which neighbor u_e (or v_n). As Patankar [1] explains, the determination of these coefficients is made by considering the diffusive and convective transport into the control volume.

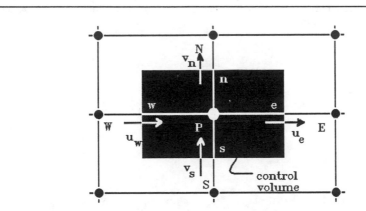

Figure 1.   Illustration of the staggered grid on which the velocity, pressure, and temperature fields are computed.

The solution of the velocity field requires the simultaneous solution of the pressure field to which it is coupled. In compressible flows, the pressure field may be determined using a thermodynamic constituitive equation of state; however, for incompressible flows, the continuity equation must serve as the link between pressure and momentum. In the SIMPLE family of solvers as in the PISO method, corrections to the velocity and pressure are defined as the difference between the actual value to be calculated minus an approximate, known, value and may be expressed as:

$$u_e{}' = u_e - u_e{}^* \tag{4}$$

$$v_n{}' = v_n - v_n{}^* \tag{5}$$

$$p' = p - p^* \tag{6}$$

An explicit relation between the velocity and pressure is established using the discretized momentum equations (equation 2 and 3) and omitting some terms :

$$u_e = u_e^* + d_e (p_P' - p_E')$$

(7)

$$v_n = v_n^* + d_n (p_P' - p_N')$$

(8)

where (for example):

$$d_e = A_e / (a_e + \rho \Delta V / \Delta t)$$

(9)

and A_e is the area of the interface associated with u_e.

These expressions are substituted into the discretized continuity equation to produce a finite differenced Poisson equation in terms of pressure corrections and approximate velocities.

$$a_P p_P' = \Sigma \, a_{NB} \, p_{NB}' + b; \qquad\qquad NB = E, W, N, S$$

(10)

where:

$$a_E = \rho_e A_e d_e$$

(11)

$$a_P = \Sigma \, a_{NB}$$

(12)

$$b = (\rho_P^0 - \rho_P) \Delta V / \Delta t + \rho_w u_w^* A_w - \rho_e u_e^* A_e + \rho_s v_s^* A_s - \rho_n v_n^* A_n$$

(13)

The SIMPLE-type methods update both the pressure and velocity by iterating between equations 4 - 13 until convergence is reached.

For the SIMPLER algorithm, this updating procedure can be described as:

1. Explicitly-determined "psuedo-velocities" are defined.

$$\underline{u}_e = \{ \Sigma \, a_{nb} u_{nb} + b_e \} / (a_e + \rho \Delta V / \Delta t)$$

(14)

$$\underline{v}_n = \{ \Sigma \, a_{nb} v_{nb} + b_n \} / (a_n + \rho \Delta V / \Delta t)$$

(15)

2. When these psuedo-velocities are substituted for u_e^*, u_w^*, v_n^*, and v_s^* in equation 13 and p is substituted for p' in equation 10, the resulting modified equations, 7 - 13 can be used to predict absolute pressure implicitly.

3. Equations 2 and 3 are solved implicitly with the updated pressure found in step 2 to determine a new value of velocities, u_e^* and v_n^*.

4. With u_e^* and v_n^*, equation 10 is solved implicitly for the pressure correction, p'.

5. The velocities in equations 7 and 8 are updated explicitly with the pressure corrections.

6. The procedure is repeated until convergence has been reached for the time step.

If only the steady state solution is desired, the need for accuracy at each intermediate time is eliminated, and the above-described cycle is performed only once for each time step (stage 6 is ignored). Jang [4] and Van Doormaal and Raithby [5] discuss the relation between the steady state and the transient algorithms in detail. Within each cycle for either the transient or the steady state algorithms, iteration is required at stage 3 to solve the velocity field implicitly and at stages 2 and 4 to solve the pressure field implicitly. The procedure is marched forward until convergence is reached for the steady state velocity and pressure fields. With the transient solution, however, the procedure is iterated until convergence at each time step (step 6 is included).

On a computer with a serial architecture, several iterative solution options exist for computing the implicitly-defined variables in the stages 2, 3, and 4; among them, Gauss-Seidel, tridiagonal matrix line-by-line, alternating direction implicit, and successive over-relaxation are probably the best known. Since each of these matrix solution methods require recursion, none of them, unless specially modified to eliminate recursion, can exploit the speed-up of pipelined processing found with parallel architectures.

To maximize the central processing unit's pipelining capabilities, a computer algorithm should be vectorized wherever possible. A vector, in this sense, refers to a set of data, each of which can undergo a series of instructions independent of the remaining data elements in the vector, without requiring recursive operations. As an example, a matrix problem can be solved as a vector using the point-Jacobi method but not with the Gauss-Seidel method. In the either of these methods, the nth variable of the nth equation is expressed explicitly in terms of the other variables of the equation, an initial guessed value is assigned to the set of variables, and the variables are updated iteratively until convergence. The methods differ in the individual iteration procedures each employs. The point-Jacobi method updates all the variables with the values from the previous iteration; while the Gauss-Seidel method updates the variables from the most recent updates available. For scalar processing, the Gauss-Seidel method is about twice as fast as and requires only half the computer storage of the point-Jacobi method [6]. Gauss-Seidel cannot be used for vector processing, however, since in each iteration, the unknown variables are updated with respect to other variables which themselves have been updated in that iteration. With the point-Jacobi method, on the other hand, the unknown variables are updated solely in terms of the known values of the variables from the previous iteration. When processed on a pipelined architecture, the vectorized version of the point-Jacobi method is over ten times as fast as the scalar Gauss-Seidel technique.

As employed with the SIMPLER procedure, the point-Jacobi method is suitable for vectorization since it does not depend on recursion; furthermore, the variables on the boundary as well as those in the interior can be grouped for dynamical storage in the computer's memory. Thus assembled, the variables can all be treated as one long vector

F

and can be broken by the computer into segments of 64 variables each, the maximum length of a vector which can be routed though a pipeline on the CRAY. The vector processing facilities of the computer are thereby used to their fullest capability.

While, in concept, it resembles the SIMPLER method as a means for determining the pressure and velocity fields, the PISO procedure does not require cycle iteration for each time increment to solve the transient convective heat transfer problem. Instead, to arrive directly at the next time step, it computes an initial estimation of the velocity field and then follows this calculation with two correction steps for updating the pressure and velocity fields. The procedure used with the PISO method can be summarized as follows:

1. Using the pressure, p^O, from the previous time step solution, equations 2 and 3 are solved implicitly for the predictor velocities, u_e^* and v_n^*.

2. The psuedo-velocities together with the coefficients for the pressure equation are calculated as in the first stage of the SIMPLER procedure.

3. The predictor pressure field, p^*, is calculated implicitly as described in stage 2 of the SIMPLER method.

4. A pressure correction defined as $p' = p^* - p^O$ and corrector velocities, u_e^{**} and v_n^{**}, are obtained by the applying explicit equations, 7 and 8.

5. New coefficients for the pressure equation and new psuedo-velocities:

$$\underline{u}_e = u_e^* + \{ \Sigma\, a_{nb} (u_{nb}^{**} - u_{nb}^*) \}/a_e - d_e\, (p_P^* - p_E^*) \tag{16}$$

$$\underline{v}_n = v_n^* + \{ \Sigma\, a_{nb} (v_{nb}^{**} - v_{nb}^*) \}/a_n - d_n\, (p_P^* - p_N^*) \tag{17}$$

are calculated.

6. The predictor pressure field, p^{**}, is determined implicitly as described in stage 2 of the SIMPLER method.

7. The corrector velocities, u_e^{***} and v_n^{***}, are obtained by applying the equation:

$$u_e^{***} = \underline{u}_e + d_e\, (p_P^{**} - p_E^{**}) \tag{18}$$

$$v_n^{***} = \underline{v}_n + d_n\, (p_P^{**} - p_N^{**}) \tag{19}$$

In comparison of the two methods, it is interesting to note that stages 2 - 4 of the PISO method are identical to the SIMPLER method; however, instead of iterating stages 2 through 4 to convergence, the PISO method uses a higher order of pressure and velocity correction (stages 5 through 7) to arrive at the final prediction for the new time step without iteration. Issa [2] has shown by error analysis that the degree of error due to the PISO procedure is less than the discretization error of the difference equations; therefore, the seven stages do not require iteration to reach an accurate solution.

Like the SIMPLER method, the PISO method requires one implicit solution of the velocity field equations (stage 1) and two implicit solutions of pressure equation (stages 3 and 6). For this purpose in the present study, separate schemes have been developed to solve the implicit pressure problem and to solve the implicit velocity problem. In stage 1, the velocity field is solved with an ADI (alternating direction implicit) solver which is suitable for vectorization.

The details of the vectorized ADI solver will be presented in a later publication [7]. Here only the concept is presented. The ADI method was developed [8, 9, 10] for solving linear, transient, three-dimensional problems by dividing the time step into three small time steps and solving the field of unknown variables over each small time step line-by-line using the Tridiagonal Matrix Algorithm, TDMA. A different sweep direction is used over each small time step to solve the variables implicitly in terms of the approximate solutions already determined for these variables. For linear problems, this three step procedure yields a stable solution without iteration. For use with stage 1 of the PISO method, the nonlinearity of velocity in the momentum equation requires that the ADI method be used with iteration and that the time step size be restricted for stability.

On a machine with serial processing, the sweep in the x direction may performed using nested DO loops in the following sequence:

DO LOOP FOR THE Y DIRECTION
    DO LOOP FOR THE Z DIRECTION
        SOLVE TDMA FOR SCALAR VARIABLES IN THE X DIRECTION
    CLOSE DO LOOP IN Z DIRECTION
CLOSE DO LOOP IN Y DIRECTION

The preceding algorithm is organized incorrectly for vectorization. Since the TDMA uses recurrence to solve for the variables in the x-direction implicitly, the algorithm can compute the variables in a serial fashion only.

To vectorize the algorithm, the variables in each y-z plane are stored dynamically as a vector and each separate step in the TDMA is modified to process an entire dynamically-stored vector rather than one separate scalar in the x-direction. In order to solve each y-z plane as a dynamically-stored vector, the TDMA must effectively incorporate within itself the DO loops in the y and z directions. With this modification the pipe-line is kept full for a large majority of the time that the sweep in the three different directions is made and maximum vectorization is obtained.

Because the momentum equation's non-linearity requires both iteration and a limited time step for the ADI method, no significant gains in speed would be expected, on the serially-processing machine, with the ADI method over Gauss-Seidel. For example, to solve the two-dimensional, non-linear, heat conduction with freezing problem, Chuan [11] found the processing time the same whether the Gauss-Seidel or the ADI method was used; however, the ADI method did produce a two-fold speed-up for the three-dimensional solution to this problem. Unlike the Gauss-Seidel method, nevertheless, the ADI method can be vectorized as described above, and, compared with the non-vectorizable Gauss-Seidel method, a substantial speed-up is realized on a pipelined architecture.

Solving for the pressure in equation 10 presents a different challenge from solving the momentum equation. While the pressure field to be determined is linear, boundary condition information is limited and prevents the use of the ADI method. Relative, and not absolute, pressure is important to the determination of the velocity; hence, boundary conditions are not explicitly defined for the pressure field. If the ADI method is used for solving equation 10, the lack of apparent boundary conditions yield an indeterminate matrix problem. Point-by-point methods skirt this problem through the iteration required for convergence of the solution; one boundary point is defined and held constant while the other boundary points are allowed to change relative to the fixed pressure point. Since all pressure nodes but one are assigned assumed values at the initiation of each new time step, the effect of the one boundary points value diffuses, as dictated by the governing equation for pressure, with iteration. For solving the pressure field, therefore, Successive Over-Relaxation, SOR, is a particularly effective point-by-point solver. It speeds up the rate at which the boundary point's effect may diffuse into the pressure field to effect a converged solution of the pressure field's relative values in considerably less time than the Gauss-Seidel method.

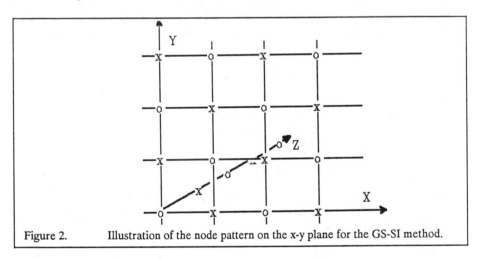

Figure 2.   Illustration of the node pattern on the x-y plane for the GS-SI method.

As a variant of the Gauss-Seidel method, however, the algorithm for SOR is serial and not suitable for vectorization; however, another Gauss-Seidel type technique is available which allows both over-relaxation and vectorization. Young [12] describes the Gauss-Seidel Semi-Iterative method (GS-SI) which was developed by Sheldon [13] and which applies the Gauss-Seidel method in a spatially leapfrog pattern. GS-SI may be over-relaxed to such a degree that it is faster than the SOR even as a serial solver. As an added bonus, the GS-SI is also vectorizable for speed-up increases on the pipe-lined supercomputer architecture. Figure 2 illustrates the x-y plane of a three-dimensional system of nodes illustrated as a pattern of alternate x's and o's. The GS-SI algorithm can calculate the pressure field by cycling between the two different points in the following way:

1. With an initial "guessed" value for all nodes, the values at the o points in terms of the x points are computed using the ordinary Gauss-Seidel method.

2. A relaxation factor is used to over-relax the newest values of all the o points

3. Steps 1 and 2 repeated for the x points.

4. The convergence criterion is checked, and the process is repeated until the solution converges.

For the purpose of vectorization, the x points and the o points could be dynamically stored as two separate vectors. For the present, however, the vectorization of the x and o points is only extended over the x direction. The reason for this partial vectorization will be considered in the results section.

RESULTS

The purpose of the research for this study was the development of a fast algorithm for solving transient, three-dimensional convective heat transfer on a supercomputer. The effort, consequently, was chronologically spent toward this end and not necessarily directed toward comparing different techniques for their efficiency and accuracy. While comparisons between several computational methods can be deduced from the study and will be presented in the present report, they should be interpreted in light of the study's over-riding objective.

Parameters common to both problems

Grid size:	10 x 10 x 10
Convergence tolerance:	10^{-5}

Parameters for internal natural convection

Grashof Number:	$Gr = 20,000$
Prandtl Number:	$Pr = 0.733$
Non-dimensional end time:	$\tau = 0.1$
Non-dimensional time step:	$\Delta\tau = 0.002$

Parameters for driven-cavity flow

Reynolds Number:	$Re = 400$
Prandtl Number:	$Pr = 1.0$
Non-dimensional end time:	$\tau = 30$
Non-dimensional time step:	$\Delta\tau = 0.5$

Table 1. Parameters for the two test cases.

In the beginning of the study, the vectorized point-Jacobi with the SIMPLER algorithm was programmed for and run on the CRAY XMP/2 using only one processor. A speed-up of one order of magnitude was gained over SIMPLER with a serial Gauss-Seidel solver run on the CRAY. With either the vectorized or the serial versions of the SIMPLER algorithm, it was noted that over 60% of the CPU time required to solve either was spent on calculating the pressure field. Vectorized versions of ADI and GS-SI, used in conjunction with SIMPLER, would have probably effected even greater speeds, but it was discovered, in the meantime, that the PISO algorithm with the point-Jacobi was far superior to the SIMPLER technique for its speed in solving transient problems. Consequently, ensuing efforts were directed at improving only the PISO method, particularly with regard to decreasing the time required to solve the pressure field at each time step.

Two convection problems are used to test the methods: the recirculating flow in a square driven cavity and internal natural convection inside a box. A summary of the details of each of these test cases is found in table 1.

Before it was vectorized, the PISO method was developed using algorithms with serial logic to yield the fastest possible algorithm for use in a scalar mode on the CRAY. For this purpose, ADI methods were developed for velocity and temperature and SOR was used to solve the pressure field. The GS-SI method was later found to be particularly efficient in lowering the time required for the pressure field solution and was subsequently substituted for SOR; a comparison between the PISO code using the GS-SI and using the SOR is seen in table 2. Note that the CPU time spent calculating the pressure in PISO is a relatively small fraction of the computation. Once it was felt that an efficient algorithm for use on a serially-processing architecture had been devised, vectorization was implemented. The scalar ADI and GS-SI methods were replaced with the vectorized versions described in the previous section. The comparison between the vectorized and scalar versions is found in table 3; table 4 describes the percentage allocation of CPU time to the different functions performed by the vectorized PISO method.

	Total # of iterations	Total CPU time
SOR	6586	151 sec.
GS-SI	5022	100 sec.
Ratio: GS-SI / SOR	0.76	0.66

Driven Cavity

	Total # of iterations	Total CPU time
SOR	2346	110 sec.
GS-SI	1363	69 sec.
Ratio: GS-SI / SOR	0.58	0.63

Internal Natural Convection

Table 2. Comparison of SOR and GS-SI (vector version) to steady state. Relaxation factor used is 1.6 in all cases.

It is noted that unlike the first algorithm described which combines the point-Jacobi method with SIMPLER, the PISO-based algorithm requires only a small fraction of the total CPU time for the pressure computation. For this reason, dynamic storage and vectorization of the full field of pressure variables at the o and x points has not been implemented since further fractional time savings in the computation of the pressure field would not contribute substantially to overall algorithm speed-up.

	Scalar version	Vectorized version	Ratio
Cavity Flow	346	100	0.29
Natural Convection	248	69	0.28

Table 3. Comparison between CPU times required for the scalar and vectorized versions of the PISO method on the CRAY XMP/2.

	Velocity	Pressure	Temperature	I/O
Cavity Flow	76%	21%	1%	2%
Natural Convection	88%	9%	0.5%	2.5%

Table 4. Relative amounts of CPU time spent for various parts of the whole algorithm

A measure of how well an algorithm is vectorized can be found by comparing its average processing speed to the maximum speed possible for the computer. For an artificially ideal computation, one CPU of the CRAY XMP can process data at the rate of 117 Million Floating Points per Second (MFLOFS). This rate is, of course, not sustainable for algorithms used in realistic problem solving but represents the maximum throughput attainable. In comparison with this ideal speed, the average speed of the vectorized SIMPLER algorithm is 40 MFLOFS while that of the vectorized PISO routine is 35 MFLOFS. The higher average speed of SIMPLER with point-Jacobi can be attributed to its larger degree of vectorization than PISO with ADI and GS-SI. For transient problems, however, the PISO method is much more stable than SIMPLER and can achieve accurate solutions with time steps which are two orders of magnitudes larger than the time steps needed for convergence with SIMPLER.

ACKNOWLEDGEMENTS

The authors wish to acknowledge the support of:

1.   CRAY Research

2.   NSF/EPSCoR

3.   Alabama Supercomputer

REFERENCES

1.   Patankar, S. V.; *Numerical Heat Transfer and Fluid Flow*; Hemisphere Publ. Co.;
       1980.

2.   Issa, R. I.; "Solution of the Implicitly Discretized Fluid Flow Equations by
       Operator-Splitting"; *J. Comp. Phys.*, vol. 62, pp. 40-65; 1985.

3.   Issa, R. I., Gosman, A. D., and Watkins, A. P.; "The Computation of
       Compressible and Incompressible Recirculating Flows by a Non-Iterative
       Scheme"; *J. Comp. Phys.*, vol. 62, pp. 66-82; 1985.

4.   Jang, D. S., Jetli, R., and Acharya, S.; "The Comparison of the PISO, SIMPLER,
       and the SIMPLEC Algorithms for the Treatment of the Pressure-Velocity
       Coupling in Steady Flow Problems"; *Num. Heat Transfer*, vol. 10, pp. 209-228;
       1986.

5.   Van Doormaal, J. P. and Raithby, G. D.; "Enhancements of the SIMPLE
       Method for Predicting Incompressible Fluid Flows"; *Num. Heat Transfer*, vol. 7,
       pp. 147-163; 1984.

6.   Stoer, J. and Bulirsch, R.; *Introduction to Numerical Analysis*; Springer-Verlag
       Publ.; 1976.

7.   Schreiber, W. C. and Chuan, C. H.; "Vectorization of the GS-SI and ADI
       Methods for Solving Convection Heat Transfer Problems on a Supercomputer";
       in preparation.

8.   Peaceman, D. W. and Rachford, H. H.; " The Numerical Solution of Parabolic
       and Elliptic Difference Equations"; *J. Soc. Ind. Appl. Math.*; vol. 3, pp. 28-41;
       1955.

9.   Douglas, J.; "On the Numerical Integration of $\partial^2 u/\partial x^2 + \partial^2 u/\partial y^2 = \partial u/\partial t$ by
       Implicit Methods"; *J. Soc. Ind. Appl. Math.*; vol. 3, pp. 42-65; 1955.

10.   Douglas, J. and Gunn, J. E.; "A General Formulation of Alternating Direction
       Methods - Part I: Parabolic and Hyperbolic Problems"; *Numerische Mathematik*,
       vol. 6, pp. 428-453; 1964.

11.   Chuan, C. H., Schreiber, W. C., and Huang, C. L.; " A Comparison of ADI,
       Line-by-Line and Gauss-Seidel in Conjunction with the Finite Volume Method
       for Solving 3-D Conduction/Freezing Problems"; *Numerical Methods in Thermal
       Problems*; vol. 6, pp. 68-78; 1989.

12.   Young, D. M.; *Iterative Solution of Large Linear Systems*; Academic Press; 1971.

13.   Sheldon, J. W.; "On the Spectral Norms of Several Iterative Processes"; *J. Assoc.
       Comput. Mach.*, vol. 6, pp. 494-505; 1959.

Transition to Chaos in a Differentially-Heated Square Cavity Filled with a Liquid Metal

A.A. Mohamad, R. Viskanta

School of Mechanical Engineering, Purdue University, West Lafayette, IN 47907, USA

ABSTRACT

A numerical investigation is performed to study the transition from steady to chaotic flow of a fluid confined in a two-dimensional square cavity. The cavity has rigid, adiabatic top and bottom connecting walls, while the vertical walls are kept at constant but different temperatures. A fluid having a Prandtl number of 0.005 fills the cavity. The transition scenario is shown by displaying the time series of local variables and streamlines. At sufficiently low Grashof numbers the flow is laminar and steady. As the Grashof number is increased the flow exhibits an oscillatory transient period, but after sufficiently long time the flow becomes steady. A further increase in the Grashof number produces oscillatory flow with a fundamental frequency, then quasi- periodic and chaotic behavior is finally evident. The flow structure is completely different than the published results for ordinary liquids such as air. For low Prandtl number fluids the nonlinearity in the momentum equations dominate the flow. Hence, the flow mechanism is different from that of ordinary liquids, where the viscosity has a greater dissipating effect.

INTRODUCTION

Thermal convection in differentially heated cavities filled with a liquid metal has applications in crystal growth from melts and in solidification of metals. For example, in metal casting the uniformity of the temperature is highly desirable for control of thermal stresses and grain sizes. Unstable flow produces spatial temperature distortions with time. Since in many applications the flow is unstable and possibly turbulent, it is essential to understand the different physical processes responsible for the transition of initially laminar flow to a turbulent one. Thermal convection flow of a liquid metal is laminar at sufficiently low Rayleigh numbers, but becomes periodic in time at a critical value, which depends on the geometric constraints and the Prandtl number.

Transient thermal convection in cavities has been studied experimentally by Hurle et al. [1], Pumplin and Bolt [2], Hart [3] and Kamotani and Sahraoui [4] using mercury and gallium as working fluids. The results showed that the temperature oscillates sinusoidally with low frequency and low amplitude. Both the frequency of the oscillation and the amplitude increase with increasing temperature difference. For numerical simulations of the transient thermal convection in differentially heated cavities of low-Prandtl number fluids, we refer to a number of papers by Jones [5], Benocci [6], Gresho and Upson [7], Winters [8], Crochet et al. [9] and Dupont et al. [10]. For an aspect ratio of one, weak secondary flow was predicted at the corners of the cavity in addition to the main circulation, but the results of Gresho and Upson [7] showed two rotating vortices inside the main circulation, in addition to the corner vortices.

The flow structure in low Prandtl number fluids due to the thermal buoyancy force is not well understood. The flow is highly nonlinear, because the inertia force dominates the flow, and viscous effects are mainly confined to the very thin boundary layers. The purpose of the direct numerical analysis is to study the transitions to various time-dependent flows. With increasing Grashof number the onset of periodic flow is calculated for a cavity having an aspect ratio of one. Power spectra of the temperature and the velocity components are examined with phase diagrams by projecting the dependent variables on a plane. We are primarily concerned with instabilities that precede aperiodic or chaotic flows, not fully turbulent flow. All of the simulations have been carried out for $Pr = 0.005$, which corresponds to the Prandtl number of liquid sodium or tin at elevated temperature.

ANALYSIS

Physical and Mathematical Model

Consider a two-dimensional, differentially heated cavity with an aspect ratio of one ($A = L/H$), where L and H are length and height of the cavity, respectively. The isothermal vertical walls are kept at constant but different temperatures, while the upper and lower boundaries are insulated. Initially the fluid is at the cold wall temperature (T_c), then suddenly the temperature of left-hand vertical wall is increased to a constant value (T_h) at times $t \geq 0$. Hence, the convective flow is generated by the thermal buoyancy force as soon as $T_h \neq T_c$. It is assumed that the fluid is homogeneous with constant properties, except where the buoyancy is concerned, i.e., the Boussinesq approximation is assumed to be valid. Scales of H, H^2/α, α/H and ΔT are used for length, time, velocity and temperature, respectively.

The conservation equations can be written as

$$\nabla \cdot \overline{V} = 0 \tag{1}$$

$$Pr^{-1} \left(\frac{\partial \overline{V}}{\partial \tau} + \overline{V} \cdot \nabla \overline{V} \right) = - Pr^{-1} \nabla P + \nabla^2 \overline{V} - Ra\, \theta \overline{j} \tag{2}$$

$$\frac{\partial \theta}{\partial \tau} + \nabla \cdot \left(\overline{V} \theta \right) = \nabla^2 \theta \tag{3}$$

In the equations (1) to (3), $\overline{V} = (i\, U + j\, V)$ is the dimensionless velocity vector; P is pressure; θ is temperature; τ is dimensionless time; $Pr = \nu/\alpha$ and $Ra = Gr \cdot Pr = g\, \beta\, \Delta T\, H^3/\nu\alpha$ are the Prandtl and Rayleigh numbers, respectively. The boundary conditions are taken as

$$\theta = 1 \qquad at \qquad \xi = 0 \tag{4}$$

$$\theta = 0 \qquad \text{at} \qquad \xi = 1 \tag{5}$$

$$\frac{\partial \theta}{\partial \eta} = 0 \qquad \text{at} \qquad \eta = 0, 1 \tag{6}$$

$$\overline{V} = 0 \qquad \text{at all the boundaries} \tag{7}$$

where $\xi = x/H$ and $\eta = y/H$.

Method of Solution

A primitive variables approach is used to solve the finite-difference forms of equations (1) - (7). The domain of interest is divided into nonuniform control volumes. The mesh is staggered in a way that the velocities are calculated at the points which lie on the faces of each control volume, while the temperature is calculated a the mid-point of the control volume (Patankar [11]). The nonlinear algebraic equations are obtained by integrating the equations over each control volume and the difference equations are solved iteratively. Fully implicit, finite-difference equations are formulated by marching the solution in time.

Central-difference discretization of the diffusive-advective flux is used for the spatial derivatives, with the truncation error $O\left(\Delta\tau, \Delta\xi^2, \Delta y^2\right)$. Grid independence of the results has been carefully tested. The results show insignificant difference between those obtained with nonuniform meshes of 51 $\times$ 51 and 61 $\times$ 61 for Gr $= 1 \times 10^7$ and Pr $= 0.005$. The dependence of the results on the time step have also been tested. However, the grid size and time steps are reduced as the Grashof number is increased to resolve the fine structure of the flow field and reduce aliasing errors from the time series of the flow field.

Table 1 lists the number of meshes and time steps used to carry the calculations for each Grashof number. All the calculations are performed implicitly with double-precision to yield sufficiently accurate solutions.

RESULTS AND DISCUSSION

It is well known that if the Rayleigh number in thermal convection problems is above the threshold value, the flow

Table 1. Meshes and time increments used in numerical calculations.

Gr	Grid numbers	Time increment
3×10^6 to 1×10^7	61 $\times$ 61	0.002
5×10^7	71 $\times$ 71	0.001
6×10^7 to 1×10^8	81 $\times$ 81	0.0005

undergoes a sequence of instabilities which lead finally to chaos. In the Rayleigh-Benard problem, Gollub and Benson [12] have identified four distinct routes to chaos. They noted that one of the routes has a time independent regime followed by bifurcation to a state where the flow oscillates with a single frequency and its harmonics. Further bifurcations lead to states with two incommensurate frequencies with several linear combinations. Then, phase locking of the two frequencies occurs before the broadband noise components appear in the spectra and begin to grow.

A multi-cellular flow is predicted as the Grashof number is increased. First, small circulations appear at the corners of the cavity, in addition to the main circulation. Thereafter, other vortices appear within the main circulation at the core of the cavity. A sequence of instabilities is quite similar to that described by Gollub and Benson [12] in their experimental study of Rayleigh-Bernard convection and by the direct numerical simulation by Paolucci and Chenoweth [13] in a differentially heated cavities for Pr = 0.71. However, the flow structure is different in many respects from that for ordinary fluids. The flow structure is similar to the numerical results of Benocci [6] and Gresho and Upson [7] for Pr = 0.005 and 0.01, respectively. These are interesting results which have not been published previously.

For Gr = 3×10^6 the damped oscillatory transient flow was predicted (Figure 1), which is the time series for U-velocity at the center of the cavity. For Gr = 5×10^6, the time dependency begins with a bifurcation from the steady state to an oscillatory periodic flow. Thermal convection in different systems with low Pr-fluids showed similar bifurcation (i.e., Hopf bifurcation) as the Grashof number exceeded a threshold value [14-16]. The power spectra analysis of the time series of velocities and temperatures at various locations indicate that the flow oscillates with one fundamental frequency of 12.2 (dimensionless frequency $1/\tau$) with harmonics at Gr = 5×10^6. Figures 2(a) and 2(b) show the time series of U-velocity at the left-hand of the cavity and the corresponding power spectra. Only the power content of the frequency of the oscillation (i.e., amplitude) is location dependent. The fundamental frequency increases to 14.6 and 16.1 as the Grashof number is increased to 8×10^6 and 1×10^7, respectively. Figures 3(a) to 3(d) show the time series and power spectra

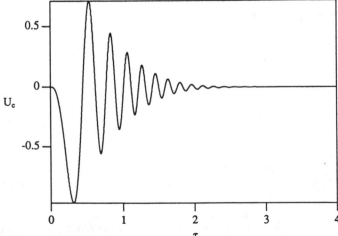

Figure 1   Time series for U-velocity at the center of the cavity, Gr = 3×10^6

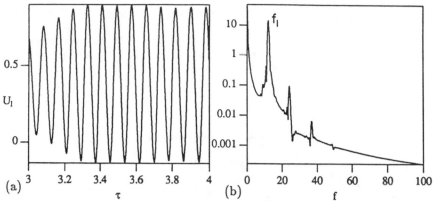

Figure 2   Time series for U-velocity at the mid-left hand of the cavity (a) and
corresponding $\log_{10}$ power spectra of velocity (b) for $Gr = 5 \times 10^6$.
(Note: In the remaining figures presented in the paper $\log_{10}$ is
deleted from the text or figure captions but is implied)

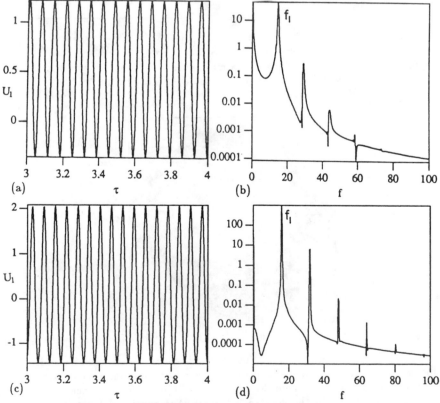

Figure 3   Time series of U-velocity, at the mid-left hand of the cavity for
$Gr = 8 \times 10^6$ (a), corresponding power spectra (b), time series of
U-velocity at the mid-left hand of the cavity for $Gr = 1 \times 10^7$ (c),
and corresponding power spectra (d)

for Gr $= 8 \times 10^6$ and Gr $= 1 \times 10^7$, respectively. The limit cycle oscillation can be seen in Figures 4(a) and 4(b), which are plots of temperature θ against U-velocity at the left-hand side of the cavity. It should be mentioned that integral quantities such as average Nusselt number at the vertical walls and global kinetic energy of the flow showed the same fundamental frequency of oscillation. Also, the oscillation of the Nusselt numbers at the vertical walls are in phase. The streamlines for Gr $= 1 \times 10^7$ (Figure 5) show weak circulations at the corners of the cavity.

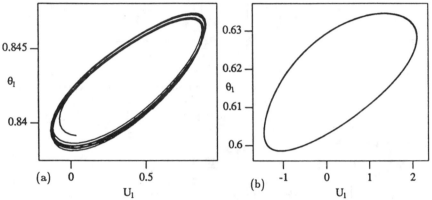

Figure 4   Limit cycle of the oscillation for Gr $= 5 \times 10^6$ (a) and 1×10^7 (b), respectively, projected onto the U—θ plane (at the mid-left hand of the cavity)

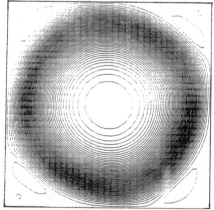

Figure 5   Streamlines for Gr $= 1 \times 10^7$, $\tau = 6$

For Gr $= 5 \times 10^7$ the flow became quasi-periodic with two incommensurate frequencies and their linear combination was evident in the power spectra of the flow field [Figure 6(a) and 6(b)]. The presence of high-order mixing components in the spectra indicates that the time dependent processes are strongly nonlinear. The evidence of nonperiodicity is shown by plotting temperature θ against U-velocity at the left-hand of the cavity (Figure 7). The time series of stream functions at the center of the cavity showed that the amplitude of the oscillation modulated with low frequency (Figure 8).

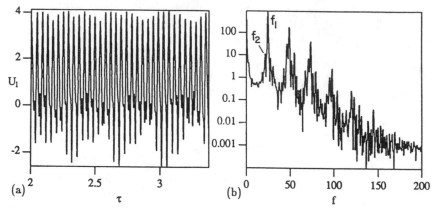

Figure 6   Time series for U-velocity at the mid-left hand of the cavity (a) and
                corresponding power spectra (b) for $Gr = 5 \times 10^7$

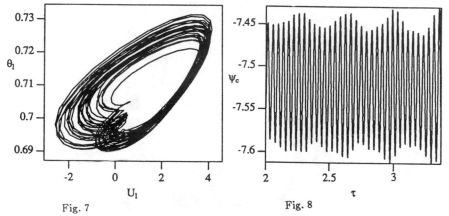

Fig. 7   Fig. 8

Figure 7   Dimensionless temperature θ against U-velocity at the mid-left
                hand of the cavity for $Gr = 5 \times 10^7$

Figure 8   Time series for streamline at the center of the cavity, $Gr = 5 \times 10^7$

Phase locking of the two frequencies occurred when the Grashof number
increased to 7×10^7. The ratio of $f_1/f_2 = 4/3$ and other peaks are harmonics
of the difference of the two frequencies. Figures 9(a) and 9(b) illustrate the times
series of the U-velocity at the mid-left hand of the cavity with its power spectra.
The flow field still reveals skew-symmetry. Weak vortices appear at the corner
of the cavity. In addition, there are three weaker cells inside the main
circulation. These three cells are arranged in an interesting manner. One cell is
at the center of the cavity while two other orbital circulations are rotating
around the center one (Figure 10).

For $Gr = 1 \times 10^8$, the power spectra revealed broad band spectrum and
high non-periodicity, even though there are sharp peaks at $f = 53.7$ and $f = 50.2$. Figures 11(a) and 11(b) show the time series of U-velocity at the mid-left
hand of the cavity with the corresponding power spectra. Also, the time series
of Nusselt numbers at both vertical walls showed a phase shift of $90°$, as

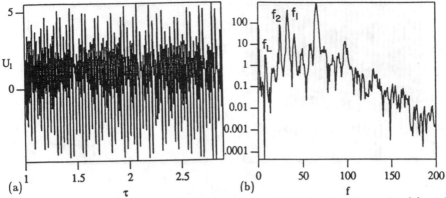

Figure 9   Time series for U-velocity at the mid-left hand of the cavity (a), and
                corresponding power spectra (b) for $Gr = 7 \times 10^7$

Figure 10   Streamlines for $Gr = 7 \times 10^7$, $\tau = 2.992$

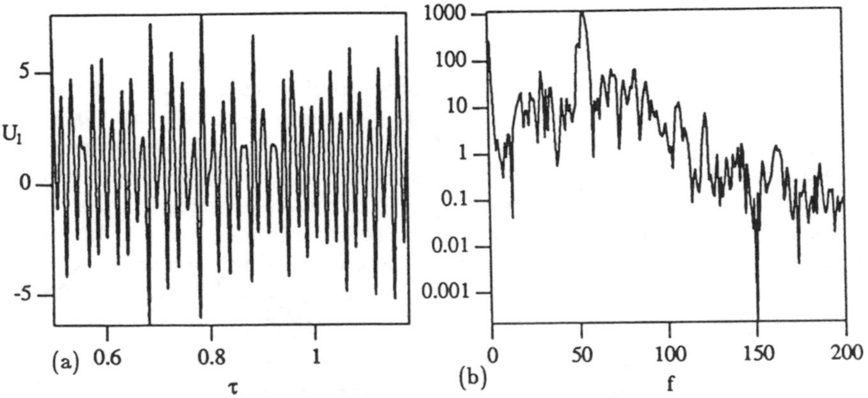

Figure 11   Time series for U-velocity at the mid-left hand of the cavity (a) and
                corresponding power spectra (b) for $Gr = 1 \times 10^8$

revealed in Figure 12. Such prediction was not found for $Gr = 7 \times 10^7$, 6×10^7 or smaller values, when the Nusselt numbers at the vertical walls are in phase for the $Gr \leq 7 \times 10^7$. Also, the skew-symmetry is lost for $Gr = 1 \times 10^8$, as shown in the streamline contours (Figure 13).

A multicellular flow is predicted as the Grashof number is increased. First, small circulations appear at the upper-right and lower-left hand corners of the cavity, then at the other corners, in addition to the main circulation. Thereafter, other vortices appear within the main circulation at the core of the cavity. This sequence of instabilities is quite similar to those described by Gollub and Benson [12] in their experimental study of Rayleigh-Bernard convection. The flow structure is different than that for ordinary fluids [16], but it is similar to the numerical results of Benocci [6] and Gresho and Upson [7] for $Pr = 0.005$ and 0.01, respectively.

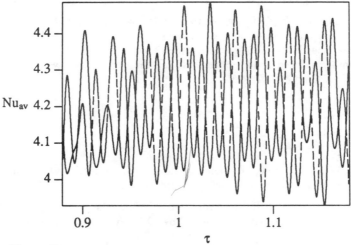

Figure 12   Time series for Nusselt number at both vertical boundaries for $Gr = 1 \times 10^8$, ——— $\xi = 0$, - - - $\xi = 1$

Figure 13   Streamlines for $Gr = 1 \times 10^8$, $\tau = 0.9995$

CONCLUSIONS

Time-dependent flows for a low Prandtl number fluid inside a differentially heated square cavity have been analyzed using direct numerical simulation. For low Grashof number $(\text{Gr} = 3 \times 10^6)$ a damped oscillatory transient flow was predicted. Further increase in the Grashof number produces a sequence of bifurcations. The route to chaos is similar to one of the routes discussed by Gollub and Benson [12] in the Rayleigh-Bernard problem, where the limit cycle oscillation bifurcates to two incommensurate frequencies (i.e., quasi-periodic) followed by frequency locking, then chaotic flow. In addition to the main circulation for $\text{Gr} = 1 \times 10^7$, four weak circulations are predicted at the corners of the cavity. For $\text{Gr} = 7 \times 10^7$ three cells are predicted inside the main circulation. These cells are arranged in such a way that one is at the center of the cavity and the other two cells are orbiting around the central one.

Increasing the Grashof number to 1×10^8 results in the loss of the skew-symmetry and one more vortex appeared at the center of the cavity. Furthermore, the flow at the corners become complex. At $\text{Gr} = 1 \times 10^8$ the Nusselt number showed a 90° phase shift at the two vertical walls, while they are in the phase for $\text{Gr} \leq 7 \times 10^7$.

REFERENCES

1. Hurle, D., Jakeman, E. and Johnson, C. Convective Temperature Oscillations in Molten Gallium, J. Fluid Mech., Vol. 64, pp. 565-576, 1974.

2. Pumplin, B.R. and Bolt, G.H. Temperature Oscillation, Induced in a Mercury Bath by Horizontal Heat Flow, J. Physics D, Vol. 9, pp. 145-149, 1976.

3. Hart, J.E. Low Prandtl Number Convection Between Differentially Heated End Walls, Int. J. Heat Mass Transfer, Vol. 26, pp. 1069-1074, 1983.

4. Kamotani, Y. and Sahraoui, T. Oscillatory Natural Convection in Rectangular Enclosures Filled with Mercury (Ed. Marto, P.J. and Tansawa, I.), pp. 2-241 to 2-245, Proceedings of the ASME-JSME Thermal Engineering Joint Conference, New York/Tokyo, 1987.

5. Jones, I.P. Low Prandtl Free Convection in a Vertical Slot, AERE Harwell Report R-10416 , 1981-1982.

6. Benocci, C. Thermohydraulics of Liquid Metals: Turbulence Modeling in Liquid Metal Free Convection, von Karman Institute Lecture Series 1983-07, 1983.

7. Gresho, P.M. and Upson, C.G. Application of a Modified Finite Element Method to the Time-Dependent Thermal Convection of a Liquid Metal (Eds. Taylor, C., Johnson, C., Smith, J.A. and Ramsey, W.), pp. 750-762, Proceedings of the 3rd International Conference on Numerical Methods in Laminar and Turbulent Flow, Pineridge Press, Swansea, U.K., 1983.

8. Winters, K. Oscillatory Convection in Crystal Melts: the Horizontal
    Bridgman Process, Theoretical Physics Division, Harwell Laboratory
    Report TP-1230, 1987.

9. Crochet, M.J., Geyling, F.T. and Van Schaftinger, J.J. Numerical
    Simulation of the Horizontal Bridgman Growth, Part I: Two Dimensional
    Flow, Int. J. Numerical Methods Fluids, Vol. 7, pp. 29-47, 1987.

10. Dupont, S., Marehal, J.M., Crochet, M.J. and Geyling, F.T. Numerical
    Simulation of the Horizontal Bridgman Growth, Part II: Three
    Dimensional Flow, Int. J. Numer. Meth. Fluids, Vol. 7, pp. 49-67, 1987.

11. Patankar, S.V. Numerical Heat Transfer and Fluid Flow, Hemisphere,
    Washington, D.C., 1980.

12. Gollub, J.P. and Benson, S.V. Many Routes to Turbulent Convection, J.
    Fluid Mech., Vol. 100, pp. 449-470, 1980.

13. Paolucci, S. and Chenoweth, D.R. Transition to Chaos in a Differentially
    Heated Vertical Cavity, J. Fluid Mech., Vol. 201, pp. 379-410, 1989.

14. Krishnan, R. A Numerical Study of the Instability of Double-Diffusive
    Convection in a Square Enclosure with Horizontal Temperature and
    Concentration Gradients (Ed. Shah, R.K.), pp. 357-368, Proceedings of the
    1989 National Heat Transfer Conference, ASME, New York, 1989.

15. Del Arco, E.C., Pulicani, J.P. and Randriamumpianina, A. Complex
    Multiple Solutions and Hysteresis Cycles Near the Onset of Oscillatory
    Convection in a $Pr = 0.0$ Liquid Submitted to a Horizontal Temperature
    Gradient, C.R. Acad. Sci. Paris, Series II, Vol. 309, pp. 1869-1876, 1989.

16. Koster, J.N. and Müller, U. Oscillatory Convection in Vertical Slots, J.
    Fluid Mech., Vol. 139, pp. 363-390, 1984.

Axisymmetric Free Convection on a Horizontal Plate with Prescribed Heat Transfer Coefficient in a Porous Medium

G. Ramanaiah, G. Malarvizhi

Department of Mathematics, Anna University, Madras 600 025, India

ABSTRACT

The axisymmetric free convection on a horizontal plate embedded in a saturated porous medium is considered assuming that the plate is subjected to a prescribed temperature or a prescribed heat flux or a prescribed heat transfer coefficient. By similarity transformation the governing equations are reduced to identical coupled equations for all the three cases with three common boundary conditions and one boundary condition depending on the thermal boundary condition imposed. It is shown that there is no need to solve the three boundary value problems independently and that the solution for one case can be used to obtain the solution for any other case by a simple algebraic method.

INTRODUCTION

The study of free, mixed and forced convections on surfaces embedded in fluid saturated porous media has received much attention during the last thirteen years because of its numerous applications in geothermal and energy related engineering problems. In these studies heat transfer coefficient has been determined assuming that the surfaces are subjected to a prescribed temperature [1–8] or a prescribed surface heat flux [9–15]. These investigators seem to have not observed that the solutions for the two cases are related to each other [16]. Further the converse problem, namely, the study of free convection when the surfaces are subjected to a prescribed heat transfer coefficient has not received sufficient attention.

In this paper we consider the axisymmetric free convection on a horizontal impermeable plate which is subjected to a prescribed heat transfer coefficient and show that the similarity solutions of free convection problems for the three cases namely, prescribed temperature, prescribed heat flux and prescribed heat transfer coefficient are dependent and the solution for one case can be obtained from the other by a simple algebraic method.

ANALYSIS

Consider the axisymmetric free convection on an impermeable horizontal plate embedded in a fluid saturated porous medium. The governing equations of free convection are

$$\frac{\partial}{\partial r} (ru) + \frac{\partial}{\partial z} (rw) = 0 \tag{1}$$

$$u = -\frac{K}{\mu} \frac{\partial p}{\partial r} \tag{2}$$

$$w = -\frac{K}{\mu} \frac{\partial p}{\partial z} + \rho g \tag{3}$$

$$u \frac{\partial T}{\partial r} + w \frac{\partial T}{\partial z} = \alpha \left[\frac{1}{r} \frac{\partial}{\partial r} \left(r \frac{\partial T}{\partial r} \right) + \frac{\partial^2 T}{\partial z^2} \right] \tag{4}$$

where z measures distance normal to the plate, r is the radial distance, u,w are the Darcy velocity components in r and z directions respectively, T the temperature, p the pressure and ρ the density. The constants K, μ, α and g denote the permeablity of the porous medium, the viscosity, thermal diffusivity and gravitational acceleration, respectively.

Let T_e be the ambient temperature. It is assumed that the plate is subjected to any one of the thermal boundary conditions:

i.   prescribed temperature (PT)
ii.   prescribed heat flux (PHF)
iii. prescribed heat transfer coefficient (PHTC)

The boundary conditions to be satisfied are :

$$w = 0 \text{ at } z = 0 \text{ and } u \to 0, \ T \to T_e \ p \to p_e \text{ as } z \to \infty \tag{5}$$

$$T - T_e = \Delta T_w(r) \tag{6a}$$

$$\text{or } -k \frac{\partial T}{\partial z} = q_w(r) \qquad \text{at } z = 0 \tag{6b}$$

or $\quad -\dfrac{r}{(T-T_e)} \dfrac{\partial T}{\partial z} = N(r) \quad$ at $z = 0$ $\qquad$ (6c)

where $\Delta T_w(r)$, $q_w(r)$ and $N(r)$ denote the plate excess-temperature (over T_e), heat flux and heat transfer coefficient. k is the thermal conductivity and p_e is the ambient fluid pressure.

Invoking the boundary layer approximations and the Boussinesq approximation,

$$\rho = \rho_e \left[1-\beta(T-T_e)\right] \qquad (7)$$

where ρ_e is the ambient fluid density and β is the coefficient of thermal expansion, Equations (3) and (4) reduce to

$$\dfrac{1}{\rho_e} \dfrac{\partial p}{\partial z} = g\beta(T-T_e) \qquad (8)$$

$$u \dfrac{\partial T}{\partial r} + w \dfrac{\partial T}{\partial z} = \alpha \dfrac{\partial^2 T}{\partial z^2} \qquad (9)$$

Introducing the similarity variables,

$$\eta = \dfrac{z}{r} R^{1/3} \quad , \quad f(\eta) = \dfrac{\psi}{\alpha r R^{1/3}} \quad , \qquad (10a,b)$$

$$P(\eta) = \dfrac{K(p_e-p)}{\alpha\mu R^{2/3}} \quad , \quad \theta(\eta) = \dfrac{T-T_e}{C} \qquad (10c,d)$$

where

$$C = c_o\, r^\lambda, \; c_o > 0, \; R = \dfrac{Kg\beta cr}{\alpha\nu} \qquad (11a,b)$$

$\nu = \mu/\rho_e$, the kinematic viscosity and ψ is the stream function, we obtain

$$u = \dfrac{1}{r} \dfrac{\partial \psi}{\partial z} = \dfrac{\alpha R^{2/3}\, f'}{r} \qquad (12a)$$

$$w = -\dfrac{1}{r} \dfrac{\partial \psi}{\partial r} = -\dfrac{\alpha R^{1/3}}{3r} \left[(\lambda+4)f+(\lambda-2)\eta f'\right] \qquad (12b)$$

where the primes denote differentiation with respect to η.

Now Equations (2), (8), (9) and the boundary conditions (5) and (6) reduce to

$$3f' + (\lambda-2)\eta\theta - 2(\lambda+1)P=0 \tag{13}$$

$$P' + \theta = 0 \tag{14}$$

$$3\theta'' + (\lambda+4)\ f\theta' - 3\lambda f'\ \theta = 0 \tag{15}$$

$$f(0) = F(\infty)= \theta(\infty) = 0 \tag{16}$$

$$\theta(0) = \frac{\Delta T_w(r)}{C} \tag{17a}$$

$$\theta'(0) = -\frac{r q_w(r)}{k C R^{1/3}} \tag{17b}$$

$$\frac{\theta'(0)}{\theta(0)} = -\frac{N(r)}{R^{1/3}} \tag{17c}$$

In the foregoing analysis use is made of the functions C and R defined by Equation (11). We shall now determine them using the thermal boundary condition prescribed at the plate.

i.   PT : If ΔT_w = const. r^m, then we set $\lambda=m$, $C=\Delta T_w$ and $R = Kg\beta(\Delta T_w)r/\alpha\nu$ so that Equation (17a) reduces to

$$\theta(0) = 1 \tag{18a}$$

In this case $q_w(r)$ and $N(r)$ are given by Equations (17b) and (17c)

ii.   PHF : If q_w=const. r^m, then we set $\lambda=(2+3m)/4$ and $kCR^{1/3}=rq_w$ which yields with the aid of Equation (11),

$$C=[\alpha\nu q_w^3\ r^2/Kg\beta k^3]^{\frac{1}{4}} \text{ and } R = [Kg\beta q_w r^2/\alpha\nu k]^{3/4}$$

and hence Equation (17b) becomes

$$\theta'(0) = -1 \tag{18b}$$

Then $\Delta T_w(r)$ and $N(r)$ are given by Equations (17a) and (17c).

iii.   PHTC : If $N(r)$ = const. r^m, then we set $\lambda=3m-1$ and $C = \alpha\nu N^3/Kg\beta r$, $R=N^3$ so that Equation (17c) becomes

$$\theta(0) + \theta'(0) = 0 \tag{18c}$$

In this case $\Delta T_w(r)$ and $q_w(r)$ are given by Equations (17a) and (17b).

Thus the free convection problem has been reduced to solving the coupled Equations (13)-(15) with the boundary conditions (16) and one of the conditions (18). The three solutions are dependent and from one solution one can obtain the other two with the aid of the following theorems:

Theorem 1.   The system of Equations (13)-(16) are invariant under the transformation group,

$$\eta^* = A\eta, \quad f^*(\eta^*) = \frac{1}{A} f(\eta), \quad P^*(\eta^*) = \frac{1}{A^2} P(\eta),$$

$$\theta^*(\eta^*) = \frac{1}{A^3} \theta(\eta) \tag{19}$$

where A is a positive constant.

Theorem 2.   If the solutions for two of the cases are same, then the solutions for all the three cases are identical.

The truth of Theorem 1 is verified by actual substitution in the concerned equations. Theorem 2 follows from the fact that any two of the conditions $\theta(0)=1$, $\theta'(0)=-1$, $\theta(0)+\theta'(0)=0$ imply the third.

Now it is a matter of simple calculation by using Theorem 1 to obtain the value of A for finding the solution of one case from the other. Table 1 shows the values of A required for the transition. For instance, if $\{f(\eta), \theta(\eta), P(\eta)\}$ is the solution for PHTC, then $\{f^*(\eta^*), \theta^*(\eta^*), P^*(\eta^*)\}$ is the solution for PT with $A = [\theta(0)]^{1/3}$.

EXACT SOLUTION

For the special case $\lambda = 2$, i.e., $\Delta T_w \sim r^2$ or $q_w \sim r^2$ or $N(r) \sim r$, Equations (13)-(15) become

$$f' - 2P = P' + \theta = \theta'' + 2(f\theta' - f'\theta) = 0 \tag{20}$$

and the boundary value problems have exact solutions,

$$f(\eta) = m (1-e^{-2m\eta}) \tag{21a}$$

$$P(\eta) = m^2 e^{-2m\eta} \tag{21b}$$

$$\theta(\eta) = 2m^3 e^{-2m\eta} \tag{21c}$$

where

$$m = (\tfrac{1}{2})^{1/3} \quad \text{for PT} \tag{22a}$$

$$= (\tfrac{1}{2})^{\frac{1}{2}} \quad \text{for PHF} \tag{22b}$$

$$= \tfrac{1}{2} \quad \text{for PHTC} \tag{22c}$$

Table 1   Values of the parameter A

→	PT	PHF	PHTC
PT	1	$[-\theta'(0)]^{\frac{1}{4}}$	$-\theta'(0)$
PHF	$[\theta(0)]^{1/3}$	1	$1/\theta(0)$
PHTC	$[\theta(0)]^{1/3}$	$[\theta(0)]^{\frac{1}{4}}$	1

Table 2   The values of $-\theta'(0)$ for PT, $\theta(0)$ for PHF and for PHTC

λ	PT	PHF	PHTC
0.5	0.9305	1.0555	1.2411
0.6472	1.0000	1.0000	1.0000
1.0	1.1615	0.8938	0.6382
2.0	1.5874	0.7071	0.2500

Table 3   The values of $f'(0)$

λ	PT	PHF	PHTC
0.5	0.8583	0.8897	0.9912
0.6472	0.9073	0.9073	0.9073
1.0	1.0144	0.9412	0.7519
2.0	1.2599	1.0000	0.5000

Table 4   The values of $P(0)$

λ	PT	PHF	PHTC
0.5	0.8583	0.8897	0.9912
0.6472	0.8263	0.8263	0.8263
1.0	0.7608	0.7059	0.5640
2.0	0.6300	0.5000	0.2500

DISCUSSION

The boundary value problems for the case of PHTC has been computed numerically for $\lambda = \frac{1}{2}$, 1 and 2 and the solutions for the other two cases have been obtained using Theorem 1.   Table 2 gives the values of $-\theta'(0)$ for PT and $\theta(0)$ for PHF and for PHTC while Table 3 and Table 4 show the values of dimensionless slip velocity $f'(0)$ and pressure drop at the plate $P(0)$.   The dimensionless temperature θ, streamwise velocity f' and pressure drop P are shown in Figures 1-3 for the three cases for comparison.   The critical value of λ for which all the three solutions are identical is found to be $\lambda_c = 0.6472$.

The unified treatment of PT, PHF and PHTC cases presented here includes the results reported by Cheng and Chau [6] for PT and [15] for PHF.   We propose to extend the analysis presented here to the cases of vertical plate and other geometries in subsequent papers.

REFERENCES

1.   Cheng, P. and Chang, I. Buoyancy Induced Flows in a Saturated Porous Medium Adjacent to Impermeable Horizontal Surfaces, Int. J. Heat Mass Transfer, Vol.19, pp. 1267-1272, 1976.

2.   Cheng, P. and Minkowycz, W.J.   Free Convection about a Vertical Plate Embedded in a Porous Medium with Application to Heat Transfer from a Dike, J. Geophys. Res., Vol.82, pp. 2040-2044, 1977.

3.   Cheng, P.   The Influence of Lateral Mass Flux on Free Convection Boundary Layers in a Saturated Porous Medium,   Int.   J.   Heat   Mass   Transfer,   Vol.20, pp. 201-206, 1977.

4.   Cheng, P.   Combined Free and Forced Convection Flow about Inclined Surfaces in Porous Media, Int. J. Heat Mass Transfer, Vol.20, pp. 807-814, 1977.

5.   Cheng, P. Similarity Solutions for Mixed Convection from Horizontal Impermeable Surfaces in Saturated Porous Media, Int. J. Heat Mass Transfer, Vol.20, pp. 893-898, 1977.

6.   Cheng, P. and Chau, W.C. Similarity Solutions for Convection of Groundwater Adjacent to Horizontal Impermeable Surfaces with Axisymmetric Temperature Distribution, Water Resour. Res., Vol.13, pp. 768-772, 1977.

7.   Merkin, J.H.   Free Convection Boundary Layers in a Saturated Porous Medium with Lateral Mass Flux, Int. J. Heat Mass Transfer, Vol.21, pp. 1499–1504, 1978.

8.   Merkin, J.H.   Mixed Convection Boundary Layer Flow on a Vertical Surface in a Saturated Porous Medium, J. Engng. Math., Vol.14, pp. 301–313, 1980.

9.   Cheng, P.   Constant Surface Heat Flux Solutions for Porous Layer Flows, Lett. Heat Mass Transfer, Vol.4, pp. 119–128, 1977.

10.   Na, T.Y. and Pop, I.   Free Convection Flow Past a Vertical Flat Plate Embedded in a Saturated Porous Medium, Int. J. Engng. Sci., Vol.21, pp.   517–526, 1983.

11.   Joshi, Y. and Gebhart, B.   Mixed Convection in Porous Media Adjacent to Vertical Uniform Heat Flux Surface, Int. J. Heat Mass Transfer, Vol.28, pp. 1783–1786, 1985.

12.   Dutta, P. and Seetharamu, K.N.   Effect of Variable Wall Heat Flux on Free Convection in a Saturated Porous Medium, Indian J. Tech., Vol.25, pp. 567–571, 1987.

13.   Merkin, J.H. and Pop, I.   Natural Convection About Two Dimensional Bodies with Uniform Surface Heat Flux in a Porous Medium, Acta Mech., Vol.70, pp. 235–242, 1987.

14.   Gorla, R.S.R. and Tornabene, R.   Free Convection From a Vertical Plate with Nonuniform Surface Heat Flux and Embedded in a Porous Medium, Transport in Porous Media, Vol.3, pp. 95–106, 1988.

15.   Merkin, J.H. and Pop, I.   Free Convection About a Horizontal Circular Disk in a Saturated Porous Medium, Warme–und Stoff., Vol.24, pp. 53–60, 1989.

16.   Ramanaiah, G. and Malarvizhi, G.   Unified Treatment of Similarity Solutions of Free, Mixed and Forced Convection Problems in Saturated Porous Media,   in Numerical Methods in Thermal Problems (Ed. Lewis,R.W. and Morgan, K.) Vol.6, pp. 431–439, Proceedings of the 6th Int. Conf. on Numerical Methods in Thermal Problems, Pineridge, Swansea, 1989.

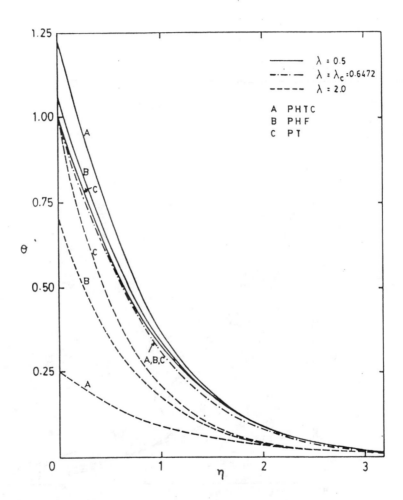

Figure 1   Profiles of dimensionless temperature θ

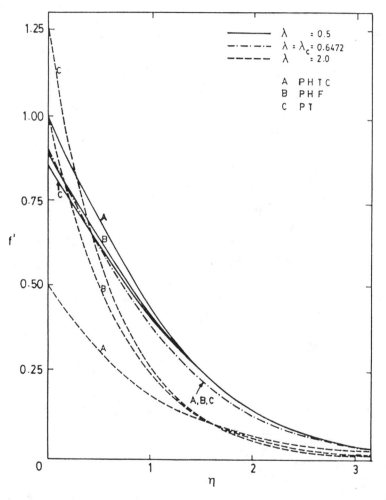

Figure 2   Profiles of dimensionless streamwise velocity f'

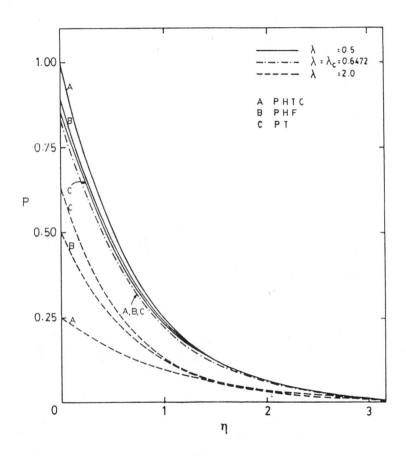

Figure 3   Profiles of dimensionless pressure drop P

Thermal Circulation in Deep Water Reservoirs

K. Onishi(*), K. Yamasaki(**)
(*) Department of Applied Mathematics,
(**) Department of Civil Engineering, Fukuoka
University, Jonan-ku, Fukuoka 814-01, Japan

ABSTRACT

Three–dimensional thermal fluid flow in reservoirs is modeled based on the finite element method. The Boussinesq approximation is considered for temperature effect to the fluid flow. Inertia and shearing stresses in the vertical direction are neglected. The Navier–Stokes equations and the heat transport equation are discretized using triangular prismatic finite elements. Two–step selectively lumping Lax–Wendroff scheme is used for time integration. The model is applied to the simulation of wind–driven thermal current in a real deep dam.

Keywords: Dam, nutrification, mathematical simulation, wind–driven current, temperature effect.

INTRODUCTION

Shallow water circulation using the mean velocity components has been often formulated by means of the finite element method for two–dimensional flow equations that are obtained by averaging the three–dimensional Navier–Stokes equations with respect to the vertical direction. The pressure distribution is assumed to be hydrostatic. The thermal energy transport in the shallow water has been modeled correspondingly in two dimensions using the mean temperature in the vertical direction.

Recent investigation on the circulation of shallow water, which is related to the protection of dams and lakes against their micro–biological pollution due to the nutrification, requires three–dimensional model which can elucidate the vertical motion of water, driven by the surface wind and temperature difference. This paper presents a finite element analysis of three–dimensional motion of water in dams with the emphasis on a simple treatment of the vertical transport of both momentum and thermal energy.

GOVERNING EQUATIONS

We shall consider a region of a dam with the rectangular coordinates x, y, z (m) directed to the east, north, and upward respectively, as illustrated in Figure 1. We choose mean level of the water surface as a reference plane $z = 0$. Let $h(x, y)$ denote the depth from the mean surface to the water bed. The elevation from the mean surface to temporary level of water surface is denoted by $\zeta(x, y, t)$ at the time t (s). The total depth from the surface to the bed is therefore given by $H = h + \zeta$.

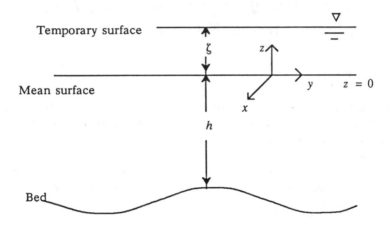

Figure 1. Vertical cross section of dam.

Density and viscosity of water depends on temperature in general. However, from the nature of problems which we shall discuss, we consider such governing equations that are based on the Boussinesq approximation: It requires that the density and viscosity are assumed to be constant in all terms of the equations except a term which describes the effect of temperature to the viscous fluid flows by the buoyancy. Accordingly, the continuity equation can be written as

$$\frac{\partial u}{\partial x} + \frac{\partial v}{\partial y} + \frac{\partial w}{\partial z} = 0 \quad , \tag{1}$$

where u, v, w are the velocity components (m/s) of the current in the x, y, z directions respectively.

Suppose that, in the vertical fluid motion in the dam, the acceleration and shear stresses are negligibly small in magnitude as compared to the pressure gradient and the gravity. Then the equations of motion can be described approximately as follows.

$$\rho_0\left(\frac{\partial u}{\partial t} + u\frac{\partial u}{\partial x} + v\frac{\partial u}{\partial y} + w\frac{\partial u}{\partial z}\right)$$

$$= \rho_0 f v - \frac{\partial p}{\partial x} - \left(\frac{\partial \tau_{xx}}{\partial x} + \frac{\partial \tau_{xy}}{\partial y} + \frac{\partial \tau_{xz}}{\partial z}\right), \tag{2}$$

$$\rho_0\left(\frac{\partial v}{\partial t} + u\frac{\partial v}{\partial x} + v\frac{\partial v}{\partial y} + w\frac{\partial v}{\partial z}\right)$$

$$= - \rho_0 f u - \frac{\partial p}{\partial y} - \left(\frac{\partial \tau_{yx}}{\partial x} + \frac{\partial \tau_{yy}}{\partial y} + \frac{\partial \tau_{yz}}{\partial z}\right), \tag{3}$$

$$0 = - \rho g - \frac{\partial p}{\partial z}, \tag{4}$$

where p is the pressure (Pa), f is the Coriolis factor given by $f = 2\omega \sin\phi$ with the angular velocity of the terrestrial rotation $\omega = 7.292\text{x}10^{\circ5}$ (rad/s) and the latitude ϕ (rad) at the location of the dam, g is the gravity acceleration = 9.81 (m/s^2); τ_{xx}, τ_{xy}, ..., τ_{yz} are the stress components (Pa), given by

$$\tau_{xx} = - \mu_h \frac{\partial u}{\partial x}, \qquad \tau_{xy} = - \mu_h \frac{\partial u}{\partial y}, \qquad \tau_{xz} = - \mu_v \frac{\partial u}{\partial z},$$

$$\tau_{yx} = - \mu_h \frac{\partial v}{\partial x}, \qquad \tau_{yy} = - \mu_h \frac{\partial v}{\partial y}, \qquad \tau_{yz} = - \mu_v \frac{\partial v}{\partial z}, \tag{5}$$

with the horizontal and vertical eddy viscosity μ_h, μ_v ($Pa.s$) respectively, and with the reference density of water ρ_0 (kg/m^3) at ordinary temperatures. The density ρ, depending on the temperature T (K) involved in the vuoyancy term of (4), is assumed to obey the equation of state

$$\rho = \rho (T). \tag{6}$$

The explicit functional form on the right hand side is given in the example.

At the surface described by the equation $z = \zeta(x, y, t)$, the boundary conditions are

$$w_S = \frac{\partial \zeta}{\partial t} + u_S \frac{\partial \zeta}{\partial x} + v_S \frac{\partial \zeta}{\partial y}, \tag{7}$$

as the surface remains through time, where the subscript s indicates the surface quantity, and

$$\tau_{Sx} = \bar{\tau}_{Sx}, \qquad \tau_{Sy} = \bar{\tau}_{Sy}, \tag{8}$$

with the stresses induced by the wind, where the bar on the variable indicates that the value is given.

At the bottom $z = -h(x, y)$ of the dam, the boundary conditions are

$$w_B = -u_B \frac{\partial h}{\partial x} - v_B \frac{\partial h}{\partial y} \quad , \tag{9}$$

as the bed is fixed with time, where the subscript B indicates the bed quantity, and

$$\tau_{Bx} = \overline{\tau}_{Bx} \, , \qquad \tau_{By} = \overline{\tau}_{By} \, , \tag{10}$$

with the frictional stresses due to the roughness of the bed.

On the surface of the embankment, it is assumed that the normal component of the velocity

$$v_n = 0 \, . \tag{11}$$

At the inlet / outlet of the dam, the boundary conditions are respectively

$$v_n = \overline{V}_{in} \, , \qquad v_n = \overline{V}_{out} \, . \tag{12}$$

The equation of energy in terms of temperature can be expressed as follows.

$$\rho_0 C \left(\frac{\partial T}{\partial t} + u \frac{\partial T}{\partial x} + v \frac{\partial T}{\partial y} + w \frac{\partial T}{\partial z} \right)$$
$$= Q - \left(\frac{\partial q_x}{\partial x} + \frac{\partial q_y}{\partial y} + \frac{\partial q_z}{\partial z} \right) \, , \tag{13}$$

where C is the specific heat ($J/kg.K$); q_x, q_y, q_z are components of the thermal flux ($J/m^2.s$) given by

$$q_x = -k_h \frac{\partial T}{\partial x} \, , \qquad q_y = -k_h \frac{\partial T}{\partial y} \, , \qquad q_z = -k_v \frac{\partial T}{\partial z} \, , \tag{14}$$

with the horizontal and vertical eddy conduction coefficients k_h, k_v ($J/m.s.K$) respectively, Q is the rate of volumetric heat source ($J/m^3.s$) due to solar radiation.

On the surface of the water, simultaneous heat transfer is expressed in the form

$$q_n = -\beta q_R + q_C + q_I + q_E \, , \tag{15}$$

where q_n is the total thermal flux ($J/m^2.s$) in the external normal direction to the surface, q_R is the radiative flux of the solar energy with the absorption coefficient β (−) on the surface, q_C ($J/m^2.s$) is the heat loss due to the forced convection of wind according to the Newton's law of cooling, q_I is the heat loss due to the Stefan–Boltzmann radiation, and qE is the heat loss due to the evapolation of water from the surface.

At the bottom of the dam we shall assume

$$q_B = 0 . \qquad (16)$$

On the surface of the embankment, it is assumed that the normal component of the thermal flux

$$q_n = 0 . \qquad (17)$$

At the inlet / outlet of the dam, the boundary conditions are respectively

$$T = \bar{T}_{in} , \qquad q_n = 0 . \qquad (18)$$

The equation (4) can be integrated to yield the solution

$$p = p_S + g \int_z^\zeta \rho \, dz , \qquad (19)$$

with the constant pressure pS on the surface. Moreover, we shall express w in terms of u and v. To this end, we integrate (1) from $-h$ to z to see

$$0 = \int_{-h}^z \left(\frac{\partial u}{\partial x} + \frac{\partial v}{\partial y} + \frac{\partial w}{\partial z} \right) dz$$

$$= \frac{\partial}{\partial x} \int_{-h}^z u \, dz - u_B \frac{\partial h}{\partial x} + \frac{\partial}{\partial y} \int_{-h}^z v \, dz - v_B \frac{\partial h}{\partial y} + w - w_B .$$

Using the boundary condition (9) we have

$$w = - \frac{\partial}{\partial x} \int_{-h}^z u \, dz - \frac{\partial}{\partial y} \int_{-h}^z v \, dz . \qquad (20)$$

Similarly, by integrating (1) on the interval $-h < z < \zeta$, we can obtain

$$\frac{\partial \zeta}{\partial t} = - \frac{\partial}{\partial x} \int_{-h}^\zeta u \, dz - \frac{\partial}{\partial y} \int_{-h}^\zeta v \, dz , \qquad (21)$$

which describes the rate of movement of the surface.

3   Finite Element Discretization

We shall apply the conventional Galerkin finite element method for the discretization of the free surface equation (21), the equations of motion (2), (3) and the equation of energy (13) in that order.

The whole domain V; $(x, y) \in A$, $-h \le z \le \zeta$ is divided into a finite number of small triangular prisms. Each prism is further subdivided into L smaller prismatic elements, as shown in Figure 2 (a), with equal depth $\Delta z_i =$ $(\zeta_i + h_i)/L$ at the node i of the projected triangular element ∇ on the xy-plane.   Each prismatic element e depicted in Figure 2 (b) has six nodes designated by a pair of numbers (α, δ). The first number α indicates the node numbers of associated triangular element, the second number δ indicates stratum number in the z direction.

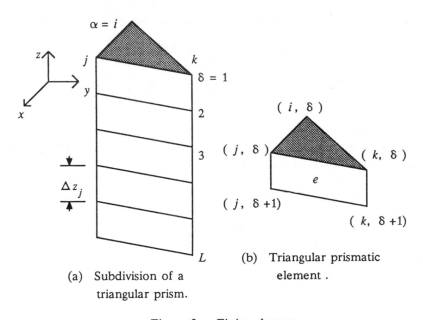

(a)   Subdivision of a
triangular prism.

(b)   Triangular prismatic
element .

Figure 2.   Finite element.

The elevation ζ is approximately expressed using the triangular linear interpolation functions $\phi\alpha$ in the element ∇ as

$$\zeta(x, y, t) \ = \ \sum_{\alpha}\phi_{\alpha}(x, y)\, \zeta_{\alpha}(t) \, , \tag{22}$$

with the nodal elevations $\zeta\alpha$ $(\alpha = i, j, k)$.

The velocity components u, v, w are approximately expressed using the six−node interpolation functions N_j in the triangular prismatic element as

$$u = \sum_{j=1}^{6} N_j(\,\xi,\,\eta,\,\zeta\,)\; u_j(\,t\,)\;,$$

$$v = \sum_{j=1}^{6} N_j(\,\xi,\,\eta,\,\zeta\,)\; v_j(\,t\,)\;, \tag{23}$$

$$w = \sum_{j=1}^{6} N_j(\,\xi,\,\eta,\,\zeta\,)\; w_j(\,t\,)\;,$$

with the nodal velocity components $u_j(t)$, $v_j(t)$, $w_j(t)$.

The temperature T is approximately expressed using the interpolation functions N_j in the triangular prismatic element as

$$T = \sum_{j=1}^{6} N_j(\,\xi,\,\eta,\,\zeta\,)\; T_j(\,t\,)\;, \tag{24}$$

with the nodal temperature $T_j(t)$, depending on the time.

For the discretization with respect to the time, we consider the element equations and apply the two–step Lax–Wendroff scheme with second–order of accuracy to the equations. To reduce the computational complexity, the selectively lumping technique is used.

NUMERICAL EXAMPLES

Traveling waves in a shallow channel.   We consider a straight channel with the constant depth $h = 20$ m, as shown in Figure 3. Initially the water in the channel is assumed to be motionless and the temperature is assumed uniformly to be at 20 °C. Along the mouth indicated by AB of the channel, the water elevation of the incident wave of the form

$$\zeta = A \sin(\,2\pi\,t/T_\zeta\,) \tag{25}$$

is specified with the amplitude $A = 0.5$ m and the period $T_\zeta = 1$ $hour$. The temperature on the boundary CD is suddenly reduced at 10 °C. The rest of the boundary; AD, BC is assumed to be impervious and thermally insulated. The purpose of computation is to obtain calculated elevation record at the monitoring station indicated by P, and to observe the vertical distributions of velocity and temperature.

For *long waves* traveling in shallow water, it is known that the wave velocity c (m/s) and the wave length λ (m) are given by the formulas

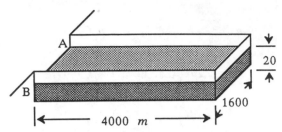

(a) Shallow rectangular channel.

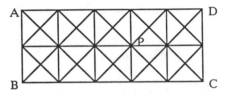

(b) Triangulation.

Figure 3.   Long traveling waves in a channel.

$$c = \sqrt{gh} \quad , \qquad \lambda = c\,T \; . \tag{26}$$

Hence it turns out that $c = 14\ m/s$ and $\lambda = 50.4\ km$ in our case.

The physical constants used in the computation are; $\nu_h = 30.0\ (\ m^2/s\ )$, $\nu_\nu = 0.001\ (\ m^2/s\ )$, $\rho_0 = 999.10\ (\ kg/m^3\ )$, $k_h = 5\times10^6\ (\ J/m.s.K\ )$, $k_\nu = 10^2\ (\ J/m.s.K\ )$, $C = 4.1855\times10^3\ (\ J/kg.K\ )$. The equation of state is given by $\rho(\theta) = 999.85497 + 0.058170 - 0.007840^2 + 0.0000400^3\ (\ kg/m^3\ )$ with the temperature in Celsius $\theta = T{-}273.16$.

The Coriolis forces are neglected. With the components W_x, W_y of the wind velocity ($\ m/s\ $) $10\ m$ high above the water surface, the shearing stresses on the surface in the boundary condition (8) are given according to Iwasa et al.[1] as   follows.

$$\bar{\tau}_{Sx} = \gamma_s^2 \rho_a\, W_x\, W \quad ,$$
$$\bar{\tau}_{Sy} = \gamma_s^2 \rho_a\, W_y\, W \quad , \tag{27}$$

with the speed of the wind W ($\ m/s\ $) given by

$$W = \sqrt{W_x^2 + W_y^2} \quad , \tag{28}$$

and with the coefficient $\gamma_s^2 = 1.3\times10^{-3}$ (–), and the density of air $\rho_a = 1.2250\ (kg/m^3\ )$. The shearing stresses in (10) are given similarly as

$$\bar{\tau}_{Bx} = \gamma_B^2 \rho_0 \, u_B \sqrt{u_B^2 + v_B^2} \quad ,$$

$$\bar{\tau}_{By} = \gamma_B^2 \rho_0 \, v_B \sqrt{u_B^2 + v_B^2} \quad , \tag{29}$$

with the coefficient $\gamma B^2 = 2.6 \times 10^{-3}$ (–).

The heat source Q in (13) is assumed to have the form

$$Q = (1 - \beta) \, q_R \exp[-\lambda(\zeta - z)] \, , \tag{30}$$

where λ is the diminishing coefficient ($1/m$). We shall assume that $\lambda = 0.1 \sim 0.5$ ($1/m$). With the cloudiness n (–), the radiative flux qR and each term on the right hand side of the boundary condition (15) are given as follows.

$$q_R = (1 - \alpha) \, \Theta \, , \tag{31}$$

$$q_C = \eta(W) \, \rho(\theta_S) \, (T_S - T_A) \, , \tag{32}$$

$$q_I = \epsilon\sigma(T_S^4 - 0.937 \times 10^{-5} T_A^6 (1.0 + 0.17 \, n^2)) \, , \tag{33}$$

$$q_E = \gamma(W) \, \rho(\theta_S) \, (E(\theta_S) - rE(\theta_A))(L + C T_S) \, , \tag{34}$$

where α is the albedo (–), Θ is the sunlightness (J/m^2s), η is the coefficient of convective heat transfer ($J.m/kg.s.K$), TS is the surface temperature (K), TA is the atmospheric temperature 0.15 m high above the surface, ϵ is the emittivity (–), σ is the Stefan–Boltzmann constant (W/m^2K^4), η is the coefficient of the Newton cooling, γ is the coefficient of evapolation heat transfer (m^2s/kg), $E(\theta)$ is the saturation vapor pressure ($mmHg$) at the temperature θ in Celsius, r is the relative humidity (–), and L is the latent heat in the evapolation of water (J/kg).

We shall assume that $\alpha = 0.4$ - 0.5, $\beta = 0.5$, $\Theta = 767.4$ - 802.3 (J/m^2s) in Fukuoka , $\eta(W) = 0.0829 + 0.0498W$, $TA = 273.16 + 20$ (K), $\epsilon = 0.97$, $\sigma = 5.67032 \times 10^{-8}$ (W/m^2K^4), $n = 0 \sim 1$, $\gamma(W) = 0.000308 + 0.000185W$, $E(\theta) = 4.47110 + 0.39626\theta + 0.00363\theta^2 + 0.00045\theta^3$ ($mmHg$), $r = 0 \sim 1$, and $L = 2.438 \times 10^6$ (J/kg), see Morikita et al.[2].

The time increment is roughly determined as $\Delta t = 10$ s, based on the Courant–Friedrichs–Lewy condition for the stability.

Figure 4 shows calculated elevation and surface temperature during 8 periods of the incident wave. The amplitude of the incident wave is modulated about 5 times in magnitude at the points P with the resonant period of about 4 hours. Figure 5 shows vertical distribution of velocity and temperature at the point P at the time $t = 8$ *hours*.

(a)   Incident wave on AB.

(b)   Elevation at P.

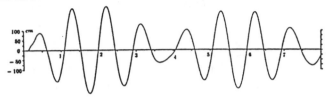

Figure 4. Calculated elevation induced in the channel.

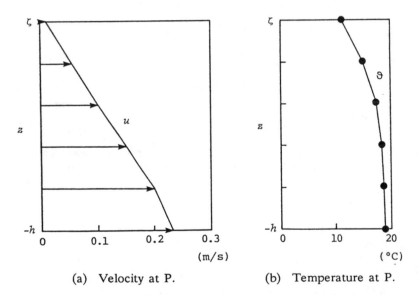

(a) Velocity at P.   (b) Temperature at P.

Figure 5.   Calculated vertical distribution.

Wind driven thermal current in the Shimouke Dam.   Figure 6 shows the outline and finite element subdivision of the dam. The outline is defined by 431 nodes on the boundary. The typical side length of the triangulation is 0.1 kilometers. The finite element mesh consists of 1143 nodes and 1555 triangular finite elements. Figure 7 shows 240 internal nodes where the depth was measured, and the contour of the depth plotted by the linear interpolation.

The depth ranges from 0 to 75.2 m, with $z = 0$ taken at EL = 336 m above the sea level.

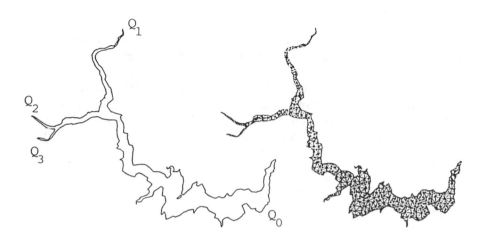

(a) Outline   (b) Finite element mesh.

Figure 6.   Shimouke dam.

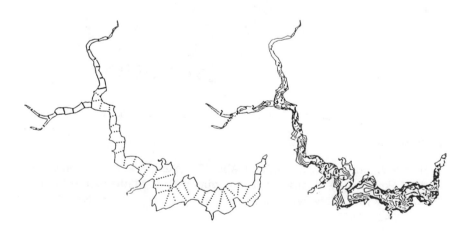

(a) Internal nodes.   (b) Depth contours.

Figure 7. Topology of the Shimouke dam.

We are interested in the flow in December, because it is the starting month of the outbreake of algae *Peridinium*. The mean elevation during Decembers from the year 1970 to 1988 is EL = 324.49, which implies ζ = $-$ 11.51 m. The mean influxes during the period are Q_1 = 2.16 (m^3/s) from the river *Taioi*, Q_2 = 1.19 (m^3/s) from the river *Kawahara*, and Q_3 = 1.15 (m^3/s) from the river *Uenoda*. The mean efflux in the corresponding period is Q_0 = 4.50 (m^3/s) out of the dam site. This implies that V_0 = 0.0316 (m/s), V_1 = -1.08 (m/s), V_2 = -0.595 (m/s), V_3 = -0.575 (m/s), respectively. The mean temperatures of the incoming water in Decembers during the period from 1985 to 1988 are θ_1 = 6.8 ($°C$) at the river *Taioi*, θ_2 = 8.1 ($°C$) at the river *Kawahara*, and θ_3 = 7.0 ($°C$) at the river *Uenoda*. The temperature of water in the dam during the corresponding period is about 8.5 - 14.5 ($°C$) and the mean atmospheric temperature is 5.3 ($°C$) in the city *Hita* near the dam. Figure 8 shows a sample of calculated distribution of mean flow velocity in the dam

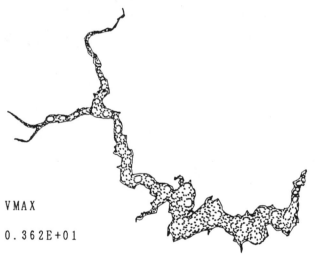

VMAX

0.362E+01

Figure 8.   Calculated mean flow velocity.

REFERENCES

1. Iwasa Y., Inoue, K., Liu S., and Abe, T.: Numerical Simulation of Flows in the Lake Biwa by Means of a Three–dimensional Mathematical Model. *in Japanese*, Kyoto University Bosai Institute Annual Report No. 26, B–2, pp.531–542 , 1987.

2. Morikita, Y., Hata, T., and Miura, S.: Numerical models on behaviors of cold–turbid water and the eutrophication in a reservoir (*in Japanese*), Japanese Public Works Research Institute, Dam Department, Water Resources Engineering Division, No.2443, ISSN 0386–5878, pp.1–313, 1987.

Numerical Prediction of Natural Convection Heat Transfer between a Cylindrical Envelope and an Internal Concentric Slotted Hollow Cylinder, with Confirmation of Experimental Results

M. Yang, W.Q. Tao, Z.Q. Chen

Department of Power Machinery Engineering, Xi'an Jiaotong University, Xi'an, Shaanxi 710049, China

ABSTRACT

In this paper a numerical prediction has been made for the natural convection heat transfer between a cylindrical envelope and internal concentric slotted hollow cylinder using the finite difference method with both colocated and staggered grid arrangements. The computed average equivalent thermal conductivities agree with experimental results quite well. The maximum relative difference between the conductivities with the two grid arrangements is less than 1%.

INTRODUCTION

The two-dimensional natural convection heat transfer between a cylindrical envelope and its inner concentric slotted hollow cylinder is of great importance to cylindrical enclosed isolated-phase busbars, which are used for transmitting large electric current in modern power plants. The enclosed busbar consists of two concentric metal cylinders, the inner hollow cylinder(the busbar) and the envelope. For the case of natural cooling of the busbar, the heat generated in the busbar from the Joule heating is first transferred from the busbar to the outer cylindrical envelope via natural convection and radiation, and then from the outside surface of the envelope to the ambient. To enhance the natural convection heat transfer between the heated inner cylinder and the outer envelope, the inner cylinder is often slotted along its mid-vertical plane. A schematic view of the busbar cross-section is shown in Fig. 1.

A number of studies, both experimental and numerical, have been conducted on laminar natural convection between two horizontal concentric cylinders maintained at constant, but different temperatures, such as references [1] , [2] . The natural convection between slotted busbar and the envelope is, however,

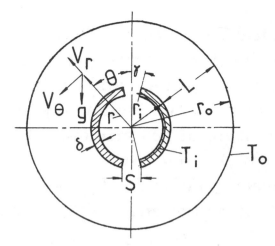

Fig. 1 Cross-section of enclosed slotted busbar

much more complicated, and very few works about this problem
have been published to-date. Kuleek [3] proposed an approximate
method of determining the total heat transfer rate for the enclosed slotted busbar according to a double-vortex flow pattern.
G. X. Wang et al. conducted an experimental study for a specific
configuration and provided an experimental correlation [4]. H.
L. Zhang et al. performed an experimental investigation on laminar natural convection between a cylindrical envelope and its
inner concentric octagonal slotted cylinder [5]. To the authors' knowledge, no any numerical study related to this problem has been published. The existence of the internal islands
(i.e. the two halves of the busbar) complicates the numerical
solution procedure. One purpose of this paper is to formulate
a mathematical model and to develop a solution procedure for
this problem, and compare the numerical results with the experimental ones provided in [4].

    In recent years in the finite difference approach with pri-
mitive variables, the colocated grid scheme is gradually be-
coming popular. Rhie [6], Peric et al. [7] have successfully obtained solutions of fluid flow problems using this grid arrangement. But there are still unexplored areas to which one can
turn attention. One of such areas is the coupled conduction and
convection problem, i. e. the conjugated problem. In the present study the heat conduction in the busbar and the fluid flow
in the rest part of the solution domain are solved simultaneously, thus the problem is of conjugated type. The second purpose
of the present study is to perform numerical computation using
both the colocated and staggered grids and to examine the characteristics of the colocated grid for the conjugated problems.

MATHEMATICAL FORMULATION AND NUMERICAL ALGORITHM

The analysis is based on the following assumptions:
(1)The fluid is of Boussinesq type of Prandtl number of 0.7; (2)
The envelope is at constant temperature T_o, and the points on
the mid-left arc in the busbar are at constant temperature T_i;
(3)The fluid flow and heat transfer are laminar and in steady
state; (4)Each half of the busbar consists of two concentric
arcs and two radial flanks. Thus in the polar coordinates the
dimensionless governing equations can be presented as

$$V\frac{\partial U}{\partial R} + \frac{U}{R}\frac{\partial U}{\partial \theta} = -\frac{\partial P}{\partial \theta} + Pr\left(\frac{\partial^2 U}{\partial R^2} + \frac{1}{R}\frac{\partial U}{\partial R} + \frac{1}{R^2}\frac{\partial^2 U}{\partial \theta^2}\right) - S_\theta \tag{1}$$

$$V\frac{\partial V}{\partial R} + \frac{U}{R}\frac{\partial V}{\partial \theta} = -\frac{\partial P}{\partial \theta} + Pr\left(\frac{\partial^2 V}{\partial R^2} + \frac{1}{R}\frac{\partial V}{\partial R} + \frac{1}{R^2}\frac{\partial^2 V}{\partial \theta^2}\right) - S_r \tag{2}$$

$$V\frac{\partial \phi}{\partial R} + \frac{U}{R}\frac{\partial \phi}{\partial \theta} = \frac{K_s}{K_f}\left(\frac{\partial^2 \phi}{\partial R^2} + \frac{1}{R}\frac{\partial \phi}{\partial R} + \frac{1}{R^2}\frac{\partial^2 \phi}{\partial \theta^2}\right) \tag{3}$$

$$\frac{1}{R}\frac{\partial U}{\partial \theta} + \frac{\partial V}{\partial R} - \frac{V}{R} = 0 \tag{4}$$

where

$$S_\theta = -\frac{UV}{R} + Pr\left(-\frac{U}{R^2} + \frac{2}{R^2}\frac{\partial V}{\partial \theta} - Ra\,\phi\,Sin\theta\right) \tag{5a}$$

$$S_r = \frac{U^2}{R} + Pr\left(-\frac{V}{R^2} - \frac{2}{R^2}\frac{\partial U}{\partial \theta} + Ra\,\phi\,Cos\theta\right) \tag{5b}$$

Due to the symmetry, only half of the solution domain needs to
be considered. The corresponding boundary conditions are:

$$r = r_i - (3/4)\delta, \quad \gamma \leqslant \theta \leqslant \pi - \gamma, \quad \phi = 1$$

$$r_i - \delta \leqslant r \leqslant r_i, \quad \gamma \leqslant \theta \leqslant \pi - \gamma, \quad U = V = 0$$

$$\theta = 0 \text{ or } \pi, \quad \frac{\partial V}{\partial \theta} = \frac{\partial \phi}{\partial \theta} = 0, \quad U = 0 \tag{6}$$

$$r = r_o, \quad U = V = 0, \quad \phi = 0$$

In the energy equation the ratio of K_s/K_f is equal to 1
for the fluid region. For the solid region of the slotted busbar the ratio is taken as 43.8 to simulate the experimental conditions of reference [4]

The finite difference formulations for equations (1)-(4)
may be derived using the control-volume approach with the staggered or the colocated grids. For the staggered grid, the discretization procedure and the resultant equations are well-known
[8], and will not be enumerated here. Only for the colocated
grid, a brief description of the discretization procedure is
given below.

The colocated grid arrangement for the polar coordinates is
shown in Fig. 2. In contrast with the staggered, in the coloca-

ted grid all variables are stored at the same locations. Therefore, only one control volume is needed. The discretization
procedure is similar to that with staggered grid. The forms of
the discretized U, V, and Φ equations are fully identical, so
only the discretized U-momentum equation is enumerated. It
reads

$$a_p U_p = \sum (a_{nb} U_{nb}) + b + (1 - \alpha) a_p U_p^* - \Delta R_p (P_e - P_w) \qquad (7a)$$

For simplicity, set

$$B = \left[\sum (a_{nb} U_{nb}) + b + (1 - \alpha) a_p U_p^* \right] / a_p \qquad (7b)$$

then we have

$$U_p = B_p - \Delta R_p (P_e - P_w) / a_p \qquad (7c)$$

In the above equations P_e and P_w are the pressures at the control
volume interfaces, which need to be determined by linear interpolation between the pressures of two neighbouring grid points.

The discretized continuity equation may be written as

$$(\rho U \Delta R)_e - (\rho U \Delta R)_w + (\rho VR \Delta\theta)_n - (\rho VR \Delta\theta)_s = 0 \qquad (8)$$

where the velocities at the control volume interfaces, U_e, U_w,
U_n and U_s, must be interpolated. By using equation (7c) at nodes
P and E, we have

$$U_p = B_p - \left[\Delta R (P_e - P_w) / a_p \right]_p \qquad (9)$$

$$U_E = B_E - \left[\Delta R (P_e - P_w) / a_p \right]_E \qquad (10)$$

According to Peric et al.[7] , for determining the interfacial
velocity U_e, following interpolation formula is used

$$U_e = \bar{B}_e + \Delta R_e (P_p - P_E) / (\bar{a}_p)_e$$

where $\bar{B}_e$ is the linear interpolation of B_p and B_E in equations
(9) and (10), and $(\bar{a}_p)_e$ is that of $(a_p)_p$ and $(a_p)_E$. Set $\bar{d} = \Delta R / \bar{a}_p$, we have

$$U_e = \bar{B}_e + \bar{d}_e (P_p - P_E) \qquad (11)$$

Similarly

$$U_w = \bar{B}_w + \bar{d}_w (PW - P_p) \qquad (12)$$

$$V_n = \bar{B}_n + \bar{d}_n (P_p - PN) \qquad (13)$$

$$V_s = \bar{B}_s + \bar{d}_s (P_S - P_p) \qquad (14)$$

In equations (13) and (14), $\bar{d}_n = (R \Delta\theta)_n / (\bar{a}_p)_n$, and $\bar{d}_s =$

$(R \Delta \theta)_s / (\bar{a}_p)_s$.

To deal with the linkage between the velocity and the pressure, the continuity equation is transformed into a pressure corection equation using a SIMPLE-like algorithm. The details of the derivation may be found in [8] . Only the resulting pressure correction equation is given below:

$$A_p P_p' = A_E P_E' + A_W P_W' + A_N P_N' + A_S P_S' - S \tag{15}$$

where P' is the pressure correction and

$$A_p = A_E + A_W + A_N + A_S$$

$$A_E = \bar{d}_e \, \rho_e \, \Delta R_e$$

$$A_W = \bar{d}_w \, \rho_w \, \Delta R_w \tag{16}$$

$$A_N = \bar{d}_n \, \rho_n (R \Delta \theta)_n$$

$$A_S = \bar{d}_s \, \rho_s (R \Delta \theta)_s$$

$$S = (\rho U^*)_e \Delta R_e - (\rho U^*)_w \Delta R_w + (\rho V^*)_n (R \Delta \theta)_n$$

$$\qquad\qquad - (\rho V^*)_s (R \Delta \theta)_s \tag{17}$$

In equation (17) U^* and V^* are the resulting velocities based on a guessed pressure field P^*.

In discretization of the convection-diffusion terms in the momentum and energy equations, the power-law scheme was used. The algebraic equations were solved by the successive line underelaxation method incorporated with the block-correction technique. The prescribed values of velocity and temperature in the

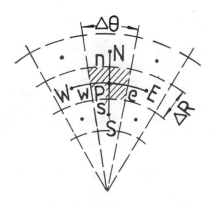

Fig. 2 Colocated grid in polar coordinates

busbar were obtained by introducing very large source terms into the discretized equations. The basic idea of this practice
was proposed in [8] . In addition, to ensure that the zero velocity could prevail throughout the solid busbar, a very small
number (say, 10^{-16}) was assigned to d_e, d_w, d_n and d_s.

The convergence criterions used in this study are

$$|F - F^*| < 10^{-3}$$

and

$$S_{max} < 10^{-4}$$

where F stands for one of the variables U, V, P and ϕ , and S_{max}
is the maximum value of the residuals in equation (15).

Attention is now turned to the calculation of the Rayleigh
number . The defining equation of Rayleigh number is

$$Ra = g\beta(T_i - T_o)L^3/(a\nu) \tag{18}$$

The average heat transfer characteristics is expressed in terms
of the mean equivalent thermal conductivity which is defined as

$$K_{eq} = \frac{Q\ln(r_o/r_i)}{2K_f(\pi - 2\gamma)(T_i - T_o)} \tag{19}$$

Due to the symmetry, equation (19) may be transformed into

$$K_{eq} = \frac{1}{\pi - 2\gamma}\int_0^\pi \frac{r_o}{T_i - T_o} \ln\frac{r_o}{r_i} \left.\frac{\partial T}{\partial r}\right|_{r=r_o} d\theta \tag{20}$$

RESULTS AND DISCUSSIONS

The solution domain was discretized with a 26(θ)x23(r) grid
shown in Fig. 3. A preliminary computation was performed to ensure the grid independence of the numerical results. It is
found that the maximum difference in the computed mean equiva
lent thermal conductivities of a 26x23grid and a 42x 42 grid is
less than 2% in the Rayleigh number range of 10^3 to 10^6. Since
this difference is small, the 26x23 grid is used for most computations of the study.

The geometric parameters used in this study are

$$r_o/r_i = 2.6, \quad \delta/2r_i = 0.09, \quad S/2r_i = 0.092$$

These specific values are taken from reference [4] , so that the
present numerical results and the experimental ones are compa-

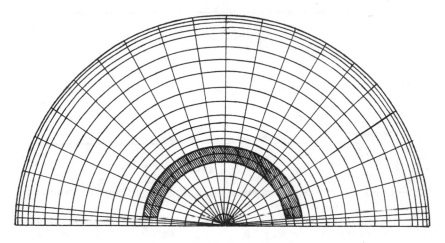

Fig. 3 Grid distribution

rable. The experimental correlations provided by [4] are

$$K_{eq} = 1 + (0.0894Ra^{0.3478})^{7.884} \quad 1/7.884, \quad 10^2 < Ra \leq 4.5 \times 10^4 \quad (21)$$

$$K_{eq} = 0.181Ra^{0.281}, \quad 4.5 < Ra \leq 4 \times 10^6 \qquad (22)$$

The maximum derivation of above correlations from the experimental data is 5.7%.

The comparison of the numerical results with the correlations are shown in Table 1 and Fig. 4. It may be seen from the table that the relative differences between the numerical prediction and the experimental correlation are in the order of 10%. With the experimental data scatter in mind, the agreement should be considered quite well.

Table 1 Mean equivalent thermal conductivity

Ra	10^3	10^4	3×10^4	10^5	1.3×10^5	1.5×10^5	5×10^5	10^6
Numerical (colocated)	1.202	2.489	3.535	5.002	5.391	5.615	7.855	9.410
Experimental	1.085	2.201	3.225	4.599	4.951	5.154	7.229	8.783
Relative difference,%	10.74	13.08	9.61	8.76	8.89	8.94	8.66	7.13
Numerical (staggered)	1.206	2.500				5.646		9.478

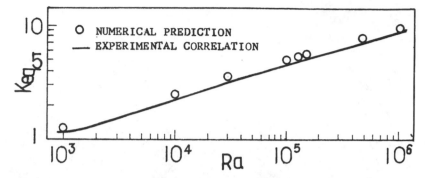

Fig. 4 Comparison between numerical and experimental results

Attention is now turned to the streamlines and isotherms. These are presented in Fig. 5 for Ra=1.5x10^5. It can be found by comparison with the photographs of flow visualization provided in [4] , that the predicted streamlines are fully agreeable with the flow visualization results. From Fig. 5 it can be seen that the flow in the envelope is mianly controlled by two vortices: one is at the annular space between the slotted busbar and the envelope; another is a larger vortex encircling the above-mentioned one. The larger vortex in formed as follows. A fluid stream goes into the space enclosed by the two halves of the busbar through the lower slot. This cold stream does not turn around immediately to meet the inner surface of the busbar. Rather, it goes up for a short distance, and then splits into two parts which turn around and flow to the vicinity of the two halves of the busbar. Hereafter, the two streams goes up along the inner surfaces of the busbar and gradually heated. They meet at the top of the space and rush out through the upper slot. Meeting with the envelope, the stream again splits into two parts, each of which turns around and goes down along the curved surface of the envelope. When the stream is coming near the bottom of the annulus, it turns around again, thus completing a close circulation of the stream. It is found that this flow pattern is more or less the same for the cases of Ra = 10^4-10^6.

Finally, attention is focused on the comparison of the results with colocated and staggered grids. The same problem was also computed with the staggered grid . Except the grid arrangement and the interpolation principle, all other numerical methods were identical for the two grid schemes. In the last row of Table 1, the computed mean equivalent thermal conductivity with staggered grid has been presented. It shows that the maximum relative difference between the equivalent conductivities with colocated grid and staggered grid is less than 1%. For most region of the computational domain, the differences between the computed velocities with these two grid arrangements are very small. However, near the slots of the busbar, where

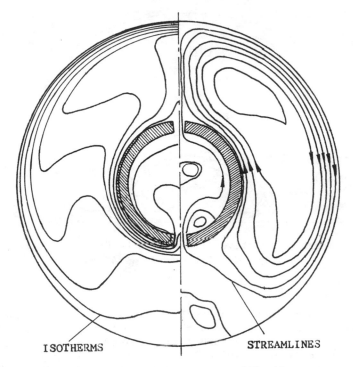

ISOTHERMS   STREAMLINES

Fig. 5 Predicted Streamlines and isotherms
(Ra = 1.5×10^5)

the velocity gradients are large, the differences are much
higher. If a relative difference, $\mathcal{E}$, is defined as $\mathcal{E}$ =
$|(F_c - F_s)/F_{max}|$ x100%, where the subscripts c and s represent
the colocated grid and staggered grid, respectively, the relative difference may be as high as 15% on a few nodes near the
upper slot. It was considered that refining the grid might reduce the differences. Computations with a 42x42 grid for both
colocated scheme and staggered scheme were then performed for
Ra=1.5×10^5. It is found that the relative differences between
the mean equivalent thermal conductivities of the two grids are
less than 0.1%. In most region of the solution domain, the velocity relative differences are in the order of 0.2%, with a
maximum value of 5%. Contours of constant relative differences
between the results of the two grid schemes on the 42x42 grid
are presented in Fig. 6. It can be seen there that the maximum
difference occurs at the nodes near the slot mouth.

As far as the solution procedures are compared, the colocated grid presents some inconvenience for conjugated problem with
internal isolated solid island. In this case for the colocated
grid the pressure values on the island boundary must be interpolated in each iteration, while for the staggered grid this
interpolation is not needed.

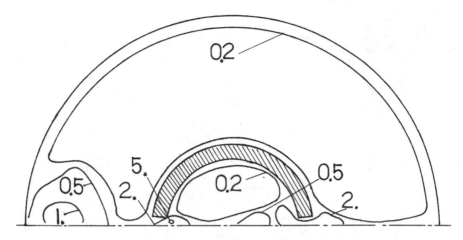

Fig. 6 Contours of relative differences between colocated and staggered solutions on 42x42 grid

CONCLUSIONS

    In the present work numerical prediction of natural convection heat transfer between a slotted hollow cylinder and its concentric envelope was performed with the finite volume approach in the range of $Ra=10^3 - 10^6$. The colocated grid was, seemingly first in the literature, successfully used to solve the conjugated heat transfer problem with internal isolated island. The numerical results of the average heat transfer characteristics and the flow patterns agreed with the experimental ones quite well. For a 42x42 grid, the relative field value difference between the colocated solution and staggered solution were mostly in the order of 0.2%, with a maximum value of 5%, occuring at the locations with the greatest velocity gradient.

NOMENCLATURE

a	= thermal diffusivity of fluid
a_{nb}, a_p	= coefficients in discretized momentum equation
A_{nb}, A_p	= coefficients in pressure correction equation
b	= constant term in discretized momentum equation
B	= temporary variable defined by equation (7b)
$d_{n,e,w,s}$	= coefficients in equations (11) - (14)
F	= general variable
g	= gravitational acceleration
K_{eq}	= equivalent thermal conductivity
K_f	= thermal conductivity of fluid
K_s	= thermal conductivity of solid
L	= width of annulus
p	= pressure
P	= dimensionless pressure, $p/[\rho (a/L)^2]$
Q	= heat transfer rate between envelope and busbar

r   = radius
R   = dimensionless radius, r/L
Ra   = Rayleigh number, $g\beta(T_i - T_o)L^3/(a\nu)$
S   = width of slot
S_r, S_θ = source terms in discretized momentum equations
S_{max} = maximum rasidual in pressure coreection equation
T   = temperature
U   = dimensionless velocity in θ-direction, $v_\theta L/a$
v_r, v_θ = velocity components in r- and θ- directions, respectively
V   = dimensionless velocity in r-direction, $v_r L/a$

Greek symbols

α = underrelaxation factor
β = volume expansion coefficient
γ = half the angle subtended by the width of slot
δ = thickness of busbar
θ = angle of polar coordinates
ν = kinematic viscoty
ρ = density
ϕ = dimensionless temperature, $(T - T_o)/(T_i - T_o)$

Superscript

*   = guessed or previous value

Subscripts

i   = inner
o   = outer

ACKNOWLEDGMENT

This work was supported by the National Educational Committee of China.

REFERENCES

1. Kuehn, T. H. and Goldstein, R. J., An Experimental and Theoretical Study of Natural Convection in the Annulus between Horizontal Concentric Cylinders, J. Fluid Mech., vol. 74, pp. 695-719, 1976
2. Date, A. W., Numerical Prediction of Natural Convection Heat Transfer in Horizontal Annulus, Int. J. Heat Mass Transfer, vol. 29, pp. 1457-1464, 1986
3. Kuleek, P. V., Heating and Cooling of Busbars for large Power Generators, Electric Power Plant(in Russian), no, 10, pp. 39-43, 1964
4. Wang, G. X., Wu, Q. J. and Wang, Q. J., An Experimental Study of Natural Convection in the Cylindrical Envelope with Two Halves of Circular Busbar, in: Proceedings of the 1986 Engineering Thermophysics Conference(in Chinese), Science Press,

Beijing, 1988
5. Zhang, H. L. Wu, Q. J. and Tao, W. Q., Experimental Study of
   Natural Convection Heat Transfer between Cylindrical Envelope
   and Internal Concentric Heated Octagonal Cylinder with or
   without Slotts, submitted to the Journal of Heat Transfer
   for publivation
6. Rhie, C. M., A Numerical Study of the Flow Past an Isolated
   Airfoil with separation, Ph. D. Thesis, University of Illinois, Urbana-Champaign, 1981
7. Peric, M., Kessler, R. and Scheuerer, G., Comparison of Finite-Volume Numerical Mehtods with Staggered and Colocated
   Grids, Computers & Fluids, vol. 16, pp. 389-403, 1988
8. Patankar, S. V., Numerical Heat Transfer and Fluid Flow,
   Hemisphere, Washington, D. C., 1980

Solution Comparison of Three Coupled Fluid Flow and Heat Transfer Problems with Staggered and Colocated Grids

M. Yang, W.Q. Tao, Z.Q. Chen
Department of Power Machinery Engineering, Xi'an Jiaotong University, Xi'an, Shaanxi 710049, China

ABSTRACT

In this paper solution comparison has been made for three coupled fluid flow and heat transfer problems with three versions of
the finite difference approach: the staggered grid and two vari-
ants of the colocated grid. For the problem studied, the numerical results of the three versions agree with each other fairly well. Special computation has been performed to demonstrate
the practical ability of the colocated grid to eliminate the
checkerboard pressure field.

INTRODUCTION

When fluid flow and heat transfer problems are to be solved
by finite difference approach in primitive variables, one often
uses a staggered grid. In the staggered grid the velocity components and pressure are located at displaced or staggered locations [1] . Although the staggered grid is able to eliminate
the checkerboard pressure field, it introduces some inconvenience. When this grid arrangement is to be extended to curvilinear
non-orthogonal coordinates or to 3-D coordinates, the inconvenience becomes more severe. Besides, when one wants to incorporate the multigrid technique in the solution algorithm to accelerate the convergence rate of solving algebraic equations, the
staggered grid has obvious disadvantages because different interpolation operators must be applied to different variables.
In recent years, efforts have been devoted to develope a non-
staggered or colocated grid. In this grid arrangement, all variables are stored at the same grid point, therefore, the solution procedure is simpler than that of the staggered grid. In
1981 Rhie [2] successfully gained the solution of a fluid flow
problem in a 2-D non-orthogonal coordinates using the colocated
grid arrangement. Since then, several other authors have also

succeeded in using the colocated grid for the numerical solutions of fluid flow problems. Peric et al. [3] conducted numerical computations using the two grid arrangements for three
problems of imcompressible 2-D fluid flow. They concluded that
the colocated grid has no disadvantages relative to the staggered version. However, Majumdar [4] found that the converged
solution resulted from Peric's method is dependent on the under-
relaxation factor used for velocity. He proposed a remedy for
it. To the authors' knowledge, all the examples solved by the
colocated grid in the published literature are fluid flow problems, no any examples of coupled fluid flow and heat transfer
problems have been provided. The purpose of the present work
is to present a solution comparison of three coupled fluid flow
and heat transfer problems by using Peric's method, Majumdar's
remedy and the standard staggered grid, and to observe whether
the colocated grid is able to eliminate the checkerboard pressure field if an initial pressure field with checkerboard character is assigned.

DISCRETIZATION EQUATIONS AND SOLUTION ALGORITHM

Governing equations
     Consider 2-D, steady state, laminar and coupled fluid flow
and heat transfer problems. In the Cartesian coordinates, the
governing equations can be presented as

$$\frac{\partial(\rho UU)}{\partial x} + \frac{\partial(\rho UV)}{\partial y} = -\frac{\partial P}{\partial x} + \frac{\partial}{\partial x}\left(\mu \frac{\partial U}{\partial x}\right) + \frac{\partial}{\partial y}\left(\mu \frac{\partial U}{\partial y}\right) + S_u \tag{1}$$

$$\frac{\partial(\rho VV)}{\partial x} + \frac{\partial(\rho UV)}{\partial y} = -\frac{\partial P}{\partial y} + \frac{\partial}{\partial x}\left(\mu \frac{\partial V}{\partial x}\right) + \frac{\partial}{\partial y}\left(\mu \frac{\partial V}{\partial y}\right) + S_v \tag{2}$$

$$\frac{\partial(\rho UT)}{\partial x} + \frac{\partial(\rho VT)}{\partial y} = \frac{\partial}{\partial}\left(\frac{k}{c_p}\frac{\partial T}{\partial x}\right) + \frac{\partial}{\partial}\left(\frac{k}{c_p}\frac{\partial T}{\partial y}\right) \tag{3}$$

$$\frac{\partial(\rho U)}{\partial x} + \frac{\partial(\rho V)}{\partial y} = 0 \tag{4}$$

In the above equations, the fluid properties, except the density , are assumed to be constant.

Finite difference momentum equations with the colocated grid
     The finite difference forms of equations (1) - (3) are now
derived using the control volume method with a colocated grid
shown in Fig. 1. In contrast with the staggered grid, in the
colocated grid all variables are stored at the same location.

Therefore, only one set of control volume is needed. The details of derivation may be found in [1] or [3] . The forms of
the discretization equations for U, V, and T are fully identical,
so only the discretized U-momentum equation is enumerated. It
reads

$$a_p U_p = \sum(a_{nb}U_{nb}) + b + (1 - \alpha)a_p U_p^0 - \triangle y_p (P_e - P_w) \tag{5a}$$

For simplicity, set

$$B = \left[\sum(a_{nb}U_{nb}) + b + (1 - \alpha)a_p U_p^0\right] \Big/ a_p \tag{5b}$$

then we have

$$U_p = B_p - (\triangle y_p/a_p)(P_e - P_w) \tag{6}$$

For the momentum equations, the pressures at the interfaces of control volume will be determined by linear interpolation between the pressures of its two neighboring grid points.

Interpolation formulas for velocity components at interfaces

As shown in Fig. 1, in the colocated grid the velocities are calculated at the same grid points as those for pressure and temperature. Thus the velocities at the control volume interfaces must be interpolated. By using equation (6) at nodes P and E, we have

$$U_p = B_p - \left[\triangle y(P_e - P_w)/a_p\right]_p \tag{7}$$

$$U_E = B_E - \left[\triangle y(P_e - P_w)/a_p\right]_E \tag{8}$$

According to Peric et al.[3] , for calculating the interfacial velocity U_e, following interpolation formula is used

$$U_e = \overline{B_e} + \triangle y_e/(\overline{a_p})_e(P_p - P_E)$$

where $\overline{B_e}$ and $(\overline{a_p})_e$ are the linear interpolations of B_p, B_E and $(a_p)_p$, $(a_p)_E$, respectively. Set $\overline{d} = \triangle y/\overline{a_p}$, we have

$$U_e = \overline{B_e} + \overline{d_e}(P_p - P_E) \tag{9}$$

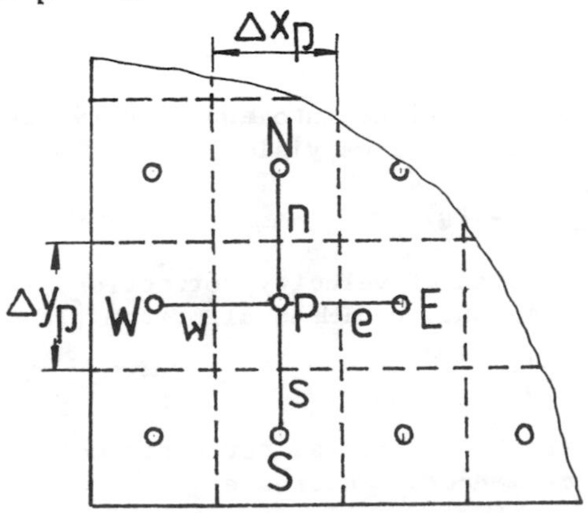

Fig. 1 Colocated grid

Similarly,

$$U_w = \overline{B}_w + \overline{d}_w(P_W - P_p) \tag{10}$$

$$V_n = \overline{B}_n + \overline{d}_n(P_p - P_N) \tag{11}$$

$$V_s = \overline{B}_s + \overline{d}_s(P_S - P_p) \tag{12}$$

From equations (9) - (12), it can be seen that the velocities at the control volume interfaces depend on the pressure difference of two neighboring grid points, which is a successful practice of the staggered grid. Equations (9) - (12) are called momentum interpolation of interfacial velocity.

Discretized continuity equation and pressure correction equation
The discretized continuity equation may be written as

$$(\rho U_e - \rho U_w)\triangle y_p - (\rho V_n - \rho V_s)\triangle x_p = 0 \tag{13}$$

To deal with the linkage between the velocity and the pressure, the continuity equation is transformed into a pressure correction equation using a SIMPLE-like algorithm. In a iterative solution procedure of the resulted equations, the velocities U_e^o, U_w^o, V_s^o and V_n^o at four interfaces of control volume P can be obtained from the initial guessed values or the previous iteration. The velocities U^* and V^*, computed with these guessed values or the results of the previuos iteration, will in general not satisfy the continuity equation (13), unless the iteration is converged. Let U' and P' be the corrections of U and P, respectively, and B' be the correction of B, then we have

$$U = U^* + U'$$

$$P = P^* + P'$$

$$B = B^* + B'$$

Substituting these equations into equation (9) and noting that U^* satisfies equation (9), we yield

$$U_e' = B_e' + \overline{d}_e(P_p' - P_E')$$

where B' is a function of velocity corrections and will be neglected, as this is done in SIMPLE algorithm [1] , thus we have

$$U_e' = \overline{d}_e(P_p' - P_E') \tag{14}$$

The velocity correction formulas for other components can be derived in the same manner. These are

$$U_w' = \overline{d}_w(P_W' - P_p') \tag{15}$$

$$V_n' = \bar{d}_n(P_p' - P_N') \tag{16}$$

$$V_s' = \bar{d}_s(P_S' - P_p') \tag{17}$$

Substituting $(U_e^* + U_e')$, $(U_w^* + U_w')$, $(V_n^* + V_n')$ and $(V_s^* + V_s')$ into equation (13), after rearrangement we obtain following pressure correction equation

$$A_p P_p' = A_E P_E' + A_W P_W' + A_N P_N' + A_S P_S' - S \tag{18}$$

where

$$A_p = A_E + A_W + A_N + A_S$$

$$A_E = d_e \rho_e \triangle y_e$$

$$A_W = d_w \rho_w \triangle y_w$$

$$A_S = d_s \rho_s \triangle x_s \tag{19}$$

$$A_N = d_n \rho_n \triangle x_n$$

$$S = \rho_e U_e^* \triangle y_e - \rho_w U_w^* \triangle y_w + \rho_n V_n^* \triangle x_n - \rho_s V_s^* \triangle x_s \tag{20}$$

Majumdar's remedial method

The colocated grid presented above has an undesirable feature, i. e., its converged results are, in some extent, dependent on the underrelaxation parameters. According to Majumdar's analysis [4] , this is because when an iterative algorithm is used and underrelaxation technique is incorporated, the equations for dertermining interfacial velocities, equations (9) - (12), are mixed expressions, which are made up of α portion of the momentum interpolation, and $(1-\alpha)$ portion of linear interpolation. Therefore, the converged solution of the interfacial velocity does not converge to the desired value of the full momentum interpolation. To eliminate this inconsistency in interpolation of the interfacial velocity, Majumdar proposed a remedial method. To distinguish, we call the method presented above as CR, Majumdar's remedy as CM and the standard staggered grid as SS. In CM the equations for calculating the interfacial velocities are rewritten as follows

$$U_e = \bar{B}_e + \bar{d}_e(P_p - P_E) + (1-\alpha)\left[U_e^0 - f_x^+ U_p^0 - (1 - f_x^+)U_E^0\right] \tag{21}$$

$$U_w = \bar{B}_w + \bar{d}_w(P_W - P_p) + (1-\alpha)\left[U_w^0 - f_x^- U_p^0 - (1 - f_x^-)U_W^0\right] \tag{22}$$

$$V_n = \bar{B}_n + \bar{d}_n(P_p - P_N) + (1-\alpha)\left[V_n^0 - f_y^+ V_p^0 - (1 - f_y^+)V_N^0\right] \tag{23}$$

$$V_s = \bar{B}_s + \bar{d}_s(P_S - P_p) + (1-\alpha)\left[V_s^0 - f_y^- V_p^0 - (1 - f_y^-)V_S^0\right] \tag{24}$$

All other calculation formulas in CM are the same as those in

CR.   In equations (21) – (24)f_x^+, f_x^-, f_y^+ and f_y^- are geometric factors. Taking f_x^+ as an example,

$$f_x^+ = eE/(Pe + eE) \tag{25}$$

In the next section, the results of three test problems will be presented. The three problems are: natural convection in a square cavity, flow and heat transfer in a 2-D sudden expansion and natural convection in a horizontal annular gap. For the third problem the polar coordinates was used.

RESULTS AND DISCUSSIONS

The computations for the three problems were conducted as follows: using CR, CM and SS for the first problem, and CM and SS for the second and third problems. The power-law scheme was used for the convection-diffusion term. The successive line underrelaxation method incorporated with the block-correction technique was used for solving the algebraic equations of the three problems. The iteration is considered to be converged if the following two conditions are both satisfied

1. $S_{max} < 5 \times 10^{-5}$ \hfill (26)

2. $e = |\phi - \phi^o| < 10^{-3}$ \hfill (27)

where S_{max} is the maximum value of the residuals in each control volume and ϕ stands for one of the variables U, V, T and P.

For comparison between the solutions of CR, CM and SS, the following relative difference is defined

$$\varepsilon = \left| \frac{\phi_A - \phi_B}{\phi_R} \right| \times 100\% \tag{28}$$

where ϕ_A and ϕ_B represent one of the solutions of CR, CM and SS, while ϕ_R is a reference field value.

Problem 1: natural convection in a square cavity
     This is a standard test problem in numerical heat transfer [5] . The statement of the problem is as follows: consider the 2-D flow of a Boussinesq fluid of Prandtl number 0.71 in an upright square cavity described in non-dimensional terms by $0 \leqslant x \leqslant 1$, $0 \leqslant y \leqslant 1$ with y vertical upwards. Assume that both components of the velocity are zero on all the boundaries, that the boundaries at y = 0 and 1 are insulated, $\partial T/\partial y = 0$, and that T = 1 at x = 0 and T = 0 at x = 1. Calculate the flow and thermal fields for Rayleigh number, $\beta g \Delta T H^3 / a\nu$, of 10^4. The geometry and the predicted streamlines are shown in Fig. 2.

Part of our results (with 20x20 grid) is listed in Tables 1 and 2. For comparison, the results in [6] and the bench mark

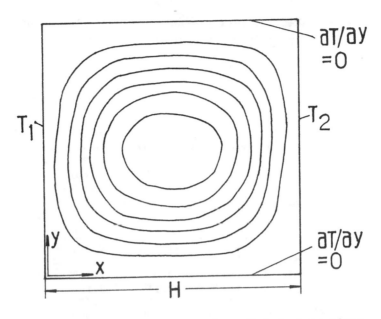

Fig. 2 Geometry of problem 1 and predicted streamlines

solution [5] are also provided. In these tables the dimensionless velocities $\widetilde{U}$ and $\widetilde{V}$ are defined as follows

$$\widetilde{U} = UH/a, \quad \widetilde{V} = VH/a \tag{29}$$

If the value 19.171 of $\widetilde{V}_{max}$ (Table 1) is taken as the reference value ϕ_R for both U and V, then from Table 2 we can
observe following features: (1) Increasing the underrelaxation
factor in CR, the relative difference $\mathcal{E}_{(CR-CM)}$ decreases, (2)
Both $\mathcal{E}_{(CR-SS)max}$ ($\alpha = 0.8$) and $\mathcal{E}_{(CM-SS)max}$ are less than
1%, and $\mathcal{E}_{(CR-SS)max}$ ($\alpha = 0.4$) is less than 1.5%, (3) $\mathcal{E}_{(CM-SS)}$
is less than $\mathcal{E}_{(CR-SS)}$, (4) Compared with the bench mark solution, the results obtained from the staggered grid and the colocated grid have the same order of accuracy.

Table 1 Comparison of heat transfer characteristics of problem 1

Solu-tions	Nu	Nu$_{max}$	y	Nu$_{min}$	y	$\widetilde{U}_{max}$	y/H x/H=.5	$\widetilde{V}_{max}$	x/H y/H=.5
[5]	2.238	3.527	0.143	0.586	1.0	16.178	0.823	19.643	0.119
[6]	2.242	3.554	0.133	0.617	0.933			19.450	0.133
CR	2.311	3.812	0.125	0.580	0.975	16.231	0.833	19.327	0.125
CM	2.303	3.805	0.125	0.579	0.975	16.113	0.825	19.164	0.125
SS	2.304	3.800	0.125	0.579	0.975	16.117	0.825	19.171	0.125

Table 2 Comparison of dimensionless velocityies of problem 1

y/H (x/H=0.5)	$\widetilde{U}$			
	CR		CM	SS
	α =0.4	α =0.8	α=0.65	α =0.65
0.975	5.208	5.233	5.232	5.237
0.875	15.069	15.177	15.189	15.287
0.775	15.189	15.302	15.323	15.439
0.675	10.719	10.790	10.809	10.892
0.575	4.610	4.656	4.661	4.706
x/H (y/H=0.5)	$\widetilde{V}$			
	CR		CM	SS
	α =0.4	α =0.8	α =0.65	α =0.65
0.025	9.323	9.354	9.367	9.377
0.125	19.082	19.137	19.164	19.327
0.225	13.463	13.554	13.570	13.683
0.325	6.935	7.058	7.062	7.103
0.425	2.439	2.518	2.517	2.530

Problem 2: flow and heat transfer in a 2-D sudden expansion
     Consider the 2-D,laminar flow and heat transfer in a sudden expansion. The wall temperature is constant(300K), the fluid inlet temperature is equal to 500K. To make the fluid flow and heat transfer coupled, the fluid density in the convection terms of the governing equations are assumed to be equal to T_w/T. Body force is neglected. Calculate the velocity and temperature fields for the inlet Reynolds number, $\rho_{in}U_{in}h/\mu_{in}$, of 20. The fluid prandtl number is 0.7. The computation domain and the predicted streamlines are shown in Fig. 3.

     For this problem, the dimensionless velocities $\widetilde{U}$ and $\widetilde{V}$ are defined as

$$\widetilde{U} = U/U_{in}, \qquad \widetilde{V} = V/U_{in}$$

If $\widetilde{U}_{in}$ is taken as the reference value, then the maximum difference between the solutions of CM and SS is less than 1%(with a 20x20 grid).

Problem 3: natural convection in an annular gap
     The third problem is the 2-D, laminar natural convection heat transfer of Boussinesq fluid of Prandtl number 0.7 in an annular gap. The inner cylinder temperature T_i is higher than the outer one T_o. The ratio of the gap width $L(= D_o - D_i)$ to the inner cylinder dyameter D_i is 0.8. Calculate the flow field and the average equivalent conductivity for Rayleigh number,

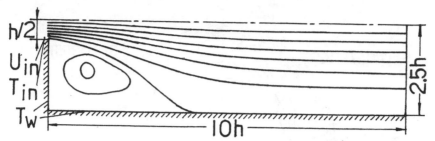

Fig. 3 Geometry of problem 2 and predicted streamlines

$\beta g \triangle TL/a\nu$, of 2×10^4.

The average equivalent conductivity is defined as

$$K_{ei} = \frac{1}{\pi} \int_0^\pi \frac{r_i}{T_o - T_i} \, \mathrm{Ln}\left(\frac{r_o}{r_i}\right)\left(\frac{\partial T}{\partial r}\right)_{r=r_i} d\theta$$

at the outer cylinder

$$K_{eo} = \frac{1}{\pi} \int_0^\pi \frac{r_o}{T_o - T_i} \, \mathrm{Ln}\left(\frac{r_o}{r_i}\right)\left(\frac{\partial T}{\partial r}\right)_{r=r_o} d\theta$$

The dimensionless velocities $\widetilde{U}$ and $\widetilde{V}$ are defined as

$$\widetilde{U} = UL/a, \qquad \widetilde{V} = VL/a$$

A grid with 20x20 points was used. The predicted streamlines and isotherms are shown in Fig. 4. In table 3 the values of the average equivalent condictivity are compared with the results provided by Küehn and Goldstein [7] . It can be seen that the agreement between the results obtained by different methods is very good. As far as the field values are concerned, the maximum relative difference between the two solutions, $\mathcal{E}$ (CM-SS), is about 2%(the reference value is $\widetilde{U}_{max}$).

Table 3 Comparison of average results of problem 3

K_e	Present results		Kuehn and Goldstein	
	CM (20x20)	SS (20x20)	Numerical (16x19)	Experimental Pr=0.706,Ra=2.11 10^4
K_{ei}	2.4080	2.4043	2.405	2.34
K_{eo}	2.4080	2.4043	2.394	2.34

Comparison of the convergence rate
    Comparison in iteration times was made for problem 2 with different underrelaxation factors. It is found that the con-

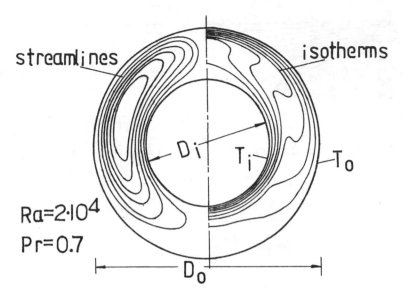

Fig. 4 Predicted isotherms and streamlines of problem 3

vergence rates of CR and SS are almost identical and sometimes the convergence rate of CR is even slightly faster.

Ability to eliminate the checkerboard pressure field
        To demonstrate the ability of colocated grid to eliminate checkerboard pressure field, special computation test was performed for problem 1, by taking (1) a checkerboard field as the initial pressure field, (2) the sum of the checkerboard field and the converged result as the initial distribution. Numerical computations reveal that the final converged solutions are the same as those obtained previously. Part of the test results for case (2) is shown in Fig. 5. In the computation the initial values of other variables were the corresponding converged results of the previous solutions.

CONCLUDING REMARKS

        Based on the numerical computations for the three coupled fluid flow and heat transfer problems, it may be concluded that the results obtained from CR, CM and SS agree with each other fairly well, especially for the heat transfer characteristics. The maximum relative difference in the local velocity values is about 1 - 2% of an appropriate reference value. The colocated scheme is able to eliminate the checkerboard pressure field. The CR solution is, in some extent, dependent on the underrelaxation factors, in an order of 1% relative to the value of itself. In order to demonstrate general applicability of the colocated scheme, attention may be turned on more complicated cases, such as flows at large Rayleigh numbers, highly swirling flows, etc.

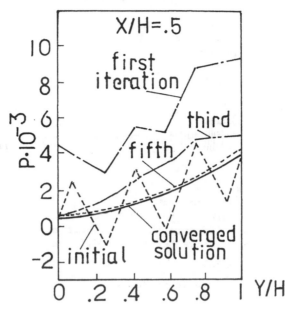

Fig. 5 Change of pressure field during iteration

NOMENCLATURE

a	=	thermal diffusivity
a_{nb}, a_p	=	coefficients in discretized momentum equation
$A_{N,E,W,S,P}$	=	coefficients in pressure correction equation
b	=	constant term in discretized momentum equation
B	=	term defined by equation (5b)
C_p	=	specific heat
d	=	ratio of $\triangle y$ to a_p or $\triangle x$ to a_p
D	=	diemeter
$f_x^+, f_{\bar{x}}^-, f_y^+, f_{\bar{y}}^-$	=	interpolation factors
g	=	gravitational acceralation
h	=	width of channel inlet, Fig. 3
H	=	height of square cavity, Fig. 2
k	=	thermal conductivity
K_e	=	equivalent thermal conductivity
L	=	width of annular gap
p	=	pressure
Pr	=	Prandtl number
r	=	radius
Ra	=	Rayleigh number
Re	=	Reynolds number
S	=	residual of continuity equation
S_u, S_v	=	source terms of discretized momentum equations
S_{max}	=	maximum residual of continuity equation
T	=	temperature
U, V	=	velocity components
x, y	=	Cartesian coordinates

Superscripts

o	= previous iteration
*	= temporary solution
—	= mean
~	= dimensionless
/	= correction

Subsctpts

i	= inner
in	= inlet
o	= outer
w	= wall

Greek symbols

α	= underrelaxation factor
β	= volume expansion coefficient
Δ	= difference in temperature or length
ε	= relative percentage difference
θ	= angle
μ	= dynamic viscosity
ν	= kinematic viscosity
ρ	= density
ϕ	= general variable

ACKNOWLEDGMENT

This work was supported by the National Educational Committee of China.

REFERENCES

1. Patankar, S. V., Numerical Heat Transfer and Fluid Flow, Hemisphere, Washington, D. C., 1980
2. Rhie, C. M., A Numerical Study of the Flow Past an Isolated Airfoil with Separation, Ph. D. Thesis, University of Illinois, Urbana-Champaign, 1981
3. Peric, M., Kessler, R. and Scheuer, G., Comparison of Finite Volume Numerical Methods with Staggered and Colocated Grids, Computers & Fluids, vol. 16, pp. 389-403, 1988
4. Majumdar, S., Role of Underrelaxation in Momentum Interpolation for Calculation of Flow with Nonstaggered Grids, Numer. Heat Transfer, vol. 13, pp. 125-132, 1988
5. De Vahl Davis, G. and Jones, I. P., Natural Convection in a Square Cavity : a Comparison Exercise, Int. J. Num. Meth. Fluids, vol. 3, pp. 227-248, 1983
6. Jones, I. P., A Comparison Problem for Numerical Methods in Fluid Dynamics: the 'Double-Glazing' Problem, in Numerical Methods in Thermal Problems, Ed. R. W. Lewis and K. Morgan, Pineridge Press, Swansea, U. K., pp. 338-348
7. Kuehn, T. H. and Goldstein, R. J., An Experimental and Theoretical Study of Natural Convection in the Annulus between Horizontal Concentric Cylinders, J. Fluid Mech., vol. 74, pp. 695-719, 1976

Boundary-Domain Integral Method for Mixed Convection

Z. Rek, P. Skerget, A. Alujevic

Faculty of Engineering, Smetanova 17, 62000 Maribor, Yugoslavia

ABSTRACT

Two dimensional mixed convection heat transfer is considered by the boundary-domain integral method. New numerical scheme for the kinetic part is introduced, which proved to be very stable even for higher Reynolds (Re), Peclet $(Pe=Re \cdot Pr)$, Grasshoff (Gr) and Rayleigh $(Ra=Gr \cdot Pr)$ number values. Comparison between new and old scheme is made for two different cases: Convection in a square cavity and convection in a channel. Theory is supported with the case dealing with injection of hot fluid into cold fluid at rest.

INTRODUCTION

Boundary-domain integral method has proved to be very useful method for solving problems in fluid mechanics and heat transfer. With transformation the system of partial differential equations into a set of boundary-domain integral equations, invoking discretisation, in many cases very stable and good conditioned numerical scheme may be obtained. This is valid for laminar flow of viscous incompressible fluid and natural convection. In case of forced convection, we found that numerical scheme for vorticity transport is not good for energy transport, although the equations are very similar. For higher values of Re or Gr numbers, the numerical scheme becomes unstable. To avoid this problem, a new scheme was introduced.

GOVERNING EQUATIONS

Time dependent laminar flow of viscous incompressible fluid and energy transport are determined with laws of conservation:

- mass

$$\vec{\nabla} \cdot \vec{v} = 0, \tag{1}$$

- momentum

$$\frac{\partial \vec{v}}{\partial t} + (\vec{v} \cdot \vec{\nabla})\vec{v} = \nu \triangle \vec{v} - \frac{1}{\rho_0}\vec{\nabla}p + [1 - \beta(T - T_0)]\vec{g}, \tag{2}$$

- energy

$$\frac{\partial T}{\partial t} + (\vec{v} \cdot \vec{\nabla})T = a\triangle T. \tag{3}$$

Material properties: ρ_0 density at reference temperature T_0, β coefficient of volume thermal expansion, ν kinematic viscosity, $a = \dfrac{\lambda}{\rho_0 c}$ diffusivity, λ thermal conductivity, c specific heat. The buoyancy forces are included with Bussinesques approximation for volume forces

$$\rho\vec{g} = \rho_0[1 + \beta(T - T_0)]\vec{g}$$

System of equations is closed, unknown are velocity $\vec{v}$, temperature T, pressure p. They are computed from the system of equations (1), (2), (3) for given boundary and initial conditions.

System of boundary-domain integral equations is

flow kinematics

$$c(\xi)v_x(\xi) + \int v_x(S)q^{*E}(\xi, S)\, d\Gamma \;=\; \int v_y(S)q^{*E}(\xi, S)\, d\Gamma$$
$$- \int \omega(s)q_x^{*E}(\xi, s)\, d\Omega$$

$$c(\xi)v_y(\xi) + \int v_y(S)q^{*E}(\xi, S)\, d\Gamma \;=\; -\int v_x(S)q^{*E}(\xi, S)\, d\Gamma$$
$$+ \int \omega(s)q_y^{*E}(\xi, s)\, d\Omega \tag{4}$$

flow kinetics

$$c(\xi)\omega(\xi, t_F) + \nu\int\int \omega(S, t)q^{*P}(\xi, t_F; S, t)\, d\Gamma dt$$

$$= \nu\int\int q^{\omega}(S, t)u^{*P}(\xi, t_F; S, t)\, d\Gamma dt$$

$$- \int\int[\omega(S, t)v_n(S, t) + \beta g_t T(S, t)]u^{*P}(\xi, t_F; S, t)\, d\Gamma dt \tag{5}$$

$$+ \int\int[\omega(s, t)\vec{v}(s, t) + \beta(g_y, -g_x)T(S, t)]\vec{\nabla}u^{*P}(\xi, t_F; s, t)\, d\Omega dt$$

$$+ \int \omega(s, t_{F-1})u^{*P}(\xi, t_F; s, t_{F-1})\, d\Omega$$

energy transport

$$c(\xi)T(\xi, t_F) \; + \; a \int\!\!\int T(S,t)q^{*P}(\xi, t_F; S, t)\, d\Gamma dt$$

$$= \; a \int\!\!\int q^T(S,t)u^{*P}(\xi, t_F; S, t)\, d\Gamma dt$$

$$- \int\!\!\int T(S,t)v_n(S,t)u^{*P}(\xi, t_F; S, t)\, d\Gamma dt$$

$$+ \int\!\!\int T(s,t)\vec{v}(s,t)\vec{\nabla}u^{*P}(\xi, t_F; s, t)\, d\Omega dt$$

$$+ \int T(s, t_{F-1})u^{*P}(\xi, t_F; s, t_{F-1})\, d\Omega \qquad (6)$$

SET OF DISCRETE EQUATIONS FOR FLUID FLOW

With discretisation of boundary Γ into boundary elements and domain Ω into cells, using method of collocation, we get the following system of equations for unknown boundary nodal values:

kinematics

$$\left([D_x^E]n_x + [D_y^E]n_y\right)\{\omega\} \; = \; \left([H_t^E]n_x - [H^E]n_y\right)\{v_x\}$$

$$+ \; \left([H^E]n_x + [H_t^E]n_y\right)\{v_y\} \qquad (7)$$

kinetics

$$[G^P]\{q^\omega\} = [H^P]\{\omega\} \; + \; \frac{1}{\nu}[G^P]\{\omega v_n - \beta T g_t\}$$

$$+ \; \frac{1}{\nu}[D_x^P]\{\omega v_x - \beta T g_x\}$$

$$+ \; \frac{1}{\nu}[D_y^P]\{\omega v_y + \beta T g_y\}$$

$$+ \; \frac{1}{\nu}[B]\{\omega\}_{F-1} \qquad (8)$$

Where: $\vec{n} = (n_x, n_y)$ unit normal; $[H^E], [H^P]$ matrix of nodal contribution of integral of elliptic or parabolic fundamental solution over the boundary; $[H_t^E]$ matrix of nodal contribution of integral of the elliptic fundamental flux in tangential direction over the boundary; $[G^P]$ matrix of nodal contribution of integral of the parabolic fundamental flux in normal direction over the boundary; $[D_x^E], [D_y^E], [D_x^P], [D_y^P]$ matrix of nodal contribution of integral of the elliptic or parabolic fundamental flux in x or y direction over the domain; $[B]$ matrix of nodal contribution of integral of the parabolic fundamental solution at a previous time step; $\{\}$ nodal values vector; $\{\}_{F-1}$ nodal values vector at a previous time step.

Velocity components v_x, v_y and vorticity ω for nodes in domain are computed explicitly from equations (4) and (5).

These two numerical schemes are stable for higher Re numbers, if velocity vector is given for boundary and the only unknown of the kinetics is vorticity flux.

SET OF DISCRETE EQUATIONS FOR ENERGY TRANSPORT — *SCHEME A*

Unknown boundary values of temperature or thermal flux are computed similarly as vorticity flux, eq. (8)

$$
\left[[H^P], -[G^P] \right] \left\{ \begin{array}{c} \{T\} \\ \{q^T\} \end{array} \right\} = \left[-[H^P], [G^P] \right] \left\{ \begin{array}{c} \{\overline{T}\} \\ \{\overline{q}^T\} \end{array} \right\} - \frac{1}{a} \left([G^P]\{Tv_n\} \right.
$$
$$
\left. + [D_x^P]\{Tv_x\} + [D_y^P]\{Tv_y\} - [B]\{T\}_{F-1} \right) \quad (9)
$$

$\{\overline{T}\}$ and $\{\overline{q}^T\}$ are known values of temperature or thermal flux. Values for temperature in the domain are obtained explicitly from equation (6).

Only heat diffusion from boundary is included in system matrix , while heat convection is treated as nonlinear right hand side term due to the products of terms $\{Tv_n\}$, $\{Tv_x\}$, $\{Tv_y\}$. This scheme is unstable for higher Pe numbers. Conditional numbers ($cond_E(A) = \|A\|_E \cdot \|A_{-1}\|_E$, and subscript $_E$ means Euclidean norm) for system matrix are also worse than at flow kinetics. We try to avoid this defectiveness by introducing a new scheme.

SET OF DISCRETE EQUATIONS FOR ENERGY TRANSPORT — *SCHEME B*

In these case, the convection from boundary is included in system matrix. We can do it, because the velocity is known on the boundary Γ (boundary conditions)

$$
\left[[H^P] + \tfrac{1}{a}\left([E^P] - [F_x^P] - [F_y^P] \right), -[G^P] \right] \left\{ \begin{array}{c} \{T\} \\ \{q^T\} \end{array} \right\}
$$
$$
= \left[-[H^P] - \tfrac{1}{a}\left([E^P] - [F_x^P] - [F_y^P] \right), -[G^P] \right] \left\{ \begin{array}{c} \{\overline{T}\} \\ \{\overline{q}^T\} \end{array} \right\} \quad (10)
$$
$$
+ \frac{1}{a} \left([D_x^P]\{Tv_x\} - [D_y^P]\{Tv_y\} - [B]\{T\}_{F-1} \right)
$$

New matrices are $[E^P]$ matrix of nodal contribution of integral of the parabolic fundamental solution and normal velocity component over the boundary; $[F_x^P]$, $[F_y^P]$ matrix of nodal contribution of integral of the parabolic fundamental flux in x or y direction and v_x or v_y velocity component over the domain for boundary nodes. By doing this, the numerical scheme becomes stabler.

EXAMPLE - Forced convection in a cavity

Let's have a square cavity with dimensions 1×1, and with cold fluid $(T = 0)$ at rest. At time $t = 0$ we start to drag upper plate with velocity $v_x = 1$. Left side is heated to $T = 1$, while right side is kept at $T = 0$. Other walls are adiabatic, see Figure 1.

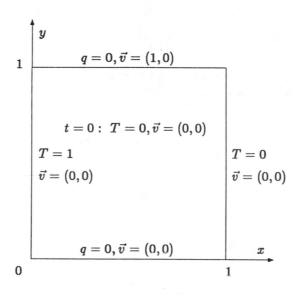

Figure 1: Problem geometry with initial and boundary conditions

This is the case of heat transfer with forced convection. Fluid carries away heat from the hot left hand wall to the cold right hand wall due to the upper plate movement. Natural convection is not included in this case.

To find a solution of the problem, we must first make a discrete model. Boundary is discretised into 16 isoparametric parabolic elements with 16 boundary nodes, while domain is discretised with 16 nine nodded Lagrangean cells with 49 internal nodes.

Case is computed for various values of Re number and $Pr=1$ by both schemes. Time step is $\Delta t = 0.5s$. Conditional numbers and convergence are shown in Table 1.

Flow kinematics is a good conditioned problem. Conditional number for flow kinetics depends on time step, and if it is chosen carefully, we can obtain good conditioned system matrix. Energy transport was always badly conditioned case, regardless on time step.

Re	cond_E			convergence			
	flow kinematics	flow kinetics	energy transport	SCHEME A		SCHEME B	
				flow	energy	flow	energy
1	45	142	277	3	4	3	4
10	45	73	269	5	14	5	14
20	45	60	271	conv.	diver.	7	24
50	45	46	282	conv.	diver.	12	44
100	45	39	309	conv.	diver.	20	90
200	45	35	365	conv.	diver.	35	147

Table 1: Conditional numbers for system matrix and convergence in time steps of schemes for different Re number values

Introducing new scheme was a required procedure, because it turns to be stabler than the old one. Conditional number did not change because normal components of velocity vector vanish. Due to this fact, matrix $[E^P]$ is absent from the system matrix, where there are only $[F_x^P]$ and $[F_y^P]$.

EXAMPLE - Forced convection in a channel

As second example gives flow in a channel, where velocity field is known (Poiseuille's flow). Temperature of the inlet flow is zero. Bottom wall is adiabatic, while upper wall is heated to a unit temperature. At outlet, the zero heat flux is prescribed, see Figure 2.

The discretised model is the same as in the previous case, except in x-direction where it has to be extended by factor of 2. Number of boundary elements and cells is unchanged. Time step is increased to $2s$. While velocity field is known, we have to compute energy transport only. Conditional numbers and convergence for this case are given in Table 2.

In this case, the old scheme is practically useless, because for $Pe=10$ its results are diverging. We achieve a stability with SCHEME B. Heat convection from the boundary is included in the system matrix. Conditional numbers are slightly worse, but we can compensate this by adjusting the time step.

EXAMPLE - Injection of hot fluid into cold fluid at rest

In the channel, we have cold fluid with $T = 0$. At time $t = 0$ the injection of the fluid is starting, where temperature field is linear, with $Re=250$, $Pr=1$, see Figure 3. At inlet, the parabolic velocity field is prescribed,

Pe	$cond_E$		convergence	
	SCHEME A	SCHEME B	SCHEME A	SCHEME B
1	224	233	4	4
10	206	315	diver.	6
20	203	415	diver.	6
50	207	678	diver.	6
100	209	1059	diver.	7
200	224	1762	diver.	8
400	—	3154	diver.	11
800	—	6041	diver.	16

Table 2: Conditional numbers and convergence in time steps for forced convection in the channel

while at outflow, the free traction condition is assumed. Upper and lower walls have temperature $T = 0$. Left and right hand sides are assumed to be isolated. Discretised model has 48 isoparametric parabolic boundary elements and 128 nine nodded Lagrangean cells. This gives 96 boundary nodes and 487 internal points.

The development of velocity field is shown in Figure 4. Temperature fields for various instants of time are given on Figure 5.

CONCLUSION

Introduction of new scheme for energy transport has proved to be very successful, since stability of scheme becomes much better. Convection from the boundary is also included in the system matrix, and not only the diffusion.

Due to great stability of SCHEME B, we can also use it in flow kinetics, while boundary values for vorticity are unknown. This happens when $v_n, \dfrac{\partial \omega}{\partial n}$ or $v_t, \dfrac{\partial \omega}{\partial n}$ are given as boundary conditions.

Given examples confirm superiority of the new scheme against the old one.

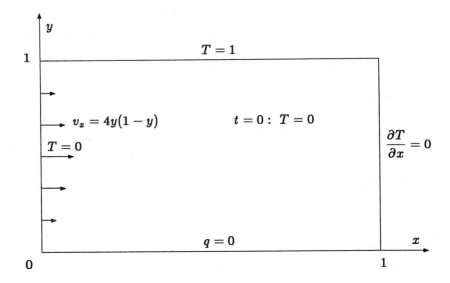

Figure 2: Geometry of the channel with boundary and initial conditions

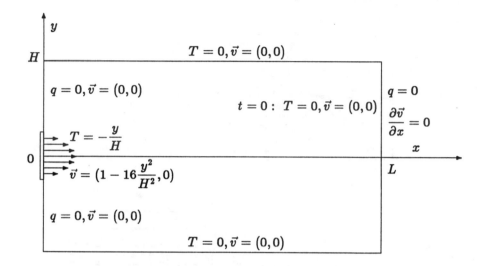

Figure 3: Injection, geometry with initial and boundary conditions $(H = 1, L = 4)$

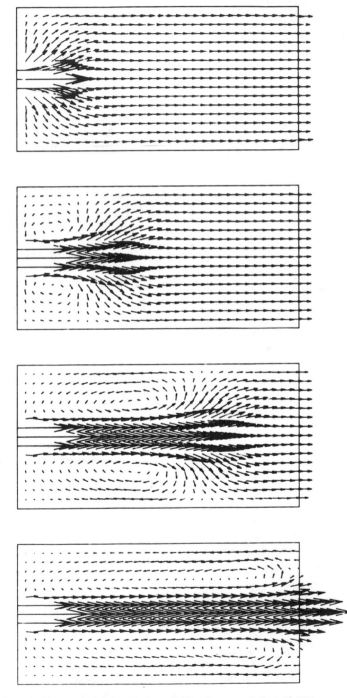

Figure 4: Injection, velocity fields for $t = 0.5, 2, 5, 10s$

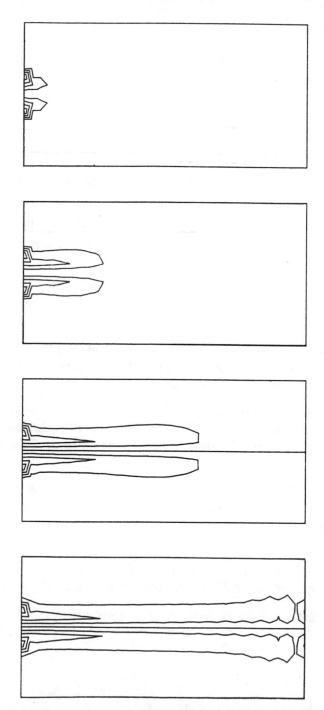

Figure 5: Injection, temperature fields $\Delta T = 0.25$, for $t = 0.5, 2, 5, 10s$

References

[1] Brebbia C.A., Telles J.C.F., Wrobel L.C.: **Boundary Element Methods-Theory and Applications**, Springer-Verlag, New York, 1984.

[2] Brebbia C.A., Skerget P.: "Time Dependent Diffusion Convection Problems Using Boundary Elements", **3rd Int. Conf. on Numerical Methods in Laminar and Turbulent Flows**, Seatle, 1983.

[3] Rek Z.: **Laminar and Turbulent Flow by Boundary Elements**, M. Sc. Thesis, University of Maribor, 1990.

[4] Rek Z., Skerget P.: "Boundary Integral Method for Time Dependent Diffusion-Convective Problems", **Mechanical Engineering Journal**, Vol. 35, No 1-3, pp. 9-12, 1989.

[5] Schnipke R.J., Rice J.G.: "A Finite Element Method for Free and Forced Convection Heat Transfer", **Int. J. Num. Meth. Eng.**, Vol. 24, pp. 117-128, 1987.

[6] Skerget P., Alujevic A., Rek Z.: "Boundary Element Method for Recirculation Fluid Flow", **UIT, 3rd Congresso Nazionale sul Transporto di Calore**, Palermo, 1985.

[7] Skerget P., Alujevic A.: "Computing Temperature and Velocity Fields in Laminar Fluid Flows by Boundary Element Method", **UIT, 2nd Congresso Nazionale sul Transporto di Calore**, Bologna, 1984.

[8] Skerget P., Alujevic A., Kuhn G., Brebbia C.A.: "Natural Convection Flow Problems by BEM", **9th Conference on Boundary Element Methods**, Vol. 3, 1987.

[9] Skerget P., Alujevic A., Brebbia C.A., Kuhn G.: "Natural and Forced Convection Simulation Using the Velocity-Vorticity Approach", **Topics in Boundary Element Research**, Vol. 5, Ch. 4, 1989.

[10] Skerget P., Kuhn G., Alujevic A., Brebbia C.A.: "Time Depended Transport Problems by BEM", **Advances in Water Resources**, Vol. 12, No. 1, pp. 9-20, 1989.

[11] Wu J.C., Rizk Y.M., Sankar N.L.,: "Problems of Time Dependent Navier-Stokes Flow", **Developments in Boundary Element Methods**, Vol. 3, Ch. 6, 1984.

Prediction of the Effect of Electrohydrodynamic (EHD) Enhancement on Fluid Flow and Heat Transfer

S.Y. Wang(*), M.W. Collins, P.H.G. Allen

Thermo-Fluids Engineering Research Centre, The City University, London, England

(*) on leave of absence from Beijing, China

ABSTRACT

In the present paper, predictions of the effect of electrohydrodynamic (EHD) forces on flow and heat transfer in low Reynolds number laminar mixed convection in a vertical channel are reported. The HARWELL code FLOW-3D, Release 2 is used, and a comparison between the predicted and experimental data is made. The theoretical results are in good agreement with the experimental results.

INTRODUCTION

It has been known for several decades that, under suitable circumstances, an electric field can give considerable enhancement of heat and mass transfer. During these years, much experimental and theoretical research has been undertaken[1].

As science and technologies develop, there are some possibilities of using EHD (electrohydrodynamic) enhancement in industrial processing. EHD seems an attractive tool for enhancing convective heat transfer, particularly in low Reynolds number flow of a weakly conducting liquid through a narrow space where the application of any of the conventional passive enhancement methods is neither easy nor effective.

Theoretical research work on the effect of EHD on flow and heat transfer are focused on flow stability. Two sorts of models have been studied previously according to accepted versions about the origin of the distribution of net space-charge, "conductivity models" and "mobility models"[2]. However, relatively little has been published on the theoretical prediction of the effect of EHD on flow and heat transfer.

The present paper reports a predictive study of the effect of EHD on flow and heat transfer in low Reynolds number laminar mixed natural and forced convection in a channel between two vertical planes using the HARWELL code FLOW-3D Release 2, and a comparison between the predictions and experimental data[3] is made. The measurement technique used was a differential one of especially high accuracy[4].

THEORETICAL MODEL

In the experimental apparatus,there are two vertical channels with different section sizes. Transformer oil flows through these channels. One wall of each channel is heated by an electric current flowing through an earthed copper winding, insulated with oil impregnated kraft paper. Heat is transferred from the wall to the transformer oil. The alternating (50Hz) high voltage is applied at the other wall, thus setting up an AC electric field.

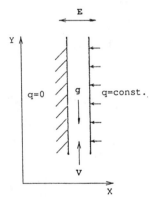

Fig.1 Illustration of Physical Model

For making a comparison between the results of experiment and theoretical predictions, the present analysis is based on the following physical model:

(1)   The flow is two-dimensional, incompressible, steady for the case of mixed natural and forced laminar convection (i.e. Boussinesq approximation for density);

(2)   The duct is vertical with upward direction of the flow;

(3)   At the entrance of the duct, the temperature and the velocity are uniform;

(4)   One wall of the duct is with constant heat flux, the other wall being adiabatic;

(5)   Except the permittivity, the properties of the fluid are constant;

(6)   The fluid is a dielectric fluid with uniform composition;

(7)   Viscous dissipation and electrical loss are neglected;

(8)   The electric field is uniform.

The physical model is illustrated in Fig.1.

The only modification to the hydrodynamical equations of mixed convection is a body force term **Fe**, which represents the response of the fluid to the electric field. The law of conservation of momentum requires

$$\rho \mathbf{V} \cdot \nabla \mathbf{V} = -\nabla p + \rho \beta \mathbf{g}(T-T_a) + \mathbf{Fe} - \mu \nabla^2 \mathbf{V} \qquad (1)$$

while the incompressibility condition is

$$\nabla \cdot \mathbf{V} = 0 \qquad (2)$$

The   energy equation is required

$$\rho c (\mathbf{V} \cdot \nabla T) = k \nabla^2 T \qquad (3)$$

Where **V** is the velocity, T the temperature, T_0 a reference temperature, p the pressure, **g** the gravitational acceleration, ρ the mass density, μ the dynamic viscosity, k the thermal conductivity, c the specific heat, and β is the volumetric coefficient of expansion.

A generally preferred expression of **Fe** is[5]

$$\mathbf{Fe} = \rho_e \mathbf{E} - \tfrac{1}{2} E^2 \nabla \mathcal{E} + \nabla [\tfrac{1}{2} \rho E^2 (\partial \mathcal{E}/\partial \rho)_T] \quad (4)$$

here ρ_e is the free electric charge density, $\mathcal{E}$ the dielectric permittivity, and **E** the electric field.

In order, these forces may be identified as Coulombic, dielectrophoretic and electrostrictive. Since curl grad $\equiv 0$ the last term produces no net circulatory effect in a hydrodynamically bounded system. For insulated duct walls:

$$\rho_e \mathbf{E} = 0 \qquad (5)$$

For uniform component liquid, $\mathcal{E}$ depends only on the mass density and the temperature[6], so

$$\nabla \mathcal{E} = (\partial \mathcal{E}/\partial T)_\rho \nabla T + (\partial \mathcal{E}/\partial \rho)_T \nabla \rho \qquad (6)$$

The predictions of the effect of electrohydrodynamic (EHD) enhancement on fluid flow and heat transfer are completed using FLOW-3D on the basis of the above model.

RESULTS AND DISCUSSION

In convective heat transfer, the nonuniformity of the temperature in the fluid produces nonuniformity of the dielectric permittivity. The electrically induced secondary flow is caused by the EHD force **Fe** which is applied in the fluid. The secondary flow results in additional changes in the velocity field and hence the temperature field in the fluid.

The profiles of the velocity and the temperature of the fluid near the exit of the channel without and with the EHD force are shown respectively in Figs.2 and 3. It can be found that the maximum velocities are near the hot wall side, either with or without the EHD force, due to the effect of the buoyancy force. The velocity profile of the flow with EHD peaks closer to the hot wall. The temperature profile of the flow with the EHD force is also slightly lower than that without EHD force. This,together with the slight increase in the gradient of the temperature near the wall results in an increase of the convective heat transfer coefficient.

These figures show that the changes of the velocity and temperature mainly occur near the hot wall side because of the greater variations there of the temperature and the electric permittivity.

The enhancement of heat transfer causes a reduction in the hot wall temperature. The distributions of the hot wall temperature without and with the EHD force are shown in Fig.4. Fig.4 shows that the effect of EHD on heat transfer is mainly in the developed region.

It can be found that the effect of EHD on the fluid flow and heat transfer is minor for conditions studied. The reasons are that the electric field strength is not sufficiently high and the temperature dependency of the electric permittivity of the transformer oil is relatively weak. The EHD secondary flow effects are small so that they cannot affect the stability of the flow.

The results of the modelling calculation agree with the results of the experiments. The results of the experiments and theoretical predictions are shown in Fig.5. In the figure, Nu_0 is the mean Nusselt number without EHD and Nu is the Nusselt number with EHD; El is the characteristic electric number,

$$El = \frac{\rho \, (\partial \varepsilon / \partial T) \, d_H^2 \, \Delta T \, E^2}{\mu^2} \qquad (7)$$

It can be found that the increase in heat transfer shows the exponential dependence on the El·Pr. (d_H is the duct hydraulic diameter; ΔT is the mean rise of duct wall over duct fluid bulk temperature and Pr is the Prandtl number.) This conclusion is in agreement with Savkar[7].

The effect on enhancement heat transfer of the electric field reduces with increase in Re[8].

It is shown that it is possible to make predictions of the effect of EHD enhancement on fluid flow and heat transfer for laminar flow but if turbulence is caused by EHD then predictions would involve using the standard k-$\mathcal{E}$ turbulence model within FLOW 3D.

CONCLUSIONS

It has been possible to make predictions for the effect of EHD on the flow and heat transfer by this computational approach. Under the conditions studied , the electrical secondary flow is rather weak, despite this, heat transfer is improved by up to about 25%. The theoretical results are in pleasing agreement with the experimental results. The deviation between the results of the predictions and the experiment is on average about 10% and not greater than 20%. The increase in heat transfer shows the exponential dependence on the El·Pr product.

ACKNOWLEDGMENT

We gratefully acknowledge the use of the AERE HARWELL CODE FLOW 3D, and the computational facilities provided by the University of London Computer Centre.

REFERENCE

(1)   Jones, T.B., Electrohydrodynamically enhanced heat transfer in liquids – a review, Advances in Heat Transfer, vol.14, pp.107-148, Academic Press, 1978.

(2)   Martin, P.J., and Richardson, A.T., Conductivity models of electrothermal convection in a plane layer of dielectric liquid, ASME J. Heat Transfer vol.106,

pp131-136,1984.

(3)   Allen,   P.H.G.,   Research   report,
unpublished,1961.

(4)   Allen, P.H.G., Heat transfer at high voltage,
Electrical Times, vol.139, pp.321-322, 1961.

(5)   Stratton, J.A., Electromagnetic theory, McGraw-
Hill, New York, 1941.

(6)   Landau, L.D., Lifshitz, E.M., and Pitaevskii,
L.P., Electrodynamics of continuous media, 2nd Edit,
Pergamon, New York, 1984.

(7)   Savkar, S.D., Dielectrophoretic effects in
laminar forced convection between two parallel
plates, Phys. Fluids, vol.14, No.12, pp.2670-2679,
1971.

(8)   Cooper, P. and Allen, P.H.G., The Senftleben
effect in cross-flow heat exchange and the part
played by electronic conduction, PhysicoChemical
Hydrodynamics, vol.4, no.2, pp.85-101, 1983.

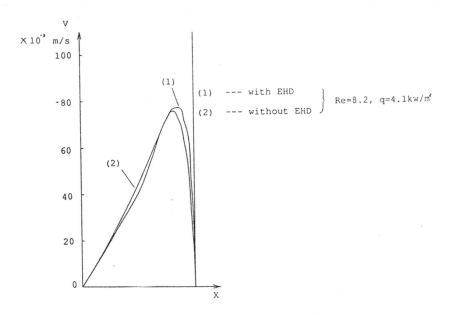

Fig.2   Profiles of velocity with and without EHD

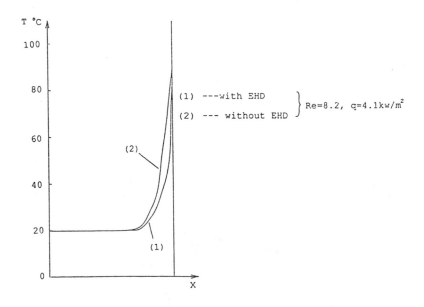

Fig.3   Profiles of temperature with and without EHD

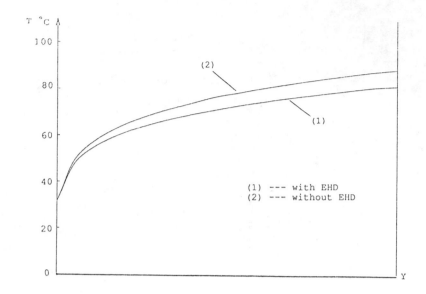

Fig.4   Computed distribution of the wall temperature
        (Re=8.2, q=4.1kw/m^2)

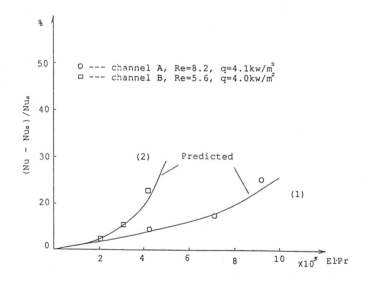

Fig.5   Increase in Heat Transfer due to EHD

Fully-Developed Non-Darcian Mixed Convection in a Horizontal Rectangular Channel Filled with Packed Spheres

F.C. Chou, W.Y. Lien

Department of Mechanical Engineering, National Central University, Chung-li 32054, Taiwan, China

ABSTRACT

This paper presents a numerical study on fully-developed non-Darcian mixed convection in a horizontal rectangular channel filled with packed spheres. The channel wall is heated with a uniform heat flux. The inertial and boundary effects are found to decrease the flow and heat transfer rate, the effects of channeling and thermal dispersion increase the heat transfer rate markedly. There exists relatively high secondary flow intensity in the region near the vertical wall due to the channeling effect, and the effect of thermal dispersion affects the flow structure and heat transfer significantly when the value of Peclet number Pe is high. The variation of Nusselt number Nu depends both on the Rayleigh number Ra and Pe, and the buoyant effect will be suppressed as Pe increasing. Moerover, even for a fixed Ra, the Nusselt number increases with the increase of Pe. These phenomena are against the fluid through a channel without packed spheres. Based on the criterion of 5% deviation of Nu from that in pure forced convection, the critical Ra for the onset of buoyant effect is found to increase almost exponentially with the increase of Pe.

INTRODUCTION

The research of heat transfer in porous media has been the subject of many recent studies due to its wide applications in various engineering systems. The great majority of previous studies dealt with the convective heat transfer in porous media confined in pure forced or pure natural convection ,however, the study of mixed convection is still limited. Heiber[1] studied the mixed convection in a porous medium involving horizontal surfaces. Combarnous and Bia [2] investigated experimentally and numerically the effect of mean flow on the onset of convection in a porous medium bounded

by isothermal planes. Cheng [3] analysed the mixed convection about
inclined surfaces in a porous medium, and later, Cheng [4] applied the
boundary-layer analysis to obtain similarity solutions for mixed convection
about a horizontal plate embeded in a porous medium with constant heat flux.
Joshi and Gebhart [5] studied the mixed convection in porous media adjacent
to a vertical uniform heat flux surface. Oosthuizen [6] reported the mixed
convection over a horizontal plate in a porous medium mounted near an
impervious adiabatic horizontal surface, and Prasad et al. [7] presented
the mixed convection in two-dimensional horizontal porous layers with
localized heating from below. All the above-mentioned studies adopted
Darcy's law to formulate the problems. Recently, Chandrasekhara and
Namboodiri [8] carried out for mixed convection about inclined surfaces in
a porous medium incorporating a non-Darcian model which included the variations of permeability and thermal conductivity, and found that these variations increased the heat transfer rate significantly. Lai and Kulacki [9]
used a non-Darcian flow model to present similarity solutions of mixed
convection for the case of constant surface heat flux from horizontal
impermable surfaces in porous media, and the inertial term is added in
Darcian model only.

Indeed few studies have been reported on the mixed convection in a
horizontal channel filled with porous media. Haajizadeh and Tien [10]
investigated analytically and numerically the mixed convective flows
through a horizontal porous channel connecting two reservoirs, and Darcy's
law was used. To the best of the authors' knowledge, a theoretical analysis
about non-Darcian effects which include high-flow-rate inertial pressure
loss, solid-boundary shear, near-wall porosity variation and thermal
dispersion on mixed convection in a horizontal porous channel has not been
studied in the literature. The aim of the present investigation is to
study numerically the mixed convection in the fully-developed region of a
horizontal rectangular channel filled with packed spheres, using the
generalized equations which consider the non-Darcian effects as described
above. The present analysis gives details of the impact of the non-Darcian
separate effects on the flow structure and heat transfer. The phenomena
which have a evident difference from a channel without packed spheres such
as high flow intensity near the vertical wall, the buoyant effect suppressed
by the effect of thermal dispersion, and the value of Nusselt number
deviated from that in pure forced convection depending on both the Rayleigh
number and Peclet number are also discussed.

THEORETICAL ANALYSIS

The present analysis introduces the volume-averaged equations which
include both global effects such as confining boundaries and local pore
effects such as dispersion. In vectorial notation, the steady three-

dimensional generalization of continuity and momentum equations in porous media are

$$\nabla \cdot \langle \vec{v} \rangle = 0 \tag{1}$$

$$(\rho/\epsilon^2)\langle \vec{v} \cdot \nabla \vec{v} \rangle = -\nabla\langle p \rangle + \rho\vec{g} - (\mu/K)\langle \vec{v} \rangle - \rho C|\langle \vec{v} \rangle|\langle \vec{v} \rangle$$
$$+ (\mu/\epsilon)\nabla^2\langle \vec{v} \rangle \tag{2}$$

where $\langle \rangle$ represents a volume-averaged quantity, $\vec{v}$ and p are the local velocity and pressure, ρ and μ the fluid density and viscosity. From the experimental results of Benenati and Brosilow [11], the porosity, ϵ, can be represented approximately by an exponential function as used by Poulikakos and Renken [12]

$$\epsilon = \epsilon_\infty [1 + a_1 \exp(-a_2 R/d)] \tag{3}$$

where ϵ_∞ is the free-stream porosity, R the transverse distance from the wall, d the partical diameter, and a_1 and a_2 are experimental parameters which depend on packing and partical size. THe values $\epsilon_\infty=0.37$, $a_1=0.43$, and $a_2=3$ were used [12]. For a liquid-saturated porous medium the permeability, K, and the flow inertial parameter, C, depend on the matrix porosity and sphere diameter which are given by the relations developed by Ergun [13]

$$K = d^2\epsilon^3/(150(1-\epsilon)^2) \tag{4}$$

$$C = 1.75(1-\epsilon)/(d\epsilon^3) \tag{5}$$

The volume-averaged, steady energy equation for porous media is (Tien and Hunt [14])

$$\rho c_P\langle \vec{v} \rangle \cdot \nabla\langle T \rangle = \nabla \cdot (k_e\nabla\langle T \rangle) \tag{6}$$

where T is the local temperature, c_P the fluid heat capacity, and k_e the effective conductivity composed of a sum of the stagnant and dispersion conductivities, $k_e=k_o+k_d$. The stagnant conductivity, k_o, depends on the porosity variation and the value of bulk stagnant conductivty, k_∞ (Zehner and Schlunder [15]). The variation of the stagnant conductivity can be written as

$$k_o = k_\infty [1 + (k_f/k_\infty - 1) \exp(-a_3 R/d)] \tag{7}$$

where a_3 is an empirical constant. The dispersion conductivity, k_d, incorporates the additional thermal transport due to the fluid's tortuous path around the solid particles. This quantity is porportional to a product of the local velocity, a constant τ_∞, and Van Driest's wall function [16]

$$k_d = \tau_\infty \, \rho \, c_P \, | \langle \bar{v} \rangle | \, l(R) \tag{8}$$

where $l(R)$ is Van Driest's wall function for thermal dispersion which given by [16]

$$l(R) = d \{ 1 - \exp[-R/(\delta d)] \} \tag{9}$$

where δ is an empirical constant. There are two different sources to show the wall effect on thermal dispersion. First, the no-slip boundary condition and variable porosity effect modify the velocity distribution near the wall. Secondly, the mixing of local fluid streams would be reduced by the presence of a wall. The values $\tau_\infty = 0.17$, and $\delta = 1.5$ were determined by Cheng and Zhu [17] by matching both the predicted temperature distribution and the Nusselt number with experimental data.

The physical configuration and the coordinate system are shown in Fig.1 for the present discussion. Consider a steady laminar flow in both hydrodynamically and thermally fully-developed region of a horizontal rectangular channel packed with fluid-saturated spheres under an axial uniform heat flux and a peripherally uniform wall temperature. The Boussinesq approximation is used to characterize the effect of free convection, and the viscous dissipation and compressibility effects in the energy equation are neglected. With the following variables

$$X = x/d, \quad Y = y/d, \quad U = \langle u \rangle/(\alpha_f/d), \quad V = \langle v \rangle/(\alpha_f/d), \quad W = \langle w \rangle/\langle \bar{w} \rangle$$

$$\theta = (\langle T_w \rangle - \langle T \rangle)/\theta_c, \quad \theta_c = q_w d/k_f, \quad Da = K_\infty/d^2, \quad Re = \rho \langle \bar{w} \rangle d/\mu \tag{10}$$

$$Pr = \mu/(\rho \alpha_f), \quad Pe = Pr \cdot Re, \quad Ra = \rho g \beta q_w d^4/(\mu \alpha_f k_f)$$

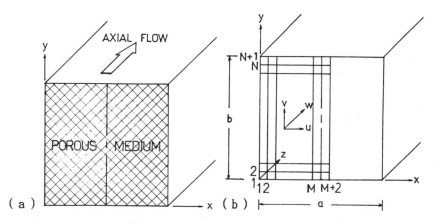

Figure 1. Physical configuration and coordinate system

and introduing the stream function and vorticity

$$U = \partial \psi / \partial Y, \quad V = - \partial \psi / \partial X \tag{11}$$

$$\xi = \partial U / \partial Y - \partial V / \partial X = \nabla^2 \psi \tag{12}$$

the governing equations can be obtained

$$\nabla^2 \psi = \xi \tag{13}$$

$$(U \partial \xi / \partial X + V \partial \xi / \partial Y)/(\epsilon Pr) + \{ U[(\partial V/\partial X)(\partial \epsilon/\partial X) - (\partial U/\partial X)(\partial \epsilon/\partial Y)]$$
$$+ V[(\partial V/\partial Y)(\partial \epsilon/\partial X) - (\partial U/\partial Y)(\partial \epsilon/\partial Y)] \}/(\epsilon^2 Pr)$$
$$= - (\epsilon d^2/K)\xi + \nabla^2 \xi + 300 \ (1-\epsilon)(U \partial \epsilon/\partial Y - V \partial \epsilon/\partial X)/\epsilon^3$$
$$- 2\epsilon Cd(||\text{a}||U/\partial Y - V \partial V/\partial X)/Pr + 1.75 \ (2-\epsilon)(U^2 \partial \epsilon/\partial Y - V^2 \partial \epsilon/\partial X)/(\epsilon^3 Pr)$$
$$+ \epsilon Ra(\partial \theta/\partial X) + Ra\theta(\partial \epsilon/\partial X) \tag{14}$$

$$(U \partial W/\partial X + V \partial W/\partial Y)/(\epsilon Pr) = \epsilon[-K_\infty(dp/dz)/(\mu \overline{w})]/Da - (\epsilon d^2/K)W$$
$$- \epsilon \ C \ d \ Re \ W^2 + \nabla^2 W \tag{15}$$

$$U \partial \theta/\partial X + V \partial \theta/\partial Y - 4Wd/De = (\partial/\partial X)[(k_e/k_f)(\partial \theta/\partial X)]$$
$$+ (\partial/\partial Y)[(k_e/k_f)(\partial \theta/\partial Y)] \tag{16}$$

where $\nabla^2 = (\partial^2/\partial X^2 + \partial^2/\partial Y^2)$, ∞ represents the bulk porous medium value, f the fluid state, — the average value computed by the Simpon's rule, and De the equivalent hydraulic diameter. The Darcian number, Da, relates the permeability to the particle size. The ratio of effective to fluid conductivity as a function of transverse distance R equals

$$k_e/k_f = k_o/k_f + k_d/k_f$$

$$= (k_\infty/k_f)[\ 1 + (k_f/k_\infty - 1)\exp(-a_3 R/d) \ ] \tag{17}$$

$$+ \tau_\infty \ Pe \ \sqrt{(U^2 + V^2)/Pe^2 + W^2} \ \{ \ 1 - \exp[-R/(\delta d)] \ \}$$

and k_o/k_f is considered as one. Equations (13) to (16) contain three independent parameters Pr, Re, and Ra. For a fluid through a channel of fixed aspect ratio, Reynolds number Re and Rayleigh number Ra govern the flow and heat transfer characteristics. The Prandtl number Pr is assigned 7.02 for water. Because of symmetry, it suffices to solve the problem in a half region of the rectangular channel such as that shown in Figure 1(b). The boundary conditions are

$$U = V = W = \theta = 0 \qquad \qquad \text{at} \quad \text{channel wall}$$
$$\tag{18}$$
$$U = \partial V/\partial X = \partial W/\partial X = \partial \theta/\partial X = 0 \quad \text{at symmetric plane } x = a/2$$

NUMERICAL SOLUTIONS

To obtain the numerical solutions to the governing equations [Equations (13)-(16)], finite-difference schemes are used. Since these equations are coupled each other, they should be solved simultaneously. The procedure for solving Equations (13)-(16) and the related boundary conditions of Equation (18) is:

1. Assign initial values for ψ, ζ, W, θ.
2. The stream function ψ at each node can be found by solving Equation (13) from the assigned valves of ζ at each node.
3. The velocity components U and V can be computed from Equation (11).
4. The boundary vorticity is then calculated from Equation (13) and the associated boundary conditions for ψ.
5. Using values of U and V from step 3, and the boundary vorticity from step 4, Equation (14) can be solved for ζ.
6. Using the same values of U and V, Equations (15) and (16) can be solved for W and θ, respectively. Check if the mean dimensionless axial velocity $\overline{W} = \langle \overline{w} \rangle / \langle \overline{w} \rangle$ is equal to 1. Otherwise, adjust the value of pressure term $- K_{\infty}(dp/dz)/(\mu \overline{w})$ in Equation (15) to meet the requirement.
7. Steps 2 to 6 are repeated until the following criterion is satified

$$\text{Error} = \sum_{ij} |(\theta_{ij}^{(n+1)} - \theta_{ij}^{(n)})/\theta_{ij}^{(n+1)} | / (M \times N) < 10^{-5} \qquad (19)$$

where M and N are the number of divisions in the x and y directions, respectively.

After the temperature field is obtained, the computation of the Nusselt number is of practical interest. The Nusselt number Nu can be evaluated on the basis of the overall energy balance

$$Nu = \overline{h} D_e / k_f = (D_e/d)(1/\theta_b) \qquad (20)$$

where b denotes the bulk quantity.

RESULTS and DISCUSSION

To illustrate the effects of buoyant force on flow and heat transfer, the isotherms and streamlines are shown. Figs.2 and 3 compare the non-Darcian separate effects on the isotherms and streamlines for $Ra=10^5$ and $Pe=10$ and 100. From Equation (11), the distance between two nearby streamlines is inversely proportional to the local speed of the secondary flow. The main eddies are drived by non-zero temperature gradients in the horizontal direction near the vertical walls. In the central region of the rectangular channel, an adversed temperature gradient in the vertical direction appears

near the bottom of the horizontal wall due to the thermal boundary condition. The adversed temperature gradient induces a second pair of counter-
rotating eddies, and isotherms are distorted due to the upward motion of
the eddies which bring the heated fluid along the central symmetric line.

(a) Boundary and inertial

    effects included

(b) Boundary, inertial, and

    channeling effects included

(c) Boundary, inertial,

    channeling, and thermal

    dispersion effects

    included

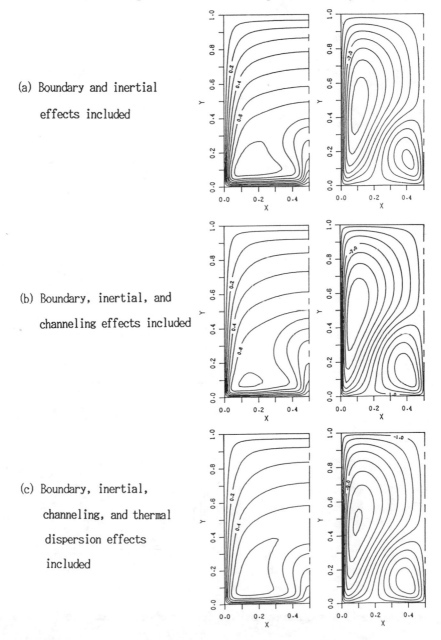

Figure 2. Isotherms and streamlines for Ra=10⁵ and Pe=10 (Δθ=0.1, Δψ=2)

(a) Boundary and inertial

effects included

(b) Boundary, inertial, and

channeling effects included

(c) Boundary, inertial,

channeling, and thermal

dispersion effects included

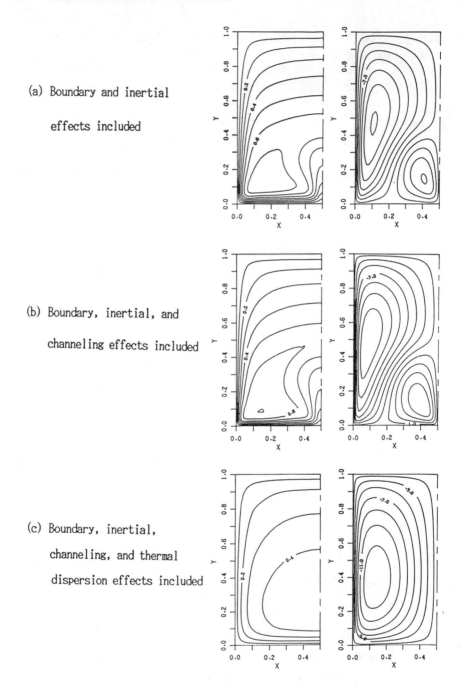

Figure 3. Isotherms and streamlines for Ra=10⁵ and Pe=100 ($\triangle\theta$=0.1, $\triangle\psi$=2)

Comparing the streamlines near the vertical wall shown in Fig.2(a) and (b), it can be seen that there exists a higher flow speed in the region near the wall when the channeling effect is included, and thus a higher local heat transfer rate is also induced. Fig.3 shows the non-Darcian separate effects for Pe=100. It can be found that if the thermal dispersion is not considered, the effect of Peclet number on heat transfer would be quite weak in the fully-developed region of a channel as comparing Fig.3(a-b) with Fig.2(a-b). The effect of thermal dispersion is relatively weak for low flow rates (Pe=10) as shown in Fig.2(c), but it significantly changes the appearances of isotherms and streamlines when the flow rates become high. In view of Fig.3(c) for Pe=100, the second pair of counter-rotating eddies in the central region near the bottom wall are nonexistent, and the isotherms and streamlines becomes more uniform. This result makes sense physically: a relatively uniform temperature distribution is caused by the violent mixing of fluid within the pores for the cases of higher flow rate , and therefore the magnitude of adversed temperature gradients decreases due to the effect of thermal dispersion and this result leads the second pair of counter-rotating eddies to be suppressed.

Figure 4 shows the variation of Nusselt number versus the Rayleigh number for the non-Darcian separate effects according to Figs. 2 and 3. The phenomena described above are observed again. Moreover, comparing the values of Nusselt number of curves a and c in Fig.4 , the different degree of the buoyant effect suppressed by the thermal dispersion will be understanded explicitly. In Fig.4(a) for Pe=10, the Nusselt number of curve c has 66.49% increment compared to curve a in pure forced convective area, and 37.52% in mixed convection (Ra=10^5). However, for Pe=100 in Fig.4(b), the Nusselt number has 259.26% increment in pure forced convection, and decreases to 58.45% in mixed convection (Ra=10^5).

Fig.5 shows the effect of Rayleigh number on the Nusselt number with the Peclet number as a parameter. The variation of Nusselt number is found to depend both on the Rayleigh number and Peclet number. This phenomenon is quite different from the fluid through a channel without packed spheres. For a fluid through a channel without porous media of fixed Prandtl number, the Rayleigh number governs the flow and heat transfer characteristics alone. But in the present study it is shown that the Nusselt number increases with the increase of Peclet number even for a fixed Rayleigh number, that is, the Rayleigh number and Peclet number govern the flow and heat transfer simultaneously. Furthermore, the Nusselt number merges into the same value for Pe=5, 10, and 30 as the Rayleigh number is larger than 5x10^4. With this information in hand, it is believed that the asymptotic representation (i.e. Nu versus Ra) in pure natural convection will be established.

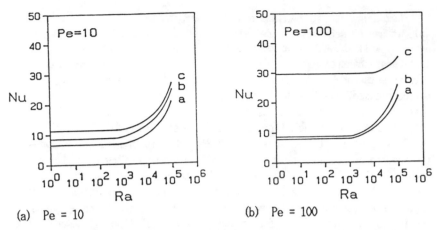

(a) Pe = 10

(b) Pe = 100

Figure 4. Relative contribution of the various effects on the variation of
Nusselt number with Rayleigh number for Pe=10 and 100
curve a: Boundary and inertial effects included
curve b: Boundary, inertial, and channeling effects included
curve c: Boundary, inertial, channeling, and thermal dispersion
effects included

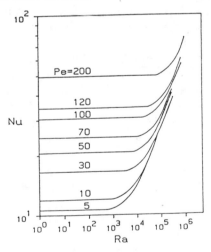

Figure 5. Nusselt number variation with Rayleigh number at various Peclet
number

Another result is found in Fig.5: The onset of buoyant effect is
delayed as the Peclet number increasing. It is interesting to discuss the
relation between the critical Rayleigh number for the onset of buoyant
effect and the Peclet number. Based on the criterion of 5% deviation of
Nusselt number from that in pure forced convection, the critical Rayleigh
number for the onset of buoyant effect is found to increase almost exponen-

tially with the increase of Peclet number as shown in Fig.6. The dependence of the critical Rayleigh number for the onset of buoyant effect on the Peclet number is a special phenomenon in porous media, and does not exist in the fluid through a channel without packed spheres.

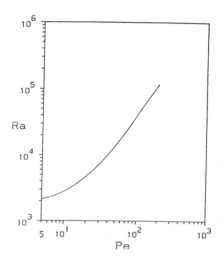

Figure 6. The critical Rayleigh number variation with Peclet number for the onset of buoyant effect based on the criterion of 5% deviation of Nusselt number from that in pure forced convection

CONCLUDING REMARKS

The problem of mixed convection in the fully-developed region of a horizontal rectangular channel filled with packed spheres incorporating non-Darcian effects is analysed by the vorticity-streamfunction formulation. The non-Darcian separate effects significantly alter the velocity and temperature profiles and make some different phenomena from the fluid through a channel without packed spheres. The boundary and inertial effects decrease the flow and heat transfer rate; there exists a quite high flow speed in the region near the vertical wall due to the channeling effect; and the effect of thermal disperaion affects the flow structure and heat transfer markedly when the Peclet number is high, moreover, the buoyant effect will be suppressed as the Peclet number increasing. The variation of Nusselt number is found to depend both on the Rayleigh number and Peclet number, and the critical Rayleigh number for the onset of buoyant effect increases exponentially with the increase of Peclet number. The asymptotic representation of Nu versus Ra in pure natural convection is believed to establish in future.

ACKNOWLEDGEMENT

The authors would like to thank the National Science Council of Republic of China for its support of the present work through project NSC79-0401-E008-16.

REFERENCES

1. Heiber, C.A. Mixed Convection above a Horizontal Surface, Int. J. Heat Mass Transfer, Vol.16, pp. 769-785, 1963.
2. Combarnous, M.A. and Bia, P. Combined Free and Forced Convection in Porous Media, Soc. Pet. Eng J., Vol.11, pp. 399-405, 1971.
3. Cheng, P. Combined Free and Forced Convection Flow about Inclined Surfaces in Porous Media, Int. J. Heat Mass Transfer, Vol.20, pp. 807-814, 1977.
4. Cheng, P. Similarity Solutions for Mixed Convection from Horizontal Impermeable Surfaces in Saturated Porous Media, Int. J. Heat Mass Transfer, Vol.20, pp. 893-898, 1977.
5. Joshi, Y. and Gebhart, B. Mixed Convection in Porous Media Adjacent to a Vertical Uniform Heat Flux Surface, Int. J. Heat Mass Transfer, Vol.28, pp. 1783-1786, 1985.
6. Oosthuizen, P.H. Mixed Convection Heat Transfer from a Heated Horizontal Plate in a Porous Medium Near an Impermeable Surface, ASME J. of Heat Transfer, Vol.110, pp. 390-394, 1988.
7. Prasad, V., Lai, F.C., and Kulacki, F.A. Mixed Convection in Horizontal Porous Layers Heated from Below, ASME J. of Heat Transfer, Vol.110, pp. 395-402, 1988.
8. Chandrasekhara, B.C. and Namboodiri, P.M.S. Influence of Variable Permeability on Combined Free and Forced Convection about Inclined Surfaces in Porous Media, Int, J. Heat Mass Transfer, Vol.28, pp. 199-206, 1985.
9. Lai, F.C. and Kulacki, F.A. Non-Darcy Convection from Horizontal Impermeable Surfaces in Saturated Porous Media, Int. J. Heat Mass Transfer, Vol.30, pp. 2189-2192, 1987.
10. Haajizadeh, M. and Tien, C.L. Combined Natural and Forced Convection in a Horizontal Porous Channel, Int. J. Heat Mass Transfer, Vol.27, pp.799-813, 1984.
11. Benenati, R.F., and Brosilow, C.B. Void Fraction Distribution in Packed Beds, AIChE J., Vol.8, pp.359-361, 1962.
12. Poulikakos, D., and Renken, K. Forced Convection in a Channel Filled with Porous Medium, Including the Effects of Flow Inertia, Variable Porosity, and Brinkman Frction, ASME Journal of Heat Transfer, Vol.109, pp. 880-888, 1987.
13. Ergun, S. Fluid Flow Through Packed Columns, Chem. Eng. Prog., Vol.48, pp. 89-94.14, 1952.
14. Tien, C.L., and Hunt, M.L. Boundary-Layer Flow and Heat Transfer in Porous Beds, Chem. Eng. Proc., Vol.21, pp. 53-63, 1987.
15. Zehner, P., and Schlunder, E.U. Thermal Conductivity of Packed Beds at Moderate Temperatures, Chem. Eng. Sci., Vol. 42,pp. 933-940, 1970.
16. Cheng, P., and Hsu, C.T. Applications of Van Driest's Mixing Length Theory to Transverse Thermal Dispersion in Forced Convection Flow Through a Packed Bed, Int. Comm. Heat Mass Transfer, Vol.13, pp.613-625, 1986.
17. Cheng, P., and Zhu, H. Effects of Radial Thermal Dispersion on Fully-Developed Forced Convection in Cylindrical Packed Tubes, Int. J. Heat Mass Transfer, Vol.30, pp. 2373-2383, 1987.

Laminar Flow Heat Transfer in Thermal Entrance Region of Circular Sector Ducts

B.T.F. Chung, R.P. Hsia

Department of Mechanical Engineering, The University of Akron, Akron, OH 44325-3903, USA

ABSTRACT

Forced convection for thermally developing and hydrodynamically developed flow in circular sector ducts is analyzed. The flow is laminar, incompressible and is subjected to uniform wall temperature in peripheral direction and uniform axial heat flux boundary condition. It is assumed that the thermal properties are constant; the axial conduction and viscous dissipation are negligible. An integral transform is employed to determine the velocity profile analytically. Energy equation is solved by employing a finite difference method. A Successive Over Relaxation (SOR) line iteration technique is utilized for solving this parabolized energy equation which is elliptic in the cross plane of the duct. The calculations of temperature profile and Nusselt number march forward axially in the thermal entrance region. Numerical solutions of developing temperature field, peripheral distribution of duct surface heat flux as well as heat transfer coefficient are presented as functions of axial distance for different open angles of sector.

INTRODUCTION

Heat transfer in noncircular ducts has received considerable attention because of their application in the design of high performance heat exchange equipment. Circular sector ducts also represent the limiting cases of internally finned tubes with equi spaced full tapered fins. Configurations of this type may be found in the cores of nuclear reactors, where fuel elements are arranged in various patterns. In the case of gas turbines or nuclear power plants which operate at extremely high temperature levels, the designer is often interested in determining the location where the wall temperature within the heat exchanger will assume its highest value. In order to calculate those hot spots, the local heat transfer coefficient has to be known.

Eckert et al. [3] considered both the thermally and hydraulically fully developed laminar flow through circular sector ducts and presented Nusselt number for the sector angle between 10 and 60 degrees. Sparrow and Haji-Sheikh [11] extended the plot of Nusselt number to 0-180 degrees. Hu and Chang [6] found their results agree well with those of Eckert et al. qualitatively with a maximum error of about 9.6 percent. Recently Lei and Trupp [7] investigated the maximum velocity location and pressure of fully developed laminar flow in circular sector ducts. The authors [8] further studied heat transfer in circular sector ducts subject to uniform axial heat flux and uniform peripheral temperature for the curved surface and the adiabatic condition for the flat surfaces. Trupp and Lau [12] and Ben-Ali et al. [2] presented analyses of laminar heat transfer in circular sector ducts with isothermal walls. All aforementioned analyses are only for hydrodynamically and thermally fully developed laminar flow. The only pertaining work for thermally developing flow was that of Hong and Bergles [5]. However, this paper is restricted to the geometry of semicircular duct.

An extensive literature review reveals that solutions for forced convection of thermally developing laminar flow in circular sector ducts with an arbitrary apex angle have not been available. Yet, in many practical applications such as flow in short ducts, temperatures are not fully developed. Therefore, the prediction of local heat transfer coefficient in the thermal entrance region of circular sector ducts becomes useful and needed.

ANALYSIS

Consideration is given to the steady state forced convection of a laminar incompressible flow in uniform straight ducts with a circular sector cross section as shown in Fig.1. The no slip, no temperature jump, constant property and a large Peclet number are assumed. Consistent with the usual boundary layer approximations, viscous dissipation and axial conduction are neglected for a large Peclet number. The system is thermally developing but hydrodynamically developed and is subject to the uniform peripheral surface temperature and uniform axial heat flux boundary condition.

Velocity Field

From Fig.1, the location within the duct cross section is described by the radial distance R from the corner and by the angular distance ϕ from the duct center line. The opening angle of the duct is 2β. The only existing flow is in axial direction and velocity field is a function of radius and angle. The dimensionless momentum equation in the cylindrical co-ordinates is given by,

$$\frac{\partial^2 w}{\partial r^2} + \frac{1}{r}\frac{\partial w}{\partial r} + \frac{1}{r^2}\frac{\partial^2 w}{\partial \phi^2} = C \qquad 1 > r > 0, \quad \beta > \phi > -\beta \qquad (1)$$

with the boundary conditions

$$w = 0 \qquad \text{for} \quad r = 1 \quad \text{and} \quad \phi = \pm\beta \qquad (2)$$

where C is a constant in term of pressure gradient and the dimensionless parameters r and w are defined as:

$$r = \frac{R}{R_0} \quad , \quad w = \frac{W}{\overline{W}} \qquad (3)$$

Following Ozisik [9], we define the Fourier integral transform and the corresponding inversion formula as

$$\overline{w} = \int_0^\beta \sqrt{\frac{2}{\beta}}\cos\nu_n\phi\, w\, d\phi \qquad (4)$$

$$w = \sum_{n=1}^\infty \sqrt{\frac{2}{\beta}}\cos\nu_n\phi\,\overline{w} \qquad (5)$$

where ν_n's are the eigenvalues with

$$\nu_n = \frac{(2n-1)\pi}{2\beta} \qquad \text{for} \qquad n = 1,2,3,\dots \qquad (6)$$

Applying the above integral transformation and the inversion formula, the velocity field can be found as

$$w = \frac{\displaystyle\sum_{n=1,3,5,\dots}^\infty (-1)^{\frac{n+1}{2}}\left(r^2 - r^{\frac{n\pi}{2\beta}}\right)\frac{\cos\frac{n\pi}{2\beta}\phi}{n(16\beta^2 - n^2\pi^2)}}{\displaystyle\frac{1}{\pi}\sum_{n=1,3,5,\dots}^\infty \frac{1}{n^2(4\beta+n)^2}} \qquad (7)$$

Detailed derivations of Eq.(7) may be found in the thesis of Hsia [4].

Temperature Field

By neglecting energy dissipation and axial conduction in the fluid, the dimensionless energy equation in cylindrical co-ordinates for steady laminar thermally developing flow of a fluid with constant properties is given by,

$$\frac{\partial^2 \theta}{\partial r^2} + \frac{1}{r}\frac{\partial \theta}{\partial r} + \frac{1}{r^2}\frac{\partial^2 \theta}{\partial \phi^2} = w\frac{\partial \theta}{\partial z} \qquad 1 > r > 0, \beta > \phi > -\beta, z > 0 \qquad (8)$$

with the inlet and boundary conditions as follows

$$\theta = 0 \qquad @ \qquad z = 0 \qquad (9)$$

$$\theta = \theta_w = const \qquad \text{for} \qquad r = 1 \quad \text{and} \quad \phi = \pm\beta \qquad (10)$$

where θ and z are defined by,

$$\theta = \frac{T - T_0}{q_w R_0 / K} \quad , \quad z = \frac{Z}{R_0 Re Pr} \qquad (11)$$

Eq.(8) is written in finite difference form using the first order forward difference in axial direction and second order central difference in radial and angular directions, we obtain

$$A_i \theta_{i-1,j}^k + B_i \theta_{i,j}^k + C_i \theta_{i+1,j}^k = D_i(\theta_{i,j-1}^k + \theta_{i,j+1}^k) + \frac{w_{i,j}}{\Delta z^k}\theta_{i,j}^{k-1} \quad (12)$$

where

$$A_i = -\frac{1}{\Delta r^2} + \frac{1}{2r_i \Delta r} \quad (13)$$

$$B_i = \frac{2}{\Delta r^2} + \frac{2}{r_i^2 \Delta \phi^2} + \frac{w_{i,j}}{\Delta z^k} \quad (14)$$

$$C_i = -\frac{1}{\Delta r^2} - \frac{1}{2r_i \Delta r} \quad (15)$$

$$D_i = \frac{1}{r_i^2 \Delta \phi^2} \quad (16)$$

$$r_i = (i-1)\Delta r \quad (17)$$

The variable increment in axial direction is given by

$$\Delta z^k = \lambda \Delta z^{k-1} \quad (18)$$

where i,j and k denote the radial, angular and axial increments, respectively and λ is the increment modulus which has to be greater than unity. The total length in the axial direction is given by

$$z^k = \frac{\Delta z^1(\lambda^{k-1} - 1)}{\lambda - 1} \quad (19)$$

A uniform step size is employed in the radial and circumferential directions. Successive Over-Relaxation (SOR) by line iteration method was utilized to solve for dimensionless temperature field $\theta_{i,j}^k$. The coefficient matrix in Eq.(12) is in tridiagonal form which permits the use of a general and efficient matrix inversion technique such as tridiagonal solver by Anderson, Tannehill and Pletcher [1]. The solution is carried out by solving the set of equations defined in Eqs.(12)-(19) for temperature $\theta_{i,j}^k$, beginning at the inlet of duct and marching downstream. The truncation error of this representation is the order of $O(\Delta r^2, \Delta \phi^2, \Delta z^k)$ where Δz^k implies the kth increment in the axial direction.

The tube wall temperature was obtained from the boundary condition and the fluid temperature gradient at the wall. A three point second order forward difference is used to calculate the temperature gradient at the wall. The integration of the temperature gradient along the duct wall in circumferential direction has to be equal to the axial heat flux, Q_w. After the temperature field is solved, the next step is to determine the heat transfer coefficient, h and Nusselt number, Nu.

An average heat transfer coefficient may be defined in the conventional manner by

$$\overline{q}_w = h(T_w - T_b)$$ (20)

where T_b is the bulk stream temperature.
The Nusselt number, hR_0/k is expressed by

$$Nu = \frac{1}{\theta_w - \theta_b}$$ (21)

where θ_b is the dimensionless bulk stream temperature

$$\theta_b = \frac{2}{\beta}\int_0^1\int_0^\beta w\theta r\,dr\,d\phi$$ (22)

All the numerical integrations in this work are carried out, by using Four-Panel Newton-Cotes formula as the following

$$\int f(\xi)d\xi \doteq \sum_{l=1,5,9,\ldots}^{m+1\,or\,n+1} \frac{2\Delta\xi}{45}[7(f_l + f_{l+4}) + 32(f_{l+1} + f_{l+3}) + 12f_{l+2}]$$

(23)

where $\Delta\xi$ is the interval between every two consecutive nodal points and m, n are the numbers of interval in radial and angular directions, respectively.   The truncation error of Eq.(23) is the order of $O(\Delta\xi^7)$ and Eq.(23) is applied twice to obtain the surface integration in Eq.(22).

RESULTS AND DISCUSSION

Based on the analyses presented above, numerical results are obtained for fluid temperature, peripheral wall heat flux and Nusselt number. They are presented graphically.

For comparison purpose, a limiting solution is considered first. The developed radial temperature profiles at center line are calculated and are compared with those of Eckert et al. [3]. A rather close agreement is observed with a difference of 3 percent. Fig.2 presents the developing temperature profiles at the center line vs. the radial distance with the axial position as a parameter. The apex angle is chosen to be 20°. Similar plots for 2β equal to 60° and 120° are shown in Figs.3-4, respectively. These figures illustrate how heat flows from the duct walls to the fluid. Noted that the temperature profiles at center line vary from a flat distribution to a concave distribution as the axial location moves toward downstream. The temperature profiles then remain the same shape when the thermally fully developed region is attained.   The minimum value of temperature distribution always appears at the same radial position for a given β. The influence of the curve wall becomes more dominant as the apex angle increases, i.e., the minimum value of temperature distribution shifts from the position near the curve wall to the position near the duct center as the apex angle increases. Fig.5 shows that small apex angle ducts will raise the flow temperature faster than that of large apex angle ducts.

It is of interest to examine the heat flux distribution along the straight and curve walls of the duct at thermal entrance region. The local heat flux q_w and the average heat flux $\overline{q}_w$ indicate the heat flow from the duct surface to the fluid around the periphery of ducts. The local heat flux q_w is obtained from the gradient of the fluid temperature in a direction normal to the duct surface and at the duct wall. In addition, $\overline{q}_w$ is the heat flow averaged over the duct periphery and it remains a constant for any axial position. Figs.6-8 illustrate the development and the distribution of the local heat flux, $q_w/\overline{q}_w$, along the straight walls of the ducts at three different apex angles. Similarly, Figs.9-11 show heat flux distribution along the angular direction normal to the curve wall of the duct. It is found that the local heat flux changes from a flat distribution to a convex distribution along the curve and straight walls as the fluid flows toward downstream. The maximum heat flux first occurs at the location near the curve wall then shifts slightly inward and maintains at the same radial location when flow moving down the stream. The local heat flux remains the same distribution along the curve wall and straight wall once thermally fully developed region is attained. The maximum values of the local heat flux along the radial direction becomes smaller if the apex angle of ducts increases. As can be examined from Figs.6-8 that the location of the maximum values of heat flux along the radial direction shifts from the curve wall to the tip (r=0) further as the apex angle of the ducts increases. Figs.6-8 reflect that the maximum values of the local heat flux along the angular direction becomes smaller if the apex angle of ducts becomes larger. Because of symmetry, the location of the maximum values of the heat flux along the angular direction remains at the center line for all apex angles of the ducts.

Since a zero inlet condition has been used here the Nusselt number approaches infinity at z=0 and decays exponentially. The developing Nusselt numbers are evaluated incrementally by marching downstream. Thus, in order to determine the entrance length, the march was terminated in the program whenever

$$\frac{\mathrm{Nu}^k - \mathrm{Nu}^{k+1}}{\Delta z^{k+1}} < 10^{-3}$$

where the Δz and superscript k denote the axial increment and marching increment, respectively.

Fig.12 reveals the developing Nusselt number as functions of the dimensionless distance from the entrance with the apex angle of the sector as a parameter. The circle on this figure represents the results of developing Nusselt number from Hong and Bergles [5] for $2\beta = 180°$ and the corresponding dash lines on the right hand side of the figure imply the developed Nusselt number from Shah and London [10]. It is found that they agree well within 2 percent. Because the thermal boundary layers meet each other faster for smaller apex angle, the smaller the apex angle of the sector, the shorter the entrance length. The Nusselt number decreases as the opening angle of circular sectors increases, because the local heat flux is lower for larger apex angle sector ducts.

CONCLUSION

Temperature distribution of the flow through circular sector ducts is obtained using a finite difference technique. Based on this result developing heat transfer coefficient and local heat flux distribution are calculated. As a limiting case, when the flow reaches the fully developed condition, the present predictions of Nusselt number and temperature agree very well with the solutions of previous investigations.

NOMENCLATURE

C	$(-1/\mu)(\partial P/\partial Z)(R_0^2/\overline{W})$;
C_P	specific heat at constant pressure;
h	average heat transfer coefficient, $\overline{q}_w/(T_w - T_b)$;
K	fluid thermal conductivity;
Nu	Nusselt number, hR_0/K;
P	pressure;
Pr	Prandtl number, $\mu C_P/K$;
q_w	rate of heat transfer per unit area of the wall;
$\overline{q}_w$	average rate of heat transfer per unit area of the wall;
Q_w	rate of heat transfer per unit length of the wall;
R, ϕ, Z	dimensional cylindrical coordinates;
r, ϕ, z	dimensionless cylindrical coordinates;
R_0	radius of the circular sector;
Re	Reynolds number, $\rho \overline{W} R_0/\mu$;
T, θ	dimensional and dimensionless local fluid temperature;
T_b, θ_b	dimensional and dimensionless local bulk stream temperature;
T_w, θ_w	dimensional and dimensionless wall temperature;
T_0	dimensional fluid entrance temperature;
T_c, θ_c	dimensional and dimensionless fluid temperature at the center line of sector ducts;
W, w	dimensional and dimensionless axial velocity;
$\overline{W}$	average axial velocity;
β	half apex angle of the circular sector;
μ	viscosity;

ρ   density;

REFERENCES

1. Anderson, D.A., Tannehill, J.C. and Pletcher, R.H. Computational Fluid Mechanics and Heat Transfer, pp. 549-550., Hemisphere Publishing Corporation, New York, N.Y., 1984.
2. Ben-Ali, T.M., Soliman, H.M. and Zaiffeh, E.K. Further Result for Laminar Heat Transfer in Annular Section and Circular Sector Ducts, *J. of Heat Transfer*, Trans. ASME, Vol. 111, pp. 1090-1093, 1989.
3. Eckert, E.R.G., Irvine, T.F. Jr. and Yen, J.T., Local Laminar Heat Transfer in Wedge-shaped Passages, Trans. ASME, vol. 80, pp. 1433-1438, 1958.
4. Hsia, R.P. Laminar Flow Heat Transfer in Thermal Entrance Region of Circular Sector Ducts, MS thesis, The University of Akron, 1990.
5. Hong, S.W. and Bergles, A.E. Laminar Flow Heat Transfer in Entrance Region of Semi-Circular Tubes with Uniform Heat Flux, *Int. J. Heat Mass Transfer*, Vol. 19, pp. 123-124, 1976.
6. Hu, M.H. and Chang, Y.P. Optimization of Finned Tubes for Heat Transfer in Laminar Flow, *J. of Heat Transfer*, vol. 95, pp. 332-338, 1973.
7. Lei, Q.M. and Trupp, A.C., Maximum Velocity Location at Pressure Drop of Fully Developed Laminar Flow in Circular Sector Ducts, *J. of Heat Transfer*, Trans. ASME, Vol. 106, pp. 1085-1087, 1989.
8. Lei, Q.M. and Trupp, A.C. Further Analyses of Laminar Flow Heat Transfer in Circular Sector Ducts, *J. of Heat Transfer*, Trans. ASME, Vol. 106, pp. 1088-1090, 1989.
9. Ozisik, M.N. Heat Conduction. Chapter 13, Integral-Transform Technique, pp. 561-563, John Wiley & Sons, Inc., New York, N.Y., 1980.
10. Shah, R.K. and London, A.L. Laminar Flow Forced Convection in Ducts, Advances in Heat Transfer, Supplement I, Academic Press, New York, 1978.
11. Sparrow, E.M. and Haji-Sheikh, A. Laminar Heat Transfer and Pressure Drop in Isosceles Triangular, Right Triangular and Circular Sector Ducts, *J. of Heat Transfer*, Trans. ASME, vol. 87, pp. 426-427, 1965.
12. Trupp, A.C. and Lau, A.C.Y. Fully Developed Laminar Heat Transfer in Circular Sector Ducts with Isothermal Walls, *J. of Heat Transfer*, Trans. ASME, Vol. 106, pp. 467-469, 1984.

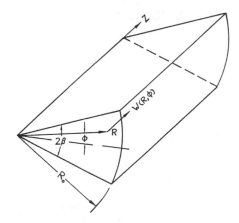

Fig1. Co—ordinate System for A Circular Sector Duct

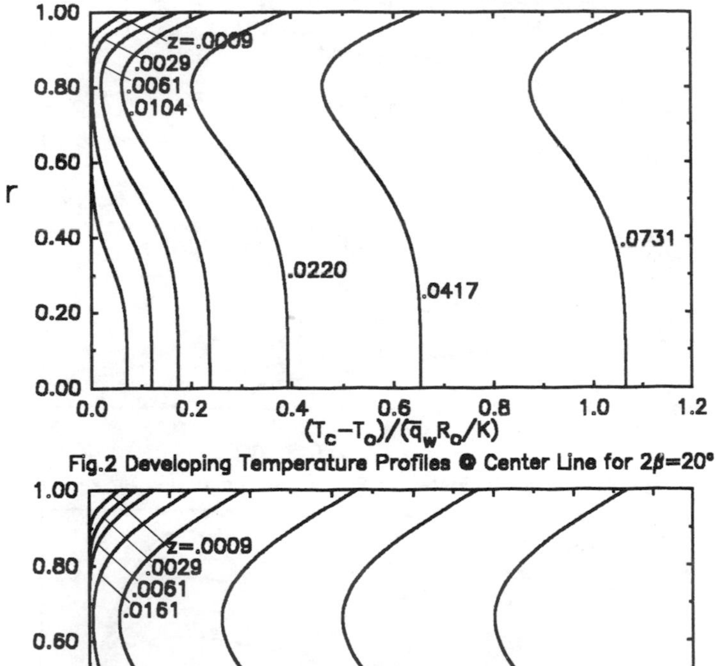

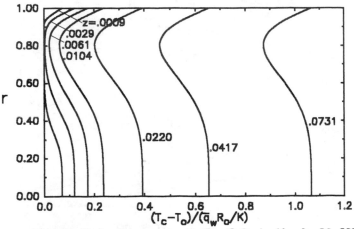

Fig.2 Developing Temperature Profiles @ Center Line for $2\beta=20°$

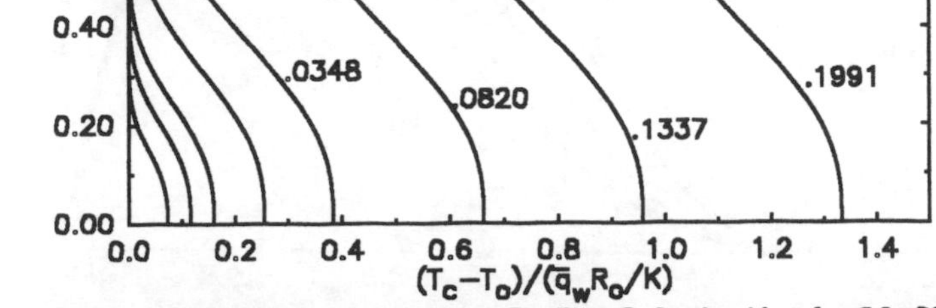

Fig.3 Developing Temperature Profiles @ Center Line for $2\beta=60°$

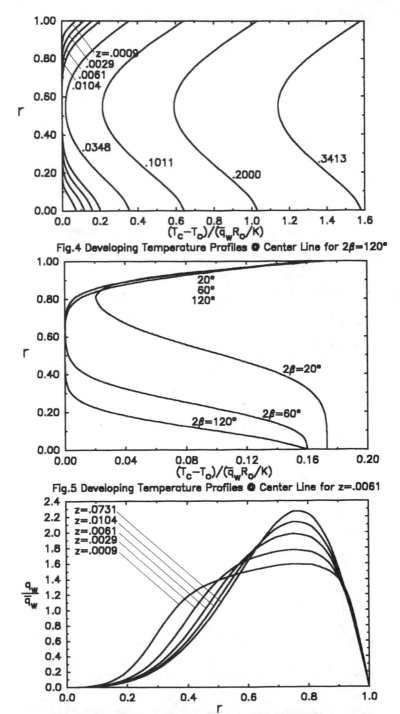

Fig.4 Developing Temperature Profiles @ Center Line for $2\beta=120°$

Fig.5 Developing Temperature Profiles @ Center Line for $z=.0061$

Fig.6 Developing Local Heat Flux in Radial Direction for $2\beta=20°$

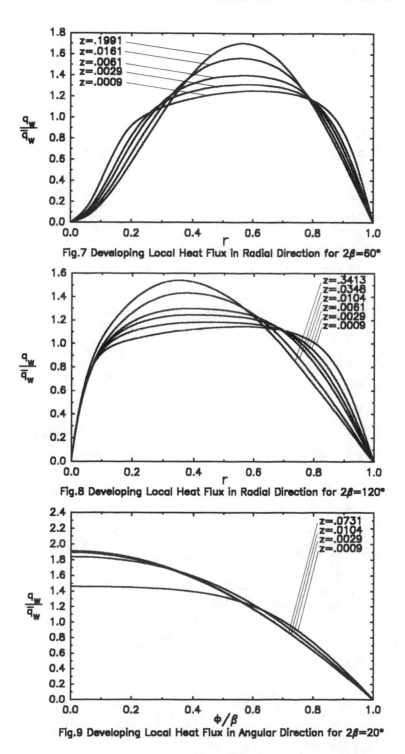

Fig.7 Developing Local Heat Flux in Radial Direction for $2\beta=60°$

Fig.8 Developing Local Heat Flux in Radial Direction for $2\beta=120°$

Fig.9 Developing Local Heat Flux in Angular Direction for $2\beta=20°$

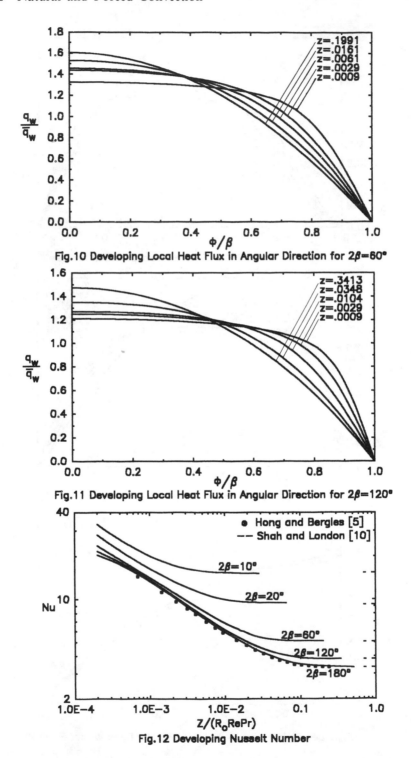

Fig.10 Developing Local Heat Flux in Angular Direction for $2\beta=60°$

Fig.11 Developing Local Heat Flux in Angular Direction for $2\beta=120°$

Fig.12 Developing Nusselt Number

Natural Convection of Gases with Property Variations and the Deviation of Heat Transfer in Calculation Caused by Ignoring the Variable Fluid Properties

D-Y. Shang(*), B-X. Wang(**)

(*) Department of Ferrous Metallurgy, Northeast University of Technology, Shenyang, Shenyang 110006, China

(**) Department of Thermal Engineering, Tsinghua University, Beijing 100084, China

ABSTRACT

The ordinary differential governing equations in considering variable fluid properties of gases are simplified to the ones when $T_w/T_\infty \to 1$ and $n_{c_p} \to 0$, which is taken as the classical Boussinesq's approximation. The corresponding expressions of the temperature gradient on the flat plate for the heat transfer are respectively summarized in considering the variable fluid properties and in ignoring the variable fluid properties. The deviations of the local heat transfer coefficient and the local Nusselt number are calculated together with the T_w/T_∞ for several gases, and then necessity of the treatment for considering variable fluid properties is pointed out.

NOMENCLATURE

a	diffusivity, m^2/s
c_p	specific heat at constant pressure, $J/(kg\ K)$
g	gravitational acceleration, m/s^2
Gr_x	local Grashof number, $gx^3(T_w/T_\infty - 1)/\nu_f^2$
$Gr_{x,\infty}$	local Grashof number, $gx^3(T_w/T_\infty - 1)/\nu_\infty^2$
n_{c_p}	temperature exponent for specific heat
n_λ	temperature exponent for thermal conductivity
n_μ	temperature exponent for dynamic viscosity
Nu_x	local Nusselt number, $a_x\ x/\lambda_f$
$Nu_{x,\infty}$	local Nusselt number, $a_x\ x/\lambda_\infty$
p	pressure, N/m^2
Pr	Prandtl number, $\mu c/\lambda$
q_x	local heat transfer rate pe unit area from wall to fluid, w/m^2
T	absolute temperature, K
w_x, w_y	velocity component in x- and y- direction respectively, m/s

W_x, W_y dimensionless velocity component in x- and y- direction respectively

Greek Sympols,
δ   boundary layer thickness, m
θ   dimensionless temperature, $(T-T_\infty)/(T_w-T_\infty)$
μ   absolute viscosity, kg/(ms)
ν   kinematic viscosity, m^2/s
ν_f kinematic viscosity at the film temperatrue $((T_w+T_\infty)/2)$,
     m^2/s
ρ   density, kg/m^3
λ   thermal conductivity, w/(m K)
λ_f thermal conductivity, at the film temperature $((T_w+T_\infty)/2)$,
     w/(m K)
a_x local heat transfer coefficient, $w/(m^2 K)$

Subscripts,
w   at wall; ∞ far from the wall surface

INTRODUCTION

In the literatures [1] and [2] we described the free convection and
heat transfer on the isothermal vertical flat plate for monatomic
and biatomic gases, as well as polyatomic gases. In both of the two
papers variable fluid property was dealt with. In this paper we
would further study the validity for the heat transfer by ignoring
the variable fluid property.

GOVERNING EQUATIONS AND THE HEAT TRANSFER EXPRESSIONS

According to the literatrue [2] the governing equations of conservation for mass, momentum and thermal energy in a two dimensional
steady laminar boundary layer on an isothermal flat plate for variable fluid properties of gases are respectively

$$\frac{\partial}{\partial x}(\rho w_x) + \frac{\partial}{\partial y}(\rho w_y) = 0 \tag{1}$$

$$\rho(w_x\frac{\partial w_x}{\partial x} + w_y\frac{\partial w_x}{\partial y}) = \frac{\partial}{\partial y}(\mu\frac{\partial w_x}{\partial y}) + \rho g\frac{T-T_\infty}{T_\infty} \tag{2}$$

$$\rho(w_x\frac{\partial(c_pT)}{\partial x} + w_y\frac{\partial(c_pT)}{\partial y} = \frac{\partial}{\partial y}(\lambda\frac{\partial T}{\partial y}) \tag{3}$$

The boundary conditions are
at y = 0 : $w_x=0$, $w_y=0$, $T=T_w$ $\tag{4}$

as y → ∞ : $w_x=0$, $T=T_\infty$ $\tag{5}$

For the equations (1)−(3) the following similarity transformations could be applied,

$$\eta = \frac{y}{x} \frac{(Gr_{x,\infty})^{1/4}}{\sqrt{2}} \tag{6}$$

$$\Theta = \frac{T-T_\infty}{T_w-T_\infty} \tag{7}$$

$$W_x = (2\sqrt{gx}(T_w/T_\infty-1)^{1/2})^{-1} w_x \tag{8}$$

$$W_y = (2\sqrt{gx}(T_w/T_\infty-1)^{1/2}(\tfrac{1}{4}Gr_{x,\infty})^{-1/4})^{-1} w_y \tag{9}$$

Then equations (1)−(3) could be transformed to dimensionless ordinary differential equations as follows respectively,

$$2W_x-\eta\frac{dW_x}{d\eta} +4\frac{dW_y}{d\eta} - \frac{1}{\varrho}\frac{d\varrho}{d\eta}(\eta W_x-4W_y) = 0 \tag{10}$$

$$\frac{V_\infty}{\nu}(W_x(2W_x-\eta\frac{dW_x}{d\eta})+4W_y\frac{dW_x}{d\eta}) =\frac{d^2W_x}{d\eta^2}+ \frac{1}{\mu}\frac{d\mu}{d\eta}\frac{dW_x}{d\eta} + \frac{V_\infty}{\nu}\Theta \tag{11}$$

$$(1+n_{c_p})Pr\frac{V_\infty}{\nu}(-\eta W_x+4W_y)\frac{d\Theta}{d\eta} =\frac{d^2\Theta}{d\eta^2} + \frac{1}{\lambda}\frac{d\lambda}{d\eta}\frac{d\Theta}{d\eta} \tag{12}$$

with following boundary conditions,

$$\eta = 0: W_x=0, W_y=0, \Theta=1 \tag{13}$$
$$\eta \rightarrow \infty: W_x=0, \Theta =0 \tag{14}$$

According to the corresponding treatment of variable fluid properties in paper [2] the equations (10)−(12) with the boundary conditions (13)−(14) could be solved numerically, and the following expressions of the temperature gradient and heat transfer for gases could be obtained by a curve matching method,

$$- \frac{d\Theta}{d\eta}\Big|_{\eta=0} =(1+0.3n_{c_p})\psi(Pr)(T_w/T_\infty)^{-m} \tag{15}$$

where $\psi(Pr)=0.567+0.186\ln Pr$ (16)

$$m =0.35n_\lambda+0.29n_\mu+0.36 \quad (\text{for } T_w/T_\infty>1) \tag{17}$$
$$m =0.42n_\lambda+0.34n_\mu+0.28 \quad (\text{for } T_w/T_\infty<1) \tag{18}$$

The local heat transfer rate per unit area from surface of the plate to gas can be calculated by fourier's law of heat conduction,

$$q_x = -\lambda_w(\frac{\partial T}{\partial y})_{y=0}$$

In term of the variable, q_x, $a_{x,\infty}$ and $Nu_{x,\infty}$ can be expressed as follows when variable fluid property is considered,

$$q_x = -\lambda_w (T_w - T_\infty) \frac{d\Theta}{d\eta}\Big|_{\eta=0} \left(\frac{1}{4} Gr_{x,\infty}\right)^{1/4} x^{-1}$$

$$\alpha_{x,\infty} = -\lambda_w \frac{d\Theta}{d\eta}\Big|_{\eta=0} \left(\frac{1}{4} Gr_{x,\infty}\right)^{1/4} x^{-1} \tag{19}$$

$$Nu_{x,\infty} = \frac{\alpha_{x,\infty} x}{\lambda} = -\frac{\lambda_w}{\lambda_\infty} \frac{d\Theta}{d\eta}\Big|_{\eta=0} \left(\frac{1}{4} Gr_{x,\infty}\right)^{1/4} \tag{20}$$

When simple power law of gas is used the $Nu_{x,\infty}$ could be expressed by

$$Nu_{x,\infty} = -\left(\frac{T_w}{T_\infty}\right)^{n_\lambda} \frac{d\Theta}{d\eta}\Big|_{\eta=0} \left(\frac{1}{4} Gr_{x,\infty}\right)^{1/4} \tag{21}$$

According to the similarity transformation in literatures [1] and [2], by means of equations (6) and (7), the corresponding factors of the physical properties in equations (10)−(12) would be expressed as follows,

$$\frac{1}{\rho} \frac{d\rho}{d\eta} = -\frac{(T_w/T_\infty - 1)\frac{d\Theta}{d\eta}}{(T_w/T_\infty - 1)\Theta + 1} \tag{22}$$

$$\frac{1}{\lambda} \frac{d\lambda}{d\eta} = \frac{n_\lambda (T_w/T_\infty - 1)\frac{d\Theta}{d\eta}}{(T_w/T_\infty - 1)\Theta + 1} \tag{23}$$

$$\frac{1}{\mu} \frac{d\mu}{d\eta} = \frac{n_\mu (T_w/T_\infty - 1)\frac{d\Theta}{d\eta}}{(T_w/T_\infty - 1)\Theta + 1} \tag{24}$$

$$\frac{\nu_\infty}{\nu} = ((T_w/T_\infty - 1)\Theta + 1)^{-(n_\mu + 1)} \tag{25}$$

Introducing the Boussinesq's approximation corresponding with the condition $T_w/T_\infty \to 1$ and $n_{c_p} \to 0$ for the equaitons (10)−(12), then, the corresponding ordinary governing equations for the free convection would be as following ones,

$$2W_x - \eta \frac{dW_x}{d\eta} + 4\frac{dW_y}{d\eta} = 0 \tag{26}$$

$$W_x \left(2W_x - \eta \frac{dW_x}{d\eta}\right) + 4W_y \frac{dW_x}{d\eta} = \frac{d^2 W_x}{d\eta^2} + \Theta \tag{27}$$

$$(-\eta W_x + 4W_y)\frac{d\Theta}{d\eta} = \frac{1}{Pr} \frac{d^2\Theta}{d\eta^2} \tag{28}$$

The boundary conditions of the equations $(26)-(28)$ are also the equations (13) and (14).

The equations $(26)-(28)$ with the boundary conditions (13) and (14) could be solved on the basis of the variation of Pr number, and the corresponding expression of the temperature gradient on the flat plate could be obtained by curve matching method as follows:

$$-\left.\frac{d\Theta|'}{d\eta}\right|_{\eta=0} = \psi(Pr) \tag{29}$$

The equation (29) is coincident with the classical boussinesq's approximation.

In ignoring the variable fluid properties the local heat transfer coefficient and local Nusselt number could be respectively expressed by

$$\alpha_x = -\lambda_w \left.\frac{d\Theta|'}{d\eta}\right|_{\eta=0} \left(\frac{1}{4}Gr_x\right)^{1/4} x^{-1} \tag{30}$$

$$Nu_x = \frac{\alpha_x x}{\lambda_f} = -\frac{\lambda_w}{\lambda_f} \left.\frac{d\Theta|'}{d\eta}\right|_{\eta=0} \left(\frac{1}{4}Gr_x\right)^{1/4} \tag{31}$$

By introducing simple power law, λ_w / λ_f can be expressed by

$$\frac{\lambda_w}{\lambda_f} = \left(\frac{T_w/T_\infty +1}{2T_w/T_\infty}\right)^{-n_\lambda} \tag{32}$$

From equation (31) and equation (32), Nu_x can be also expressed as follows:

$$Nu_x = -\left(\frac{T_w/T_\infty +1}{2T_w/T_\infty}\right)^{-n_\lambda} \left.\frac{d\Theta|'}{d\eta}\right|_{\eta=0} \left(\frac{1}{4}Gr_x\right)^{1/4} \tag{33}$$

THE COMPARISON BETWEEN RESULTS CALCULATED BY CONSIDERING VARIABLE FLUID PROPERTIES AND BY IGNORING THE VARIABLE FLUID PROPERTIES

The calculated results of $(1+0.3n_{c_p})(d\theta/d\eta)|'_{\eta=0}/(d\theta/d\eta)|_{\eta=0}$ for several gases by means of the corresponding numerical results are plotted in Fig.1.

From equation (15) combined with equation (29), the ratio could be expressed by

$$(1+0.3n_{c_p})\left.\frac{d\Theta|'}{d\eta}\right|_{\eta=0} / \left.\frac{d\Theta}{d\eta}\right|_{\eta=0} = (T_w/T_\infty)^m \tag{34}$$

It could be known that the result by equation (34) is very well coincident with the one calculated by the corresponding numerical solution.

From the equations (19) and (30), the following equation could be obtained,

$$(\alpha_x/\alpha_{x,\infty}) = (\frac{d\Theta}{d\eta}\Big|_{\eta=0}' / \frac{d\Theta}{d\eta}\Big|_{\eta=0})(Gr_x/Gr_{x,\infty})^{1/4} \qquad (35)$$

where ($a_x / a_{x,\infty}$) is the ratio of the local heat transfer coefficient for ignoring the variable fluid properties to the corresponding coefficient in considering variable fluid properties.

From equation (35) with the definitions of Gr_x and $Gr_{x,\infty}$, the equation (35) can be transformed to the following ones,

$$(\alpha_x/\alpha_{x,\infty}) = (\frac{d\Theta}{d\eta}\Big|_{\eta=0}' / \frac{d\Theta}{d\eta}\Big|_{\eta=0})(\nu_\infty/\nu_f)^{1/2} \qquad (36)$$

By means of simple power law, ($\nu_\infty / \nu_f)^{1/2}$ can be expressed by

$$(\nu_\infty/\nu_f)^{1/2} = (\frac{T_w/T_\infty+1}{2})^{-(n_\mu+1)/2} \qquad (37)$$

From equation (36) with (37), ($\alpha_x / \alpha_{x,\infty}$) can be expressed by

$$\alpha_x/\alpha_{x,\infty} = (\frac{d\Theta}{d\eta}\Big|_{\eta=0}' / \frac{d\Theta}{d\eta}\Big|_{\eta=0})(\frac{T_w/T_\infty+1}{2})^{-(n_\mu+1)/2} \qquad (38)$$

With the equations (15) and (29) the $a_x / a_{x,\infty}$ could be expressed as follows,

$$\alpha_x/\alpha_{x,\infty} = (T_w/T_\infty)^m(\frac{T_w/T_\infty+1}{2})^{-(n_\mu+1)/2}(1+0.3n_{c_p})^{-1} \qquad (39)$$

The calculated results of $(1+0.3n_{c_p})\,\alpha_x / \alpha_{x,\infty}$ from the equation (38) with the corresponding numerical solution $(d\theta/d\theta)|'_{\eta=0}$ and $(d\theta/d\eta)|_{\eta=0}$ for the several gases are plotted in Fig. 2. It could be known that the results are very well coincident with the one calculated by the equation (39).

From the equation (20) and equation (31) the ratio $Nu_x/Nu_{x,\infty}$ could be expressed as follows,

$$Nu_x/Nu_{x,\infty} = (\lambda_\infty/\lambda_f)(\frac{d\Theta}{d\eta}\Big|_{\eta=0}' / \frac{d\Theta}{d\eta}\Big|_{\eta=0})(Gr_x/Gr_{x,\infty})^{1/4} \qquad (40)$$

where ($Nu_x / Nu_{x,\infty}$) is the ratio of the local Nusselt number for ignoring the variable fluid properties to the one for considering the variable fluid properties.

By introducing the simple power law, $(\lambda_\infty / \lambda_f)$ can be expressed by

$$\lambda_\infty / \lambda_f = (\frac{T_w/T_\infty + 1}{2})^{-n_\lambda} \tag{41}$$

From equation (40) combined with definitions of Gr_x and $Gr_{x,\infty}$ and the equation (41) the ratio $Nu_x / Nu_{x,\infty}$ could be expressed by

$$Nu_x / Nu_{x,\infty} = (\frac{T_w/T_\infty + 1}{2})^{-n_\lambda} (\frac{d\Theta}{d\eta}\big|_{\eta=0}' / \frac{d\Theta}{d\eta}\big|_{\eta=0}) (\lambda_\infty / \lambda_f)^{1/2} \tag{42}$$

With equation (37), the equation (42) can also be described by

$$Nu_x / Nu_{x,\infty} = (\frac{T_w/T_\infty + 1}{2})^{-n_\lambda - (n_\mu + 1)/2} (\frac{d\Theta}{d\eta}\big|_{\eta=0}' / \frac{d\Theta}{d\eta}\big|_{\eta=0}) \tag{43}$$

With the equation (34), the equation (43) could be transformed as follows:

$$Nu_x / Nu_{x,\infty} = (T_w/T_\infty)^m (\frac{T_w/T_\infty + 1}{2})^{-n_\lambda - (n_\mu + 1)/2} (1 + C \cdot 3 n_{c_p})^{-1} \tag{44}$$

For several gases the results for ratio $(1 + 0.3 n_{c_p}) Nu_x / Nu_{x,\infty}$ by the equation (43) with the corresponding numerical solutions for the temperature gradient are shown in Fig.3. It could be known that the results are very well coincident with the ones calculated by equation (44).

DISCUSSIONS

The equations (26)−(28) are the ordinary differential governing ones which express the case both $T_w / T_\infty \to 1$ and $n_{c_p} \to 0$ for the equations (10)−(12), and which is corresponding with the classical Boussinesq's approximation.

From Figs.1−3 it can be seen that, for considering the variable fluid properties, the temperature gradient on the flat plate, heat transfer coefficient and Nusselt number would reveal very large variation with T_w / T_∞. In addition, the larger the n_μ, n_λ and n_{c_p} of gas are, the more obvious the variation is.

From Fig.2 it could be known that, if the variable fluid properties are ignored, the calculated deviation of the corresponding heat transfer would be between −0.04 and 0.03 for Ar, between −0.045 and 0.04 for air, between −0.06 and 0.05 for water vapour, and between −0.08 and 0.065 for NH_3 when $0.9 < T_w / T_\infty < 1.1$. When $0.7 < T_w / T_\infty < 1.3$, the deviation would be between −0.14 and 0.1 for Ar, between −0.15 and 0.11 for Air, between −0.2 and 0.16 for water vapour, and between −0.23 and 0.18 for NH_3. The father from the unity the ratio T_W / T_∞ deviated, the larger the deviation caused by ignoring the variable fluid properties is.

CONCLUSIONS

From the study and the discussions, it follows the following conclutions:

When $T_W / T_\infty \to 1$ and $n_{c_p} \to 0$, the ordinary differential governing equations for free convection on isothermal flat plate should be the equations (26)−(28), which is corresponding with the classical Boussinesq's approximation. The expression of temperature gradient on the flat plate for the classical Boussinesq's approximation should be expressed by the equation (29).

In ignoring the variable fluid properties of gases, the calculated deviation for the heat transfer would increase with the case when T_W / T_∞ deviates the unity. In addition, the larger n_μ, n_λ and n_{c_p} are, the larger the deviation would be.

It is necessary to consider the variable fluid properties in calculating the heat transfer fo the free convection on the isothermal flat plate, and it also follows that the validity of the Boussinesq's approximation is very limited.

ACKNOWLEDGEMENT, The project is financially supported by National Science Foundation Committee of China with grant No.5880238.

REFERENCES

1. Shang, D.Y. and Wang, B.X. Effect of variable Thermophysical Properties on Laminar Free Convection of Gas, will be published in Int. J.Heat Mass Transfer, 1990.
2. Shang, D.Y. and Wang, B.X. Effect of Variable Thermophysical Properties on Laminar Free Convection of Polyatomic Gas, submitted to Int. J.Heat Mass Transfer, 1990.

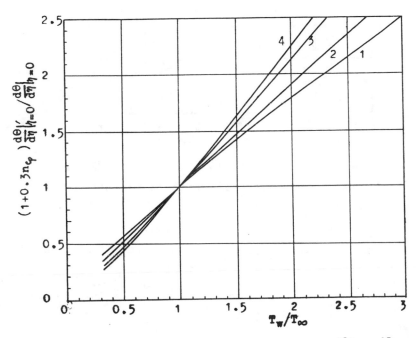

Fig. 1 The calculated results of $(1+0.3n_{c_p})\frac{d\Theta}{d\eta}\big|_{\eta=0}'/\frac{d\Theta}{d\eta}\big|_{\eta=0}$

by the corresponding numerical results

(1. air, 2. gas mixture*, 3. water vapour, 4. NH_3)

* component of the gas mixture: $co_2=0.13$,

water vapour$=0.11$ and $N_2=0.76$).

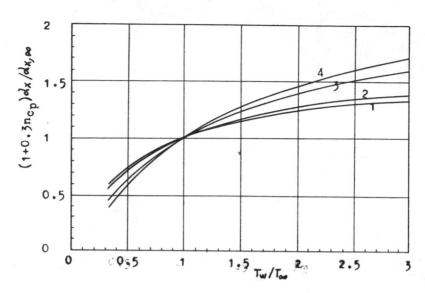

Fig. 2 The calculated results of $(1+0.3n_{c_p})d_x/d_{x,\infty}$
       from the equation (38) with corresponding
       numerical results
       (1. Ar, 2. air, 3. water vapour, 4. NH_3)

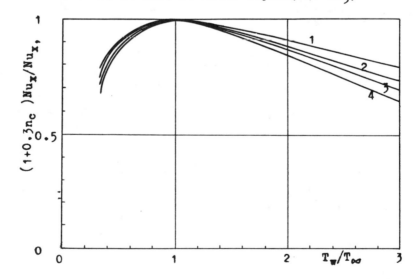

Fig.3 The calculated results of $(1+0.3n_{c_p})Nu_x/Nu_x$,
      from equation (43) with corresponding numerical
      results
      (1. air, 2. gas mixture, 3. water vapour, 4. NH_3)

Natural Convection in a Vertical Annular Enclosure Filled with a Porous Material: Adoption of the Darcy-Brinkman Model

J.C. Morales(*), U. Lacoa(*), A. Campo(**), H. Campos(*)

(*) Departamento de Mecánica, Universidad Simón Bolívar, Caracas 1080-A, Venezuela

(**) Department of Mechanical Engineering, Florida International University, Miami, Florida 33199, USA

ABSTRACT

This paper presents a numerical study of non-Darcian natural convection in a vertical annular enclosure filled with a porous material. The flow is modeled using the Brinkman-extended Darcy equations. A finite-volume based procedure was employed to numerically solve the conservation equations with the SIMPLER algorithm. A comparison between the predictive numerical results and the experimental observations demonstrate the importance of non-Darcian effects especially in enclosures having high Darcy numbers. Finally, Nusselt number curves are presented covering wide ranges of the controlling parameters.

INTRODUCTION

Buoyancy-induced convection in vertical annular enclosures containing a single-phase fluid or a porous medium has received considerable attention over the years because of its broad range of applications in engineering. In particular, problems involving vertical and slender annular enclosures filled with a porous medium have been studied and the references by Havstad and Burns [1], Reda [2], Hickox and Gartling [3] and Prasad and Kulacki [4] are representatives. These investigations have been taken in an effort to gain insight into the heat loss mechanism associated with insulated tanks, insulated oil wells and nuclear waste storage. These theoretical works were based on the traditional Darcy's law and clearly indicate the very limited effort has been made to examine the influence of other equally or more realistic parameters characterizing the porous material. For this geometrical configuration, the results reflected an interrelation of Nusselt number, Rayleigh number, Darcy number, the aspect ratio and the radii ratio of the enclosure with respect to the Darcy model.

Conversely, the situation of buoyancy-driven flows in vertical annular cavities has been extensively studied in the literature.

The inclusion of non-Darcy effects in the analysis of natural convection in porous media has become very important in recent years. In this context, Prasad et al. [5] among others have noted that in general experimental data for several combinations of fluids and solids at low Rayleigh numbers do not agree with the theoretical predictions based on the simple Darcy model. To overcome this deficiency, Tong and Sabramanian [6] and Lauriat and Prasad [7] invoked the Darcy-Brinkman model and examined the boundary effects of natural convection heat transfer in a rectangular porous cavity. It was found in both references that the Nusselt number tends to decrease with an increase in the Darcy number maintaining a fixed value of the modified Rayleigh number.

The purpose of the present work is to examine the significance of Brinkman's modification to the Darcy law on natural convection in a vertical annular cavity, where the influence of curvature has to be taken into account. Aside from a practical application, this type of flow is widely accepted for testing and evaluating numerical solution procedures. There is, however, no reported information on the heat transport behavior of porous annular layers modeled by the Darcy-Brinkman relation.

A finite-volume procedure using the SIMPLER algorithm along with a coordinate transformation is adopted in the present work to solve the coupled system of conservation equations. The effects of the dimensionless groups: Rayleigh, Ra, Darcy, Da, as well as the aspect ratio, A, and the radii ratio of the enclosure, m are examined in detail. However, due to space restrictions, a full parametric study was not carried out in this communication.

MATHEMATICAL FORMULATION

The physical system depicted in Fig. 1 consists of two adiabatic horizontal walls and two vertical isothermal concentric walls that are differentially heated. The porous medium is completely saturated with fluid and is assumed to be macroscopically isotropic, homogeneous and in local thermal equilibrium. The properties are treated as constant, except for the buoyancy term, where the classical Boussinessq approximation allows for the variation of density with temperature in the momentum equation. The thermaconvective flow induced by buoyancy forces is interpreted as steady with a two-dimensional laminar motion in the annular enclosure.

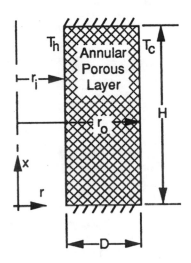

Fig.1   Annular region

In view of the foregoing description, the dimensionless equations of conservation may be written as follows:

$$\frac{\partial U}{\partial X} + \frac{1}{R}\frac{\partial}{\partial R}(RV) = 0 \tag{1}$$

$$\frac{U}{Da} = -\frac{\partial P}{\partial X} + \frac{1}{R}\frac{\partial}{\partial R}\left(R\frac{\partial U}{\partial R}\right) + \frac{\partial^2 U}{\partial X^2} + \frac{Ra}{Pr}\theta \tag{2}$$

$$\frac{V}{Da} = -\frac{\partial P}{\partial R} + \frac{1}{R}\frac{\partial}{\partial R}\left(R\frac{\partial V}{\partial R}\right) + \frac{\partial^2 U}{\partial X^2} \tag{3}$$

$$U\frac{\partial \theta}{\partial X} + V\frac{\partial \theta}{\partial R} = \frac{1}{Pr}\left[\frac{1}{R}\frac{\partial}{\partial R}\left(R\frac{\partial \theta}{\partial R}\right) + \frac{\partial^2 \theta}{\partial X^2}\right] \tag{4}$$

where the Darcy-Brinkman model has been employed for the porous medium.

The boundary conditions imposed on the governing equations are:

$$U = V = 0, \qquad on \quad the \quad walls \tag{5}$$

$$\theta = 1, \qquad R = R_i \tag{6}$$

$$\theta = 0, \qquad R = R_0 \tag{7}$$

$$\frac{\partial \theta}{\partial X} = 0 \qquad X = 0, \quad X = A \tag{8}$$

In addition, the controlling parameters are the Rayleigh number Ra, the Darcy number Da, the aspect ratio A and the radii ratio m.

Once the velocity and temperature fields are computed, the results for the total heat transfer rate across the enclosure are presented in terms of the average Nusselt number Nu, defined as

$$N = \frac{Q_{total}}{Q_{conduction}} \tag{9}$$

In this ratio, the denominator represents the heat transfer by conduction when the entire enclosure was filled with fluid alone. Accordingly, the appropriate equation for Nu may be expressed by the relation

$$N = -\frac{1}{AR_k} \int_0^A \left(\frac{1}{R} \frac{\partial \theta}{\partial R} - \theta V \right) dX \tag{10}$$

where R_k identifies the thermal conductivity ratio.

METHOD OF SOLUTION

Equations (1)-(8) are solved numerically using the finite volume approximations devised by Patankar [8], in conjunction with a suitable non-uniform grid. Accordingly, the set of conservation equations (1)-(4) belongs to the general diffusion-convection type:

$$\frac{\partial}{\partial X_j}\left(\rho U_j \phi - \Gamma_\phi \frac{\partial \phi}{\partial X_j}\right) = S_\phi \tag{11}$$

where ϕ is the dimensionless dependent variable, while Γ_ϕ and S_ϕ are the appropriate diffusion coefficient and source terms, respectively. The system of eqs. (1)-(4) represented in compact form by eq. (11) was solved iteratively using a version of a SIMPLER algorithm developed by Patankar [9] for integrating a system of elliptic partial differential equations. Correspondingly, primitive variables, U, V, P, θ were used in a staggered grid system. The computational domain was divided into rectangular control volumes with one grid point located at the center of the control volume forming the basic cell. In this regard, the variables pressure p and temperature θ were calculated at these grid points, while the velocity variables U and V were calculated at points that lie on the face of the basic cells.

To cope with enhanced accuracy, refinement of the mesh was mandatory and the grid points were positioned non-uniformly in the integration domain where the highest concentration of grid points is placed in the vicinity of the walls. Accordingly, the grid deployment was conveniently plotted according to a stretching coordinate transformation proposed by Roberts [10]. Furthermore, the resulting system of algebraic equations was solved iteratively using a line-by-line approach, in conjunction with the block correction procedure of Settari and Aziz [11] to accelerate the convergence. In addition, for each run, an overall energy balance was made for the enclosure wherein the integrated heat transfer rate through the inner cylinder must be equal to that through the outer cylinder. In general, agreement of less than 1% on the heat transfer rates were found for all runs.

To verify the numerical accuracy of the solution

procedure, results were first obtained for a Darcy
model and a rectangular cavity. Using the Nusselt
number for comparison purposes, good agreement is
observed in Fig. 2 with the results of Prasad et al.
[5].

RESULTS

To investigate properly the significance of
Brinkman's modification of the Darcy flow model, the
numerical results have been compared with the
experimental data reported by Prasad et al. [4,5] over
a wide range of Rayleigh and Darcy numbers. The
experiments were conducted with several combinations
of solids and fluids bounded by short concentric
vertical cylinders. Fig. 2 has been prepared to
examine this comparison in detail for Da = 10^{-5} and
different Ra. In this figure, it may be observed that
the predicted Nu curve reproduces the physical
behavior over the wide range of Rayleigh numbers (10^6
< Ra < 10^{10}). An increase in Ra characterizes the
heat transfer enhancement due to buoyant forces, and
consequently the Nusselt number increases with the
Rayleigh number. The slope remains almost constant

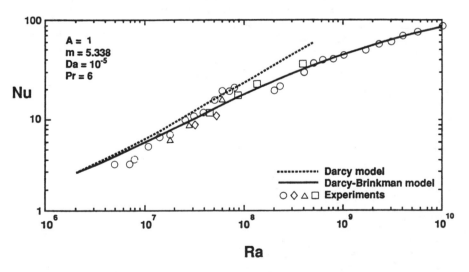

Fig.2 Comparison of the Nusselt number for Darcy and
       Darcy-Brinkman model with experiments

for the boundary layer motion in the Darcy regime,
i.e. for low values of Ra. However, when Ra is
increased beyond a certain value of Ra, the slope of
the Nusselt curve starts decreasing indicating the
departure from the Darcy regime. This is quite
evident from the curve in Fig. 2 which shows a

reduction in slope around Ra = 2×10^3. In the non-Darcian regime at high values of Ra, both viscous and inertia effects have to be incorporated in the model. The slope of the curve will further decrease and finally approach that for a single-phase fluid at higher Rayleigh numbers. In addition, the numerical results of Prasad based on a Darcy model are also induced in the figure. It should also be noted that in the non-Darcy regime, the Darcy model overpredicts the heat transfer rates; the deviation between the two Nusselt numbers increases with Ra.

Figs. 3 and 4 illustrate the variation of the Nusselt number using the Darcy-Brinkman solutions for A = 5 and m = 2. First, the trends already observed in Fig. 2 are repeated in Fig. 3 for different Da numbers. The slope of the Nusselt number curve in the non-Darcy regime decreases with an increase in Da and eventually approaches zero. This corresponds to an asymptotic convection regime where the heat transfer rate becomes independent of the Darcy number. On the contrary, the departure from the pure conduction regime starts with a combination of low Rayleigh numbers and high Darcy numbers.

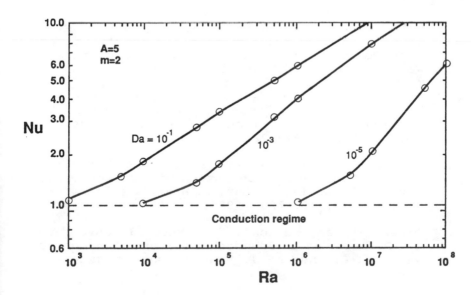

Fig.3 Variation of the Nusselt number for Darcy-Brinkman model

Additionally, Fig. 4 shows the onset of natural convection parametrized by the Rayleigh number. Here, it may be seen that the Darcy model is generally valid in the vicinity of the conduction regime.

K

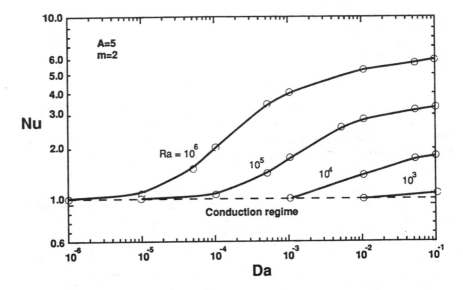

Fig.4   Variation of the Nusselt number for Darcy-
       Brinkman model

REFERENCES

1.   Havstad, M.A. and Burns, P.J. Convective Heat
     Transfer in Vertical Cylindrical Annuli Filled
     with Porous Media, Int. J. Heat Mass Transfer,
     Vol. 25, pp. 1755-1766, 1982.

2.   Reda, D.C. Natural Convection Experiments in a
     Liquid-Saturated   Porous   Medium   Bounded   by
     Vertical   Coaxial   Cylinders,   J. Heat Transfer,
     Vol. 105, pp. 795-802, 1983.

3.   Hickox,   C.E.   and   Gartling,   D.K.   A   Numerical
     Study   of   Natural   Convection   in   a   Vertical,
     Annular Porous Layer, Int. J. Heat Mass Transfer,
     Vol. 28, pp. 720-723, 1985.

4.   Prasad, V. and Kulacki, F.A. Natural Convection
     in Porous Media Bounded by Short Concentric
     Vertical Cylinders, J. Heat   Transfer, pp. 147-
     154, 1985.

5.   Prasad,   V.,   Kulacki,   F.A.   and   Keyhani,   M.
     Natural Convection in Porous Media, J. Fluid
     Mechanics, Vol. 50, pp. 89-119, 1985.

6.   Tong, T.W. and Subramanian, E. A Boundary Layer

Analysis for Natural Convection in a Porous Enclosure: Use of the Brinkman-Extended Darcy Model, Int. J. Heat Mass Transfer, Vol. 28, pp. 563-571, 1987.

7. Lauriat, G. and Prasad, V. Natural Convection in a Vertical Porous Cavity: A Numerical Study for Brinkman-Extended Darcy Formulation, J. Heat Transfer, Vol. 11, pp. 295-320, 1987.

8. Patankar, S.V. Numerical Heat Transfer and Fluid Flow, McGraw-Hill, New York, 1980.

9. Patankar, S.V. A General-Purpose Computer Program for Two-Dimensional Elliptic Situations, University of Minnesota, Minneapolis, Minnesota, USA, 1982.

10. Roberts, G.O. Computational Meshes for Boundary Layer Problems, Lecture Notes on Physics, Vol. 8, pp. 171-177, Springer-Verlag, New York, 1971.

11. Settari, A. and Aziz, K. A Generalization of the Additive Correction Methods for the Iterative Solution of Matrix Equations, SIAM J. Numerical Analysis, Vol. 10, pp. 506-521, 1973.

NOMENCLATURE

A	aspect ratio of enclosure, H/D
d	characteristic particle dimension
D	enclosure width
Da	Darcy number, κ/d^2
g	acceleration due to gravity
Gr	Grashof number, $g\beta(T_h - T_c)D^3/(\mu/\rho)^2$
H	enclosure height
m	radii ratio, r_0/r_1
k	thermal conductivity
Nu	Nusselt number, eq.(9)
p	pressure
P	dimensionless pressure, $p/(\rho v_0^2)$
Pr	Prandtl number, v/a
r	radial coordinate
r_1	inner radius
r_0	outer radius
R	dimensionless radial coordinate, r/D
R_k	thermal conductivity ratio, k_f/k_p
Ra	Rayleigh number, GrPr
T	temperature
u	vertical velocity
U	dimensionless vertical velocity, u/v_0
v	radial velocity
V	dimensionless radial velocity, v/v_0
v_0	dimensionless velocity, $\mu/(\rho D)$

x   vertical coordinate
X   dimensionless vertical coordinate, x/D

Greek Letters

α   thermal diffusivity
β   coefficient of volumetric expansion
θ   dimensionless temperature, $(T - T_c)/(T_h - T_c)$
κ   permeability
μ   dynamic viscosity
ρ   density

Numerical Simulation of the Mixed Convective Transport from a Heated Moving Plate

B.H. Kang, Y. Jaluria
Department of Mechanical and Aerospace Engineering, Rutgers University, New Brunswick, NJ 08903, USA

ABSTRACT

The conjugate mixed convection transport that arises due to the continuous movement of a heated plate in uniform forced flow in the same direction of the motion of the plate has been numerically investigated. Two important circumstances, which involve plate movement in the uniform flow in a channel and that in a uniform free stream in an extensive medium, are considered in this study. A numerical study is carried out, assuming a two-dimensional, steady, circumstance. The full elliptic equations are solved, employing finite difference and finite volume techniques. The transport in the solid material is coupled with that in the fluid through the boundary conditions and the two are solved for simultaneously. The numerical simulation of such processes is discussed, outlining the important considerations to obtain accurate results for a wide variety of practical problems that involve a moving material subjected to heat transfer. The numerical results for the flow field and the temperature distributions in the solid, as well as in the flow, are obtained. The numerical imposition of the boundary conditions is shown to be a very important aspect in the simulation. The effect of the relevant parameters that arise in the numerical scheme on the results is also determined.

INTRODUCTION

A problem of considerable practical and fundamental importance is that of a continuously moving flat plate which loses energy by convection and radiation at the surface to the ambient fluid. This circumstance is of interest in a wide variety of manufacturing processes such as hot rolling, plastic extrusion, metal forming, glass fiber drawing and continuous casting processes [2,4]. In most cases, the ambient fluid, far from the moving material, is stationary, with the fluid flow being induced by thermal buoyancy and the motion of the solid material. Therefore, the resulting flow and thermal fields are determined by these two mechanisms, buoyancy and surface motion. However, in many practical processes, the resulting flow does not provide adequate cooling rates and an additional, externally induced, forced flow is employed to increased the heat transfer rate [1]. Other heat transfer mechanisms, such as boiling with the use of sprays and radiation, are also employed for increasing the heat transfer rates [10,12].

The essential features of the two-dimensional flow induced by a long, continuously moving flat plate in a uniform forced flow are shown schematically in Fig. 1. Two forced flow circumstances are considered in this study. These concern the heated plate moving in a forced channel flow and in a forced free stream. The plate moves out of the slot in an extrusion die, or between the rollers, and proceeds at a constant speed U_s. Due to the viscous drag exerted on the fluid by the surface of the plate, the flow is aided by the plate motion if the plate speed is larger than the external flow, i.e., $U_s > U_\infty$, and the forced flow is opposed if $U_s < U_\infty$. The plate, if it is warmer than the fluid, also loses heat to the fluid. There are several important aspects in this process that need a detailed investigation. These include the effects of conjugate boundary conditions at the surface, boundary conditions upstream of the location of the emergence of the material, the channel width, and the additional free stream on the resulting thermal and flow fields and on the heat transfer. Also, of interest in this work is the numerical imposition of the relevant boundary conditions and the consequent effect on the results.

The present study is directed at a numerical simulation of the conjugate transport from a continuously moving flat plate, including the effects of thermal buoyancy and of conduction within the plate and of the non-boundary layer behavior of the flow adjacent to the surface. The material is assumed to be maintained at a given uniform temperature T_o far upstream of the point where it emerges from a slot, representing an extrusion die, furnace outlet, rollers or mold. The flow field and the temperature distributions in the solid and in the flow are computed. Numerical calculations are carried out, assuming a two-dimensional, steady, laminar flow circumstance. The full elliptic equations that govern the heat transfer and the flow are solved numerically, employing finite difference and finite volume methods. Far downstream, the flow approaches the characteristics of a boundary layer and this fact is employed in the development of the numerical method. The results obtained lead to a better understanding of the underlying physical processes and also provide inputs that may be used for the design of the relevant systems. Also considered in detail are the numerical treatment of the relevant boundary conditions, particularly far upstream and far downstream.

ANALYSIS

Consider the flow induced by a plate, which is moving with a constant velocity U_s in a uniform forced flow of velocity U_∞. The transport in the solid material is coupled with that in the fluid through the boundary conditions and the two are solved for simultaneously. The temperature of the plate is assumed to be at a uniform value T_o at some far upstream distance x_b away from the origin, i.e., at $x = -x_b$. Basically, we want to simulate the condition of uniform temperature far upstream in a furnace or an oven. When the temperature gradients are large along the plate length, the axial diffusion within the plate and the fluid must be taken into account. Therefore, the full governing equations, which are elliptic in nature, must be solved near $x = 0$. The circumstances for which the boundary layer approximations may be employed are discussed in detail by Karwe and Jaluria [8].

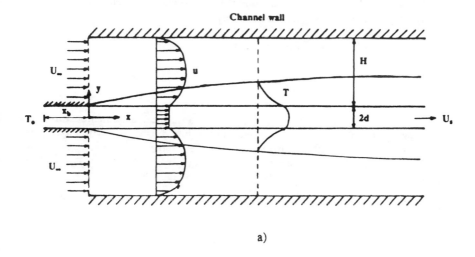

a)

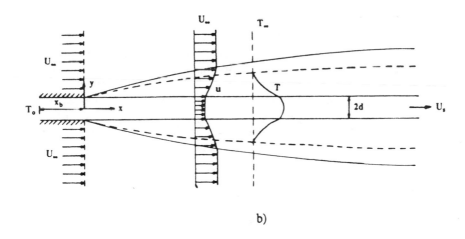

b)

Fig. 1   Schematic of the velocity and temperature profiles in the conjugate
            problem for a heated moving plate in a uniform forced flow.

            a) Plate moving in a channel flow

            b) Plate moving in a free stream

The full governing equations, including the transient terms, are nondimensionalized by employing the following transformation

$$X = x/d, \quad Y = y/d, \quad \tau = tU_s/d, \quad U = u/U_s, \quad V = v/U_s,$$

$$\Psi = \psi/U_s d, \quad \Omega = \omega d/U_s, \quad \theta = \frac{T - T_\infty}{T_o - T_\infty}, \tag{1}$$

The dimensionless equations are obtained in the vorticity-stream function formulation, by taking curl of the momentum equation and introducing fluid vorticity. This approach is particularly suitable for two-dimensional problems since the pressure is eliminated as an unknown and one less equation has to be solved [7]. For a plate which is moving vertically upward, the governing equations become

for the fluid

$$\nabla^{*2} \Psi = -\Omega \tag{2}$$

$$\frac{\partial \Omega}{\partial \tau} + \overline{V}.\nabla^* \Omega = \frac{1}{Re}(\nabla^{*2} \Omega) - \frac{Gr}{Re^2} \frac{\partial \theta}{\partial Y} \tag{3}$$

$$\frac{\partial \theta}{\partial \tau} + \overline{V}.\nabla^* \theta = \frac{1}{Re.Pr}(\nabla^{*2} \theta) \tag{4}$$

For a plate moving vertically downward, the sign of the buoyancy term is reversed. For the case of a plate moving horizontally, the thermal buoyancy gives rise to a pressure gradient normal to the plate surface [6]. This pressure gradient is aligned with the plate surface in the vertical case. The energy equation for the plate is

$$\frac{\partial \theta}{\partial \tau} + \frac{\partial \theta}{\partial X} = \frac{1}{Pe}(\nabla_s^{*2} \theta) \tag{5}$$

where the Reynolds number Re, the Peclet number Pe, the Grashof number Gr, and the Prandtl number Pr are defined as

$$Re = U_s d/\nu_f, \quad Pe = U_s d/\alpha_s, \quad Gr = \frac{g\beta(T_o - T_\infty)d^3}{\nu_f^2}, \quad Pr = \nu_f/\alpha_f \tag{6}$$

It is also noted here that the Peclet number Pe is not RePr since the Pe is based on the thermal conductivity of the solid material, not that of the fluid.

The boundary conditions on u, v and T arise from the physical considerations of the flow and heat transfer processes and are shown in Fig. 2. These are due to the no-slip conditions at the moving surface of the plate and at the channel walls, temperature and heat flux continuity between the fluid and the solid at the surface of the plate and symmetry about the x axis, which is at the midplane of the plate. The boundary conditions, in terms of the dimensionless vorticity Ω and streamfunction Ψ, were obtained by employing the appropriate transformation of the physical boundary conditions [12]. Additional boundary conditions downstream, in X, are needed because of the elliptic nature of the governing equations. This is achieved, as shown in Fig. 2, by dividing the physical domain under consideration into two regions : I) An "Elliptic" region near the slot or the die, and, II) A "Parabolic" region far downstream. Near the slot, the boundary layer approximations are not valid and, therefore, the full equations are solved in Region I. In the parabolic region, which is taken far from the slot, the boundary layer approximations are employed. To solve for the elliptic region, boundary conditions are specified at the interface between the two regions numerically.

NUMERICAL SCHEME

In all the cases considered here, the numerical calculations were carried out for half of the computational domain. The extent of the grid in the Y direction was chosen such that the ambient conditions could be specified at the edge of the domain. This extent was chosen to be at least twice the estimated boundary layer thickness at the maximum value of X in the computational domain. The transient vorticity transport equation, equation (3), was solved along with the energy equations, equations (4) and (5), in the conservative form, for the flow and the plate, respectively, using the Alternate Direction Implicit (ADI) scheme [12]. In the ADI scheme, the time step of integration is split into two halves. Integration over the first half time step is implicit in the X direction, while the second half is implicit in the Y direction. This yields two tridiagonal linear systems that are then solved by the tridiagonal matrix algorithm. The Poisson equation, equation (2), for the stream function was solved using the successive-over-relaxation (SOR) method [3]. At each time step, after advancing the solution for equations (3)-(5) by the ADI scheme, equation (2) was solved using the SOR method till the maximum change in Ψ satisfied a specified convergence criterion. For the present study, the optimum relaxation factor was found numerically to be about 1.75, and convergence was obtained only a few iterations. Three-point central differences were used to discretize all the terms except the convection terms, for which a second-order upwind differencing was used in the X direction [12]. The accuracy of such a scheme is of the order of $[\Delta X^2, \Delta Y^2, \Delta \tau^2]$.

Even though the truncation error for the ADI method is of the order of $[\Delta X^2, \Delta Y^2, \Delta \tau^2]$, for linear equations, this second order accuracy is lost when the governing non-linear equations are linearized by assuming U and V constant over a time step. In addition, the ADI method requires the boundary values of vorticity at the plate surface. The current boundary values are given as the previous time step values at these cases. This makes the method unstable for larger time steps. The Von Neumann analysis indicates unconditional stability for linear problems. For the present study, the numerical scheme was stable with the time step $\Delta \tau \leq 0.005$.

For the first time step in the elliptic region, the calculations are started with the ambient conditions at the interface between the two regions. The temperature θ within the solid is first determined and then the temperature in the flow is obtained, employing the thermal condition at the plate surface. This boundary condition is derived by using the energy balance over the control volume around the grid point lying on the surface of the plate. The upper half of the control volume lies within the fluid and the lower half lies within the plate. The new value of θ is used in the calculation of the vorticity in the flow at the same time step. The stream function equation, which requires only the values of Ψ on the boundaries and the values of the vorticity at interior points, is then solved by the SOR method to give a new value of Ψ. This new value of Ψ is used to determine the velocity components U and V and the new boundary conditions for the vorticity at the solid walls.

Using the computed values of U, V and θ at one grid point away from the interface, inside the elliptic region, the transient boundary layer equations are then solved. The modified Crank-Nicholson scheme given in [8], is used to march in the downstream direction. Thus, the values U, V and θ are updated at the interface and are then used as boundary conditions for the calculations at the next time step in the elliptic region. Such an approach has been shown to work successfully by Karwe and Jaluria [9,10]. The location of the interface between the two regions is also varied till the final solution is marginally affected by a further change, as discussed later. This leads to a computational process for calculating the temperature, vorticity, stream function, and velocity components for increasing time, starting with the initial conditions.

The calculations were stopped when [max ($\partial\theta/\partial\tau$) and max ($\partial U/\partial\tau$)] $\leq \varepsilon$, where ε is a small quantity, indicating steady-state conditions. The effect of the chosen convergence criterion ε on the steady state solution was also studied, to ensure a negligible effect on the computed results. For the present study, ε = 0.0001 was found to be adequate. Similarly other numerical parameters like time step, grid size, computational domain, etc., were varied. The number of iterations needed to reach steady state depends on the physical circumstance and on the given governing parameters, especially the Prandtl number Pr. About 30,000 iterations for the plate moving in uniform free stream, and about 25,000 iterations for the plate moving in forced flow in a channel are necessary to obtain convergence to steady state when the fluid medium is water, Pr = 7.0. However, when air is employed as the fluid medium, i.e., Pr = 0.7, about three times the number of iterations, compared to the water case, are needed to reach steady state. The calculations performed on the CDC Cyber 205 machine at JVNC/Princeton required 380K words of computer storage and 0.1 sec of CPU per iteration with vectorization. The total CPU time is about 0.85 hour for a typical computational run in which the plate is moving in a water free stream.

NUMERICAL RESULTS AND DISCUSSION

The numerical scheme has been developed to investigate the conjugate transport from the moving plate to the uniform forced flow. Numerical solutions to equations (2)-(5) were obtained for wide ranges of the governing parameters. Some of the typical results are presented here. The computed velocity field in terms of the velocity vectors and the isotherms are shown in Fig. 3 for two forced flow circumstances when a heated aluminum plate is moving vertically upward in a water channel flow and in a free water stream. A uniform flow of velocity U_∞ is taken at X = 0. The flow field is seen to develop very rapidly downstream and to yield velocity levels higher than U_∞ for both forced flow circumstances. The isotherm distributions are very similar to each other. However, the velocity profiles are seen to be very different, due to the boundary conditions at the channel wall or at the free stream flow far away from the plate. Temperature distributions across the plate are seen to be uniform due to the high thermal conductivity of the aluminum plate. It is also found that the upstream penetration of conduction effects within the aluminum plate is significant for both cases.

The effect of thermal diffusion penetrates upstream and, therefore, the relevant boundary condition must be applied numerically at $X = X_b$, where the value of X_b must be increased till the results become essentially independent of a further increase. Fig. 4 shows the temperature at $X = 0$ and $Y = 0$ for various values of X_b, when the plate surface is assumed to be adiabatic in the region $-X_b \leq X \leq 0$. The value of θ_o is specified at a uniform value of $\theta_o = 1.0$ at $X = X_b$, for $0 \leq Y \leq 1$. Therefore, the final solution is dependent on the boundary condition location $X = -X_b$ at which θ_o is specified. As seen in Fig. 4, the value of X_b has a significant effect on the computed temperature θ at $X = 0$ and $Y = 0$ with a high thermal conductivity material, such as aluminum. The effect of thermal diffusion is small enough to be negligible with Teflon cases, due to the low thermal conductivity of the material. It is also found that $X_b = 10$ is large enough to simulate the condition of $\theta = \theta_o$ far upstream even with the aluminum case.

To illustrate the effect of different boundary conditions on the plate surface in the region $- X_b \leq X \leq 0$, numerical results are presented in Fig. 5 for the adiabatic and isothermal boundary conditions at the surface upstream of $X = 0$. As seen in Fig. 5, for the isothermal boundary condition, the centerline temperature of the aluminum plate is higher than that for the corresponding adiabatic condition, implying that the upstream conduction effects are smaller in the isothermal case. However, the effects of these boundary conditions are almost negligible for a Teflon plate, due to the low thermal conductivity of the material. The results shown in Figs. 4 and 5 are important from the point of view of design of an actual system and also for the numerical simulation of the process for different materials. In many practical circumstances involving a furnace or an oven, the isothermal boundary condition is more appropriate. However, a longer length is required in this case, to bring the temperature level to a desired value. We have discussed only two of the typical boundary conditions that arise, depending upon the process under consideration. In some cases, for instance, in a process involving radiative or electrical heating, a uniform heat flux condition at the plate surface, may be applied upstream of the slot. The type of boundary condition will influence the calculated length of the cooling trough, distance between the die and the take-up spool, etc.

The channel width H/d is varied in the channel flow case and the results are compared with those in the free stream case. The effect of the channel width on the heat transfer rate is shown in Fig. 6. Also shown is the comparison with the computed heat transfer rate in free stream. The local Nusselt number Nu is defined here as

$$Nu = \frac{h.d}{K_f} = \frac{(-\frac{\partial \theta}{\partial Y})_{Y=1.0} .d}{\theta_{Y=1.0}} \qquad (12)$$

where h is the local heat transfer coefficient calculated at the surface of the plate and is given by $h = (-K_f \frac{\partial T}{\partial y}|_f)_{Y=1.0} /(T_s - T_\infty)$. As channel width H/d increases, the heat transfer rate is also increased. Mixed convection flow in vertical channels have been extensively considered and various boundary conditions are discussed by

Gebhart et. al. [5]. The heat transfer rate is increased with an increase in the Reynolds number Re_∞, which is based on the channel width H and the uniform velocity U_∞, i.e., $Re_\infty = U_\infty H/\nu_f$ [5]. For a given fluid medium and a fixed uniform velocity, Reynolds number Re_∞ is dependent upon only the channel width H. Therefore, the heat transfer rate is increased as H increases. However, when the channel width is very large, the flow approaches the external free stream case. As seen in this figure, the local Nusselt number variation for H/d = 8.5 in the channel flow case is very close to that in the free stream case.

CONCLUSIONS

A detailed numerical study of the thermal transport arising from a continuously moving plate in an uniform forced fluid flow is carried out. Numerical solutions for the flow and temperature fields, particularly in the vicinity of the die or a furnace from where the plate emerges, have been obtained. It is found that the number of numerical iterations for convergence to steady state strongly depends upon the boundary conditions, specified for the two different flows, and the properties of the fluid medium. The results obtained indicate that the penetration of the conductive effects, upstream of the point of emergence, is significant and must be treated appropriately treated in the numerical scheme. When a plate is moving in a channel flow, the effect of the channel width on the resulting heat transfer rate is significant. However, when the channel width is increased, the heat transfer rate in case of a plate moving in a channel flow approaches that in the case of a plate moving in free stream. The importance of the numerical imposition of the boundary conditions is brought out. The effect of the basic numerical parameters such as grid size, convergence criterion, etc., is also determined.

ACKNOWLEDGEMENTS

The authors acknowledge the financial support provided by the National Science Foundation, under Grant CBT-88-03049, for this work.

REFERENCES

[1] Abdelhafez, T. A. Skin friction and heat transfer on a continuous flat surface moving in a parallel free stream, Int. J. Heat Mass Transfer, vol. 28, pp. 1234-1237, 1985.

[2] Altan, T., Oh, S. and Gegel, H. Metal Forming Fundamentals and Applications, American Society of Metals, Metals Park, Ohio, 1979.

[3] Carnahan, B., Luther, H. A. and Wilkes, J. O. Applied Numerical Methods, John Wiley & Sons, New York, 1969.

[4] Fisher, E. G. Extrusion of Plastics, John Wiley & Sons, New York, 1976.

[5] Gebhart, B., Jaluria, Y., Mahajan, R. L., and Sammakia, B. Buoyancy Induced Flows and Transport, Hemisphere, New York, 1988.

[6] Jaluria, Y. Natural Convection Heat and Mass Transfer, Pergamon Press, New York, 1980.

[7] Jaluria, Y. and Torrance, K. E. Computational Heat Transfer, Hemisphere, New York, 1986.

[8] Karwe, M. V. and Jaluria Y., Thermal Transport from Heated Moving Surfaces, J. of Heat Transfer, Vol. 108, pp. 728-733, 1986.

[9] Karwe, M. V. and Jaluria Y. Fluid Flow and Mixed Convection Transport from a Plate in Rolling and Extrusion Processes, ASME J. of Heat Transfer, Vol. 110, pp. 655-661, 1988.

[10] Karwe, M. V. and Jaluria Y. Numerical Simulation of Thermal Transport Associated With a Continuously Moving Flat Sheet in Rolling or Extrusion, Vol. 3, pp. 37-45, The Proceedings of ASME 1988 National Heat Transfer Conference, Houston, 1988.

[11] Rhodes, C. A. and Chen, C. C. Thermal Radiation in Laminar Boundary Layer on Continuous Moving Surfaces, ASME, J. Heat Transfer, Vol. 96 , pp. 32-36, 1974.

[12] Roache, P. J. Computational Fluid Dynamics, Hermosa Publishers, New Mexico, 1976.

[13] Zambrunnen, D. A., Viskanta, R., and Incropera, F. P., The Effect of Surface Motion on Forced Convection Film Boiling Heat Transfer, J. Heat Transfer, Vol. 111, pp. 760-766, 1989.

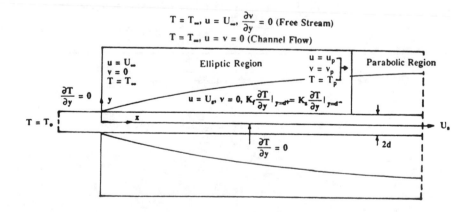

Fig. 2   Boundary conditions for a moving heated plate in a uniform forced flow, indicating the numerical boundary between the elliptic and parabolic region.

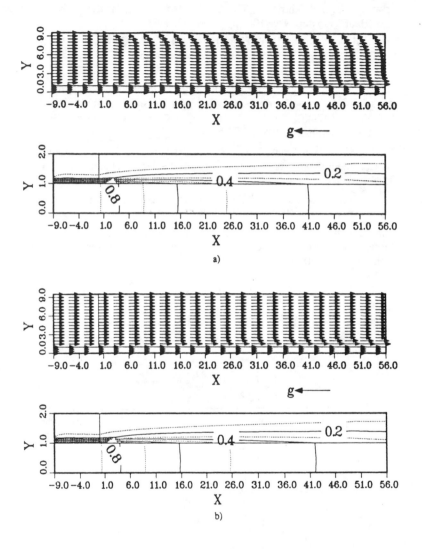

Fig. 3   Calculated velocity vectors and isotherms for typical circumstances.
            Pr = 7.0, Pe = 0.29, Re = 25, Gr = 1500, U_∞/U_s = 2.0, K_f/K_s = 0.0029
            a) Aluminum plate moving in a water channel flow
            b) Aluminum plate moving in a water free stream

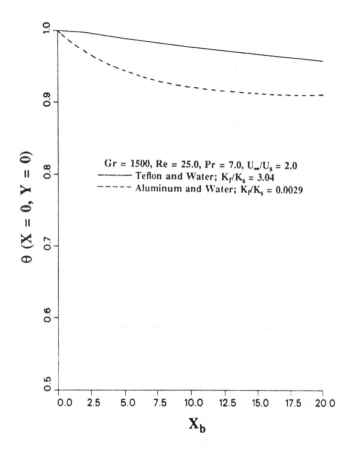

Fig. 4   The dependence of temperature at X=0 and Y=0 on the location X = −X_b upstream at which θ_o is specified for a heated plate moving vertically upward in a channel with water as the fluid, considering Teflon and aluminum as the solid material.

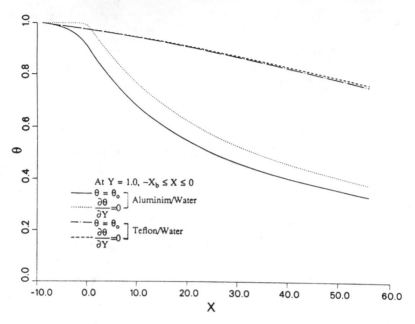

Fig. 5   Effect of the numerically imposed adiabatic and isothermal surface
boundary conditions in the region $-X_b \leq X \leq 0$, on the downstream
midplane (Y=0) temperature variation.

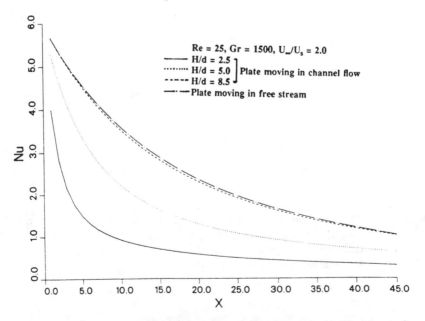

Fig. 6   Effect of the channel width on the calculated local Nusselt number,
varying with the downstream distance, for a heated aluminum plate
moving vertically upward in a uniform water flow.

Prediction of Heat Transfer from Helically Ribbed Surfaces

F.S. Henry, M.W. Collins

Thermo Fluids Engineering Research Centre, City University, London, EC1V 0HB, England

ABSTRACT

Using the AERE Harwell code FLOW3D, numerical calculations of heat transfer have been carried out for fully-developed turbulent flow through an annulus with a multi-start, helically-ribbed inner cylinder. Various rib (helix) angles were considered, and to facilitate comparison, the transverse-ribbed and smooth cases were also considered.

FLOW3D allows the use of non-orthogonal body-fitted coordinates which were necessary for the flow geometry under consideration. The code uses k-ε turbulence model, and employs wall functions at solid boundaries. As the code allows conduction and convection to be solved simultaneously, the heat was assumed to enter through the inner surface of the inner cylinder, thus avoiding the need to define explicitly either heat flux or temperature on the ribbed fluid/solid interface.

The calculations were carried out in an effort to clarify the experimental finding that helical-ribbed surfaces have better heat transfer characteristics than the corresponding transverse-ribbed case. The predictions supported the experimental finding that helical ribs with a rib angle of approximately 34 degrees have the best overall performance. It was also predicted that the 34-degree case had the lowest average ribbed-surface temperature.

While the predictions broadly agreed with the available experimental data, significant differences were noticed in detail. An attempt was made to resolve better the ribs by refining the grid. However, the resulting predictions were still significantly different from the experimental data. It was concluded that the standard k-ε turbulence model was deficient for this class of flow

1.   INTRODUCTION

Heat transfer from ribbed surfaces has been the subject of several experimental investigations. Lockett and Collins [1] included a review of past work on the subject in their investigation using holographic interferometry. Like most researchers, they studied the case of heat transfer from a transversely-ribbed plane surface. Numerical predictions of this problem are less common. Wilkes [2], using the k-ε model, predicted some of the experimental finding of Watts and Williams [3]. However, to the best of the authors' knowledge, no calculations for the cylindrical case, of any rib type, have been published.

Pirie [4] made a series of measurements of pressure drop and bulk heat transfer in the flow through annuli with inner cylinders having multi-start, helical ribs. Pirie's data showed that helical-ribbed surfaces have superior pressure drop and heat transfer characteristics when compared to surfaces with transverse ribs. However, to date, no measurements have been made of the details of the flow or heat transfer in the inter-rib channel. The heat transfer calculations to be described in the paper were carried out, in conjunction with the flow calculations described in Henry and Collins [5] to clarify the details of the heat transfer process in the inter-rib channel of cylindrical surfaces with multi-start, helical ribs.

2.   GOVERNING EQUATIONS AND THEIR NUMERICAL SOLUTION

2.1   Governing equations

The Reynolds equation for temperature can be written as:

$$U_j \frac{\partial T}{\partial x_j} = \nu \frac{\partial^2 T}{\partial x_j \partial x_j} - \frac{\partial}{\partial x_j} \overline{U_j \theta} \qquad (1)$$

where U is the mean velocity, T the mean temperature, and ν is the thermal diffusivity.

Invoking Reynolds' analogy between heat and mass transfer, the correlation between the fluctuating temperature and the velocity field is modelled as,

$$\overline{u_i \; \theta} = -\frac{\nu_t}{Pr_t} \frac{\partial T}{\partial x_i} \tag{2}$$

where Pr_t is the turbulent Prandtl number, and the eddy viscosity is computed, using the k-ε model as,

$$\nu_t = c_\mu \; k^2/\varepsilon \tag{3}$$

where c_μ is a constant, k is the turbulence kinetic energy, and ε is the rate of energy dissipation.

The equation for steady heat conduction in the solid wall can be given as,

$$\frac{\partial}{\partial x_i} \left(\lambda_s \frac{\partial T}{\partial x_i} \right) = 0 \tag{4}$$

where λ_s is the solid conductivity.

The near-wall condition on temperature is estimated by assuming there exists a universal temperature profile of similar form to that for the velocity. The near-wall temperature is then given by

$$T^+ = \frac{1}{\kappa} \ell_n \; (F \; y^+) \tag{5}$$

where

$$T^+ = \frac{(T_w - T) \; \rho \; u^* \; C_p}{Q_w}$$

$$y^+ = y \; u^*/\nu$$

and

$$u^* = (\tau_w/\rho)^{1/2}$$

T_w is the wall temperature, Q_w is the wall heat flux, C_p is the specific heat at constant pressure, F is an emperical constant, and τ_w is the wall shear stress.

As with the velocity log layer, equation 4 cannot be used as y^+ approaches zero, and it is necessary to introduce a viscous sublayer temperature profile if y^+ is smaller than some minimum value y_t^+.   The sublayer temperature profile is of the form

$$T^+ = Pr \ y^+ \tag{6}$$

where Pr is the molecular Prandtl number.   The value of F was found using the formula due to Jayatilleke [6], i.e.,

$$F = E \ exp \ [9.0 \ \kappa \ ((Pr/Pr_t)^{\frac{3}{4}} - 1)(1 + 0.28 \ exp(-0.007 \ (Pr/Pr_t))] \tag{7}$$

The matching point y_t^+ is then found iteratively from

$$y_t^+ = \frac{1}{\kappa \ Pr} \ \mathcal{l}_n \ (F \ y_t^+) \tag{8}$$

2.2 Numerical solution

The heat transfer calculations were carried out using HARWELL FLOW3D, Release 2.3, which is a general purpose code for the calculation of three-dimensional flow and heat transfer, developed at AERE Harwell.   FLOW3D uses a control volume approach, and the current version (Release 2) allows the use of non-orthogonal body-fitted coordinates on a non-staggered grid.   Body-fitted coordinates were necessary for the flow geometry under consideration.   Details of the development of FLOW3D with body-fitted coordinates have been given by Burns et al [7], and general details of FLOW3D by Jones et al [8].

The flow field was calculated separately using HARWELL FLOW3D, Release 2.1.   These calculations have been described in detail in Henry and Collins [5].   Briefly, both laminar and turbulent flows were computed.   the k-ε model together with the standard near-wall boundary treatment at solid surfaces were used for the turbulent flow calculations.

3.   PROBLEM DEFINITION AND GRIDDING SYSTEMS

3.1 Geometrical and other details

Schematics of transverse- and helical-ribbed cylinders are given in Figure 1.   Details of the five ribbed geometries considered are given in Table 1.   As can be seen, the rib (helix) angle is defined such that the transverse case has a value of zero degrees.   In all cases, the inner cylinder root diameter, D_1, was taken to be 39.44 mm, the rib height, e, to be 1.13 mm, the rib pitch to be 7.29 mm, and the inner diameter of the outer cylinder, D_2, to be 98.09 mm.   When used (i.e., in the heat transfer calculations) the inner diameter of the inner cylinder, D_o, was taken to be 36.62 mm.   The rib widths were chosen so that all geometries had the same value (0.8609 mm) in the axial direction, and hence, the same inter-rib channel width in that direction.   Further, the rib widths were chosen so that the 34-degree case matched that used by Pirie [9].   The smooth-walled annuli were assumed to have inner diameters of 39.44 mm and outer diameters of 98.09 mm.

Case	1 (transverse)	2	3	4	5
Number of starts	–	4	6	12	30
Rib angle (deg)	0	12.887	18.941	34.464	59.767
Rib width at mid height (mm)	0.8609	0.8416	0.8153	0.7100	0.4335

Table 1.   Flow channel geometry

To reduce the computational effort to a minimum, all flows were assumed to be incompressible, and fully developed.   It was further assumed that the density and molecular viscosity were not functions of temperature, thus decoupling the velocity and temperature fields.   The Reynolds number for the turbulent flows was chosen to match that used by Pirie [9].   Mass flow rate and other data used in the heat transfer calculations are given in Table 2.

Solid
  Thermal conductivity, k_s 150 W/m.K

Fluid
  Mass flow rate, m   0.45 kg/s

  Density, ρ 1.15 kg/m^3

  Viscosity, μ 1.960a$\times 10^{-5}$ Pa.s

  $R_e = \rho (D_2 - D_1) U_m / \mu$ 2.13$\times 10^5$

  Thermal conductivity of fluid, k_f 2.802$\times 10^{-2}$ W/m.K

Specific heat at constant pressure, C_p	1007.5 J/kg.K
Prandtl number = $C_p \mu / k_f$	0.705

Turbulence Model

F	3.186
y_t^+	12.546
Turbulent Prandtl number, Pr_t	0.9 or 0.7

Heat flux (constant), Q_w	61497.0 W/m^2

Table 2.   Heat Transfer Input Data

In all cases, the mass flow rate and density were kept constant at the values in Table 2. When required, the Reynolds number was changed by adjusting the viscosity.

3.2   Computational grids

The assumption of fully-developed flow required the velocity field to be periodic in the axial direction. This meant, in the transverse-ribbed case, that it was necessary to model only one rib pitch. The solution domain for the transverse case is shown schematically in Figure 1a as the hatched area ABCD. AB and CD are planes of periodicity. The flow is also axisymmetric, and hence, BC and DA are planes of symmetry.

In order to ensure that all predictions were of comparable accuracy, the transverse grid was made geometrically similar to the helical grids. That is, the grids in the radial direction were identical, and the same number of grid points was used in the axial direction of the transverse grid as was used in the angular direction of the helical grids. Also, each node point in the axial direction of the transverse grid was made to correspond to an angular node point in the helical grids.

The number of control volumes used to resolve the rib was limited by the need for the first computational point to be above the viscous sublayer. In fact only two volumes could be used over the rib height, and a similar number across the rib width, at the design Reynolds number (213,000). To reduce the total number of control volumes, the grid was expanded away from all solid surfaces.

FLOW3D treats all flows as three-dimensional, and requires dummy nodes at either end of all coordinate directions to implement boundary conditions. Consequently, the minimum number of control volumes in any one direction is three. For example, in the case of axisymmetric flows (as is the flow over transverse ribs) the grid has only one active node in the angular direction, but dummy nodes are required either side to satisfy the symmetry boundary conditions. The total number of control volumes (active and dummy) used in the transverse-ribbed case was 16×27×3.

In the helical-ribbed case it was necessary to use a non-orthogonal grid. A sample grid is shown in Figure 2. Note that the grid shown does not represent the actual grid used. In this case, the assumption of fully developed flow meant that the flow was periodic in the angular direction, and it was necessary to model only the flow in one angular segment. The segment is then bounded by planes of periodicity in the angular direction. These planes are shown schematically as lines AD and BC in Figure 1b. Only one active control volume was necessary in the flow direction for the helical-ribbed case. The nature of the fully-developed flow over the helical ribs is such that the velocity, and all other flow quantities except pressure and temperature, is constant for constant radial position in planes parallel to the ribs, for example along the lines AD and BC in Figure 1b. Fully-developed flow was generated by invoking periodic boundary conditions in the mean flow direction. Unless otherwise stated, all helical-ribbed predictions were generated on 3×27×16 grids, similar to that shown in Figure 2.

4.   BOUNDARY CONDITIONS

A constant heat flux was assumed at the inner diameter of the inner cylinder. The actual value used was chosen to give a heat input to the fluid of 5 kW/m. FLOW3D (Release 2.3) allows the conduction in the solid and the convection in the fluid to be solved simultaneously. The boundary condition between solid and fluid is treated as part of the iterative solution. The inner diameter of the outer cylinder was assumed to be an abiabatic surface. In the angular direction the temperature field was axisymmetric in the smooth-walled, and transverse-ribbed cases, and periodic in the helical-ribbed cases.

Upstream and downstream boundary conditions for the temperature field presented a similar problem to that encountered for the pressure field in the flow calculations, which have been described in Henry and Collins [5]. It can be shown that the bulk temperature, ΔT_b, increases linearly with axial distance, 1.

Specifically,

$$\Delta T_b / 1 = 2\pi \, r_o \, Q_w / C_p \, \dot{m} \qquad (8)$$

where r_o is the inner radius of the inner cylinder, Q_w is the heat flux through the inner surface, and $\dot{m}$ is the mass flow rate. The inlet and outlet boundary conditions on the temperature field can then be given as

$$T(1, J, K) = T(NI-1, J, K) - \Delta T_b \qquad (9)$$

and

$$T(NI, J, K) = T(2, J, K) + \Delta T_b \qquad (10)$$

As the smooth-walled annulus and the helical-ribbed cases have only a single active control volume in the axial direction, in these cases ΔT_b is the bulk temperature increase over one control volume. However, in the transverse-ribbed cases ΔT_b is the temperature increase over one rib pitch. It should be noted that there are NI-2 active control volumes in the axial direction. Control volumes 1 and NI are dummy volumes used to implement boundary conditions. The above boundary conditions were implemented in a modified version of the FLOW3D routine BCSPER.

5.   PREDICTIONS

5.1   Surface temperature and Stanton number

Heat transfer calculations were carried out for the smooth-wall annulus, transverse-ribbed case, and the 13-, 19-, 34-, and 60-degree helical-ribbed cases. Only turbulent flows were considered.

The heat transfer parameters of special interest in this study were the ribbed-surface temperatures and the Stanton number. The local Stanton number is defined as

$$St = \frac{h}{\rho \, c_p \, U_m} \qquad (11)$$

where

$$h = \frac{Q_w}{T_w - T_b}$$

and

$$T_b = \frac{1}{\dot{m}} \int \rho \, U \, T \, r \, dr \, de$$

Variations of local Stanton number for the transverse and helical cases are given in Figure 3. The turbulent Prandtl number was taken to be 0.9. The only direct measurements of local heat transfer coefficients for ribbed-roughened surfaces appears to be the transverse rib data of Watts and Williams [3]. Wilkes [2] performed a numerical study of the same geometry using the k-ε model. The present prediction of local Stanton number for the transverse case, given in Figure 3, matches the general trend of Wilkes' [2] predictions of local heat transfer coefficient (h = constant×St). However, as with Wilkes' [2] predictions, the predicted values show a sudden decrease as the rib is approached which is not seen in the experimental data of Watts and Williams [3]. However, this decrease does appear in the holographic interferometry data reported by Lockett and Collins [1].

The helical-ribbed predictions follow the expected trend of a decrease in the overall level of heat transfer with an increase in angle. Again as would be expected, in all cases, higher values of heat transfer are predicted to occur on the ribs than occur in the rib channel.

Predicted distributions of wall temperature are given in Figure 4. It is evident that the swirl produced by the helical ribs does reduce the wall temperature, and therefore lowers the risk of hot spots. The prediction of a nearly constant wall temperature follows the transverse rib prediction of Wilkes [2]. The difference between the maximum and minimum temperatures is predicted to be less than two degrees. It is of interest that the 34-degree helical-ribbed case, and not the 60-degree case, is predicted to have the lowest overall wall temperature.

A curve of the predicted variation of the average Stanton number with rib angle is given in Figure 5. The turbulent Prandtl number was taken to be 0.9. The numerical values are given in Table 4., together with predictions of average Stanton numbers for the same cases, but with a turbulent Prandtl number of 0.7. Predictions for the smooth-walled cases are also included.

Case	Average Stanton Number	
	$Pr_t=0.7$	$Pr_t=0.9$
Smooth	0.00265	0.00191
Transverse	0.00290	0.00234
Helical		
13	0.00296	0.00239
19	0.00298	0.00238
34	0.00279	0.00218
60	0.00207	0.00159

Table 4. Predicted Average Stanton Numbers

Inspection of Table 4 reveals that the 60-degree helical-ribbed case is predicted to have an average Stanton number that is lower than the smooth-walled case. This would seem to be physically implausible. A rough estimate of the smooth-walled Stanton number can be gained from the assumption that

$$St=f/2 \tag{12}$$

Using equation 12, and the predicted value of friction factor (0.00398) gives a Stanton number of 0.00199. Hence, it can be assumed that the smooth-walled prediction (for Pr=0.9) is reasonably accurate. Therefore, it must be concluded that the predicted Stanton numbers for all the ribbed cases are probably in error. In fact, values of untransformed Stanton number given in Pirie's [4] table of experimental results would suggest that the predictions are as much as 50% low. Wilkes [2] found predictions were as much as 40% lower than expected.

The estimate of Stanton number using equation 12 would appear to suggest that the turbulent Prandtl number should be taken to be 0.9. This is somewhat misleading as it can be shown (see for instance, Reynolds [10]) that equation 12 implies that both the molecular and the turbulent Prandtl numbers are unity. In fact, while it is assumed to be so in FLOW3D, there is experimental evidence that the turbulent Prandtl number is not a constant in wall-bounded flows. In pipe flow of air it has been found that the turbulent Prandtl number is approximately 0.9 in the wall layer and reduced to about 0.7 in the turbulent core of the flow. Hence, it would seem legitimate too use a lower value to improve the predictions. However, while significant improvements in the predicted Stanton number are seen in Table 4 for the cases with a turbulent Prandtl number of 0.70, they are still much lower than Pirie's [4] experimental values. Obviously, other factors such as grid dependency, and inaccuracies within the turbulence model must also be causing the poor predictions.

5.2 Grid refinement

Flow fields for refined grids were available for the 34-degree helical-ribbed case. Hence, it was a simple matter to consider the effect of grid refinement on the heat transfer predictions. Calculations were carried out for the coarse grid ($3 \times 27 \times 16$) and a refined grid of $3 \times 40 \times 22$ at the flow Reynolds number of one million. The turbulent Prandtl number was assumed to be 0.9.

The predicted average Stanton number for the refined grid was 0.00085 compared to 0.00064 for the coarse grid. Hence, grid refinement gave an increase in average Stanton number of approximately 33%. The unrealistically low values were due to the fact that the molecular Prandtl number was inadvertently set to 0.15. However, it can be assumed that if grid refinement were possible for the lower Reynolds number it would significantly improve the Stanton number prediction.

5.3 Performance parameter

Pirie [4] following Wilkie [11] defined the thermal performance of the ribbed surfaces using the group $St/f^{1/3}$. Predicted values of this parameter are plotted versus rib angle in Figure 6. The predicted trend follows Pirie's [4] experimental results. It can be seen that the 34-degree helical-ribbed surface is predicted to have the best performance. A similar conclusion was drawn from Pirie's [4] data.

6. CONCLUSION

Heat transfer in turbulent flow through annuli with rib-roughened inner cylinders has been modelled using FLOW3D. Where possible, the results have been compared to existing data. The predictions support the experimental finding that helical ribs with a rib angle of approximately 34 degrees have the best overall performance.   It was also predicted that the 34-degree case had the lowest average ribbed-surface temperature.

While the predictions broadly agreed with the available experimental data, significant differences were noticed in detail.   An attempt was made to resolve better the ribs by refining the grid.   However, the resulting predictions were still significantly different from the experimental data.   The differences seen between the experimental data and the predictions for turbulent heat transfer from the rib-roughened surfaces underlines the need for improvement in the k-ε turbulence model.

ACKNOWLEDGEMENT

This work was supported by a Central Electricity Generating Board Contract No. RK 1679.   The authors would also like to thank Mr H G Lyall, Drs S A Fairbairn, R J Firth, R T Szczepura, and Mr M A M Pirie at Berkeley Nuclear Laboratories, and Drs I P Jones and N S Wilkes and their colleagues at AERE Harwell for many helpful discussions.

REFERENCES

1.   Lockett, J F and Collins, M W.   "Holographic interferometry applied to rib-roughness heat transfer in turbulent flow".   To be published Int. J. Ht. Mass Trans 1990.

2.   Wilkes, N S.   "The prediction of heat transfer from surfaces roughened by transverse ribs".   UKAEA Report AERE-R 10293 1981.

3.   Watts, J and Williams, F.   "A technique for the measurement of local heat transfer coefficients using copper foil".   CEGB Report RD/B/5023/N81 1981.

4.   Pirie, M A M.   "CEGB Report RD/B/N2760 1974".

5.   Henry, F S and Collins M W.   "Prediction of flow over helically ribbed surfaces.   Submitted to Int. J. Num. Meth Fluids 1990.

6.   Jayatilleke, C L V.   "The influence of Prandtl number and surface roughness on the resistance of the laminar sublayer to momentum and heat transfer".   Prog. Heat Mass Trans. 1, p193-329, Pergamon, London 1969.

7.   Burns, A D, Jones, I P, Kightley, J R and Wilkes, N S. "Procs. 5th Int. Conf. on Numerical Methods in Laminar and Turbulent Flow".   Swansea 1987.   Pineridge Press, Swansea 1987.

8.   Jones, I P, Kightley, J R, Thompson, C P and Wilkes, N S. "UKAEA Report AERER 11825, 1985.

9.   Pirie, M A M.   "Unpublished communication".   Berkeley Nuclear Labs., CEGB, 1988.

10. Reynolds, A J.   "Turbulent flows in engineering". Wiley-Interscience 1974.

11. Wilkie, D. "Nuc. Eng. Intl., March, pp215-217, 1971.

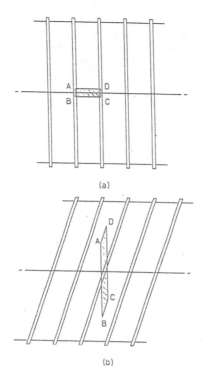

(a)

(b)

Figure 1   Schematics of ribbed surfaces.

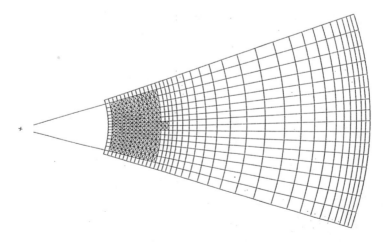

Figure 2   Grid slice for helical-ribbed case.
Heat transfer calculation.

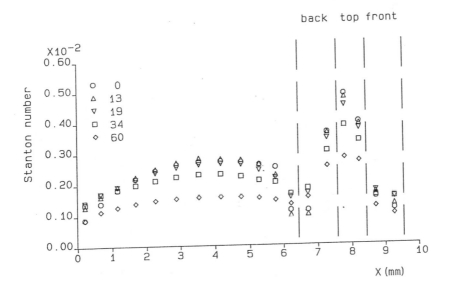

Figure 3   Predicted variation of Stanton number on the
ribbed surface of turbulent flow over
transverse and helical ribs.

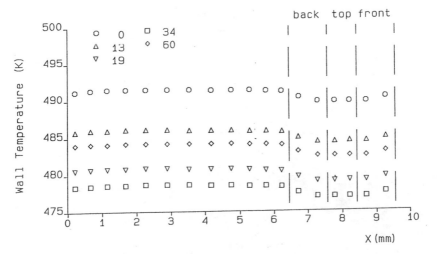

Figure 4   Predicted variation of temperature on the ribbed
surface of turbulent flow over transverse and
helical ribs.

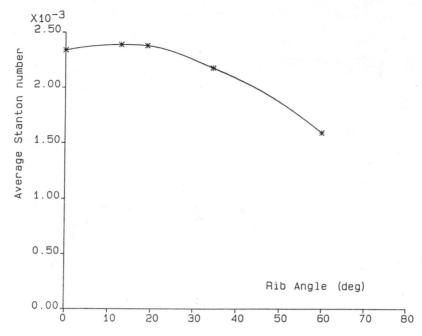

Figure 5   Predicted variation of average Stanton number
with rib angle.

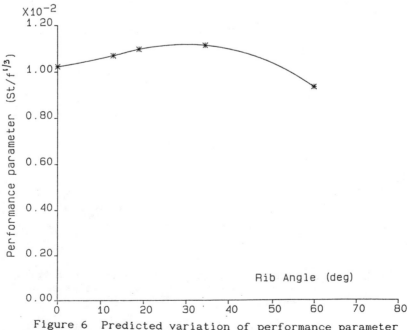

Figure 6   Predicted variation of performance parameter
$(St/f^{1/3})$ with rib angle.

Computation of the Velocity and Temperature Fields in Turbulent Channel Flows with Longitudinal Vortex Generators

J.X. Zhu, N.K. Mitra, M. Fiebig
Institut für Thermo- und Fluiddynamik,
Ruhr-Universität Bochum, 4630 Bochum 1, West
Germany

ABSTRACT

A numerical investigation of turbulent flow and heat transfer in a three-dimensional channel with a built-in vortex generator is presented. A computer program based on the SOLA-algorithm has been modified for this investigation. The standard k-ε turbulence model and wall functions are employed in the code. The results show that the turbulence energy level in the flow is elevated considerably by the longitudinal vortices and this in turn enhances the heat transfer in the channel. The mean heat transfer augmentation over the entire channel, induced by the vortex generator in the present study, amounts to 19% for a plate area which is 30 times the area of the vortex generator.

KEY WORDS: Three-dimensional channel flow, Turbulent heat transfer, Longitudinal vortices

INTRODUCTION

In gas-liquid and gas-gas heat exchangers extended surfaces in the form of fin plates are frequently used to augment the heat transfer on the gas side, see figure 1.

In a compact fin-tube heat exchanger a bank of tubes share a number of common plate fins. Generally a liquid flows through the tubes and a gas flows through the channels formed by the parallel plate fins. In order to enhance heat transfer at the fin plates, longitudinal vortices are introduced into the channel flows [1,2]. The longitudinal vortices can be produced by placing a vortex generating device in the form of wings or winglets at the channel wall. Typically a small triangular piece of the fin can be punched out of the plate and bent in such a way that this triangular piece, while remaining attached to the plate at the base, will stick out like a half-delta wing (a delta winglet) in the flow between two fins with an angle of attack to the streamwise direction, see figure 2.

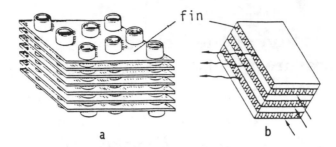

Figure 1. Schematic of a compact gas-liquid and gas-gas heat
exchanger with extended surfaces (a) finned-tube,
(b) finned-plate. The fin area may amount up to 90%
of the total heat transfer.

A longitudinal vortex will form along the side edge of the
winglet because of the pressure difference between the front
surface facing the flow and the back surface of the vortex
generator. The longitudinal vortices will spiral the flow,
thereby continuously mixing the fluid near the channel wall
and the fluid in the core region and thus increasing the heat
transfer at the wall.

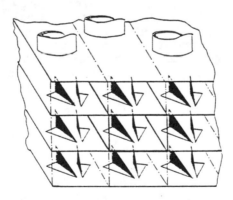

Figure 2. Configuration of a finned-tube heat exchanger with
delta winglets punched out of the fins.

The numerical simulation of the flow fields in a heat
exchanger, shown in Fig. 2, is a tremendous task. The first
step to such a task is to simulate the flow and temperature
fields in a 3D channel. Fiebig et al [3] have reported numerical investigations of flow fields, heat transfer and flow losses for laminar flows in a rectangular channel with longitudinal vortex generators. Since the gas flow is turbulent in many
practical applications of heat exchangers, the need remains
for the numerical simulation of turbulent flows in a channel
with longitudinal vortex generators. The present work addres-
ses to this need.

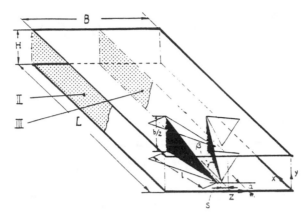

Figure 3. Computational domain: Punched winglet pair between
two fin plates.

MATHEMATICAL FORMULATION

Computational Domain

The chosen geometry for the simulation of flow and heat transfer is shown in figure 3, which can be seen as an element of
the configuration shown in figure 2. Because of the symmetry
about the planes II and III the flow simulation is performed
only for the region within these two planes in the z-direction. The dimensions of the computational domain and the vortex generator relative to the channel height H are:

$$B/2 = 3H \quad L = 10H \quad a = 0.5H \quad s = 0.5H$$
$$\beta = 25° \quad l = 2H \quad b/2 = H$$

The height of the delta winglets, which are punched out
of the plates, is equal to the channel height, so that the
vortex generators can act also as fin-spacers in the heat exchanger.

Governing Equations

The Reynolds averaged Navier-Stokes and energy equations incorporating the eddy viscosity concept are used to describe
the incompressible flow in the channel (figure 3).

Continuity equation:
$$\frac{\partial u}{\partial x} + \frac{\partial v}{\partial y} + \frac{\partial w}{\partial z} = 0 \tag{1}$$

Momentum equations:
$$\rho\frac{Du}{Dt} = -\frac{\partial p}{\partial x} + \frac{\partial}{\partial x}[2\mu_{eff}\frac{\partial u}{\partial x} - \frac{2}{3}\rho k] + \frac{\partial}{\partial y}[\mu_{eff}(\frac{\partial u}{\partial y} + \frac{\partial v}{\partial x})]$$
$$+ \frac{\partial}{\partial z}[\mu_{eff}(\frac{\partial u}{\partial z} + \frac{\partial w}{\partial x})] \tag{2}$$

$$\rho\frac{Dv}{Dt} = -\frac{\partial p}{\partial y} + \frac{\partial}{\partial x}[\mu_{eff}(\frac{\partial v}{\partial x} + \frac{\partial u}{\partial y})] + \frac{\partial}{\partial y}[2\mu_{eff}\frac{\partial v}{\partial y} - \frac{2}{3}\rho k]$$
$$+ \frac{\partial}{\partial z}[\mu_{eff}(\frac{\partial v}{\partial z} + \frac{\partial w}{\partial y})] \tag{3}$$

$$\rho\frac{Dw}{Dt} = -\frac{\partial p}{\partial z} + \frac{\partial}{\partial x}[\mu_{eff}(\frac{\partial w}{\partial x} + \frac{\partial u}{\partial z})] + \frac{\partial}{\partial y}[\mu_{eff}(\frac{\partial w}{\partial y} + \frac{\partial v}{\partial z})]$$
$$+ \frac{\partial}{\partial z}[2\mu_{eff}\frac{\partial w}{\partial z} - \frac{2}{3}\rho k] \tag{4}$$

Energy equation:

$$\rho\frac{DT}{Dt} = \frac{\partial}{\partial x}[(\Gamma + \Gamma_t)\frac{\partial T}{\partial x}] + \frac{\partial}{\partial y}[(\Gamma + \Gamma_t)\frac{\partial T}{\partial y}] + \frac{\partial}{\partial z}[(\Gamma + \Gamma_t)\frac{\partial T}{\partial z}] \tag{5}$$

Here μ_{eff} is the effective viscosity, $\mu_{eff} = \mu + \mu_t$. The turbulent viscosity μ_t is obtained from the standard k-ϵ model of Launder and Spalding [4] which relates μ_t to the turbulent kinetic energy k and its dissipation rate ϵ. The local values of μ_t, k, ϵ and the turbulent dynamic thermal diffusivity Γ_t are calculated by using the following equations:

$$\mu_t = c_\mu\rho\frac{k^2}{\epsilon}, \qquad \Gamma_t = \frac{\mu_t}{\sigma_t} \tag{6}$$

$$\rho\frac{Dk}{Dt} = \frac{\partial}{\partial x}(\frac{\mu_t}{\sigma_k}\frac{\partial k}{\partial x}) + \frac{\partial}{\partial y}(\frac{\mu_t}{\sigma_k}\frac{\partial k}{\partial y}) + \frac{\partial}{\partial z}(\frac{\mu_t}{\sigma_k}\frac{\partial k}{\partial z}) + G - \rho\epsilon \tag{7}$$

where G denotes the production rate of k which is given by:

$$G = \mu_t\left[2(\frac{\partial u}{\partial x})^2 + 2(\frac{\partial v}{\partial y})^2 + 2(\frac{\partial w}{\partial z})^2 + (\frac{\partial u}{\partial y} + \frac{\partial v}{\partial x})^2\right.$$
$$\left. + (\frac{\partial u}{\partial z} + \frac{\partial w}{\partial x})^2 + (\frac{\partial v}{\partial z} + \frac{\partial w}{\partial y})^2\right] \tag{8}$$

$$\rho\frac{D\epsilon}{Dt} = \frac{\partial}{\partial x}(\frac{\mu_t}{\sigma_\epsilon}\frac{\partial\epsilon}{\partial x}) + \frac{\partial}{\partial y}(\frac{\mu_t}{\sigma_\epsilon}\frac{\partial\epsilon}{\partial y}) + \frac{\partial}{\partial z}(\frac{\mu_t}{\sigma_\epsilon}\frac{\partial\epsilon}{\partial z}) + c_1\frac{\epsilon}{k}G - c_2\rho\frac{\epsilon^2}{k} \tag{9}$$

The standard values of the empirical constants appearing in Equations (6)-(9) suggested by Launder and Spalding [4] are chosen:
$c_\mu = 0.09$, $c_1 = 1.44$, $c_2 = 1.92$, $\sigma_k = 1.0$, $\sigma_\epsilon = 1.3$, and the turbulent Prandtl number is taken as $\sigma_t = 0.9$.

Boundary Conditions

As indicated in figure 3, the computation domain has five types of boundaries: entrance, exit, symmetry, periodicity and walls. The boundary conditions are given in this order.
Entrance
A fully developed flow profile and a uniform temperature T_o are taken at the inlet in the present study. The inlet velocity, k and ϵ profiles are obtained from a calculation for 2D turbulent duct flow.
Exit
At the outlet the streamwise gradients of all variables are set to zero:

$$\frac{\partial f}{\partial x} = 0, \qquad f = \{u,v,w,k,\epsilon,T\} \tag{10}$$

Symmetry

At the symmetry planes (II and III) the normal velocity component and the normal derivatives of all other variables are set to zero:

$$w = 0, \qquad \frac{\partial f}{\partial z} = 0, \qquad f = \{u,v,k,\epsilon,T\} \tag{11}$$

Periodicity

The stamping regions at the channel top and bottom walls are treated with periodic boundary conditions which are defined as follows:

$$f_T = f_B, \qquad \left.\frac{\partial f}{\partial y}\right|_T = \left.\frac{\partial f}{\partial y}\right|_B, \qquad f = \{u,v,w,k,\epsilon,T\} \tag{12}$$

$_T$ - top wall, $_B$ - bottom wall

Walls

Wall functions given by Launder and Spalding [4] are employed to prescribe the boundary conditions along the channel walls. The wall functions are applied in terms of diffusive wall fluxes. For the wall-tangential moment these are the wall shear stresses:

$$\tau_w = \frac{\rho U_p c_\mu^{1/4} k_p^{1/2} \kappa}{\ln(Ey^+)} \tag{13}$$

with the non-dimensional wall distance y^+ defined as:

$$y^+ = \frac{\rho y_p c_\mu^{1/4} k_p^{1/2}}{\mu} \tag{14}$$

and $\kappa \simeq 0.42$, $E = 9.0$.

The subscript p denotes the variables at the grid point adjacent to the wall. U_p, k_p and y_p are the tangential velocity at point p, the value of k for the grid point p and the distance of the point p from the wall respectively. The velocities at the wall are set to zero.

From the assumption of the local turbulence equilibrium, the production rate of the turbulent kinetic energy in the k-equation at the near wall grid points are computed from

$$G_p = \tau_w \frac{U_p}{y_p} \tag{15}$$

and the dissipation rates from

$$\bar{\epsilon} = \frac{1}{y_p}\int_0^{y_p}\epsilon dy = \frac{c_\mu^{3/4} k_p^{3/2}}{\kappa y_p} \ln(Ey^+) \tag{16}$$

The ϵ-equation should not be solved at the near wall points. In accordance to the assumption that the turbulent eddy length scale in the vicinity of the wall is linearly proportional to the wall distance

$$L_p = \frac{k_p^{3/2}}{\epsilon_p} = \frac{\kappa}{c_\mu^{3/4}} y_p \tag{17}$$

the value of ϵ at point p is obtained from

$$\epsilon_p = c_\mu^{3/4} \frac{k_p^{3/2}}{\kappa y_p} \tag{18}$$

For the temperature boundary condition, the heat flux to the wall is derived from the thermal wall function [4]

$$q_w = \frac{(T_w-T_p)\rho c_p c_\mu^{1/4} k_p^{1/2}}{Pr_t[\frac{1}{\kappa}\ln(Ey^+)+P]}$$

(19)

where the empirical P function is specified as:

$$P = \frac{\pi/4}{\sin(\pi/4)} \left(\frac{A}{\kappa}\right)^{1/2} \left(\frac{Pr}{Pr_t}-1\right)\left(\frac{Pr_t}{Pr}\right)^{1/4}$$

(20)

and T_w is the wall temperature which is a constant in the present work, $T_w=2T_o$.

NUMERICAL METHOD

A numerical procedure, based on the SOLA-algorithm [5], was modified by the authors' group for the investigation of the velocity and temperature fields in 3D laminar channel flows with longitudinal vortex generators [3]. In the present work, this procedure has been extended to solve the turbulent problem.

The SOLA-algorithm solves the time dependent Navier-Stokes equations directly for the primitive variables by advancing the solution explicitly in time.

In order to validate the modified computer program, the code has been used to simulate the well documented 3D turbulent flow experiments of Pauley and Eaton [6]. The comparison of the numerical simulation and experiments [6] will be presented elsewhere [7].

RESULTS AND DISCUSSION

Figures 4-10 illustrate calculated results obtained for a Reynolds number, based on the channel height, $Re_H=50000$ and a Prandtl number Pr=0.7. The cross sectional velocity vectors, streamwise vorticity contours, contours of the turbulent kinetic energy and the isotherms at four axial locations (x/H=1.3, 2.3, 4.3 and 9.3) are shown in figures 4, 5, 6 and 7 respectively. The streamwise vorticity ω_x is defined as usual:

$$\omega_x = \frac{\partial w}{\partial y} - \frac{\partial v}{\partial z}$$

(21)

The cross sectional velocity vectors indicate the behaviour and the extent of the vortices which diverge from the central plane of the channel as they move in the streamwise direction. As shown in figure 4, the vortices have elliptic shapes and their major axes turn gradually from vertical to lateral direction.

Figure 5 shows at cross section x/H=2.3 four zones of negative vorticity ω_x in each half of the channel. They lie at the corners of the top wall and the bottom wall and the central symmetry plane, top wall and the side boundary plane and around the delta winglet. The negative vorticity might be produced by the interaction of the secondary flow and the frictions at the walls and the vortex generators. The generation of the negative vorticity in the outer side of the winglets is attributed to the inflow of the fluid through the punched area

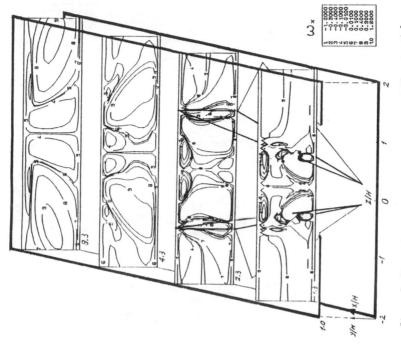

Figure 5. Streamwise vorticity contours ω_x at different axial stations.

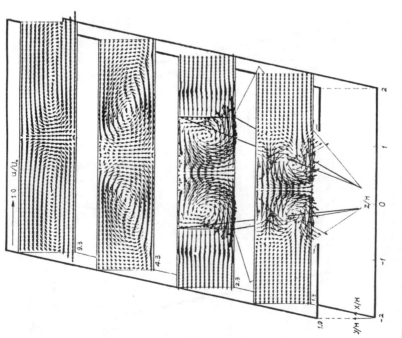

Figure 4. Cross sectional velocity vectors at different axial stations x/H = 1.3, 2.3, 4.3, 9.3 showing the formation and development of the longitudinal vortices.

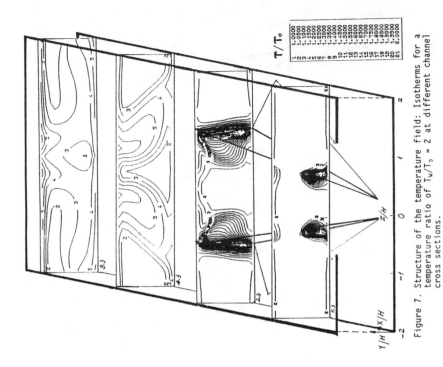

Figure 7. Structure of the temperature field: Isotherms for a temperature ratio of $T_w/T_0 = 2$ at different channel cross sections.

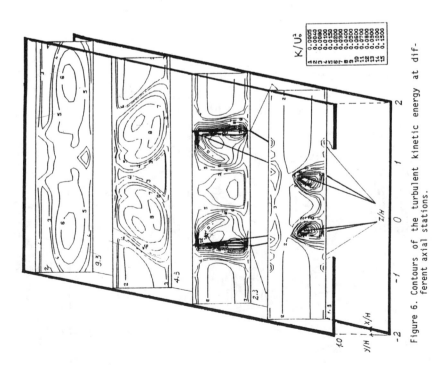

Figure 6. Contours of the turbulent kinetic energy at different axial stations.

at the top wall (see figures 3 and 8). Figure 8 shows the inflow and outflow through the punched hole at the top and bottom walls respectively. The negative vorticity region near the
central plane enlarges at the cross sections x/H=4.3 and 9.3,
where it extends from the top to the bottom area. At the same
time, the negative vorticity in the outer region is convected
by the primary vortex and pushed to the top center line area.
The positive vorticities, which represent the primary longitudinal vortices, possess an elliptic shape.

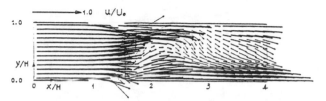

Figure 8. Streamwise velocity vectors at the axial section
z/H = 0.72, showing the in- and outflow through the
punched hole at the top and bottom walls.

Figure 6 shows in the outer region of the first cross
section, where the flow field is still not affected by the
vortex generator, a typical distribution of the turbulence
energy of a 2D channel flow. The turbulence energy increases
with proximity to the wall. Higher magnitudes of k appear
around the vortex generator, where due to the high shear the
turbulence energy generation is large. Large values of k are
also found in the core region of the vortices at subsequent
axial stations in figure 6. Such regions are away from the
channel walls, therefore the dissipation rates of k there is
low. On the other hand, the generation of k is large because
of the high velocity gradients in this region. At the last
cross section in figure 6, the usual turbulence structure in a
channel flow is totally distorted by the vortices.
        The isotherms in figure 7 reveal that large temperature
gradients are found around the vortex generators. This means,
that the heat transfer near the vortex generator is very high,
which is the consequence of the high turbulence level around
the vortex generators. This consequence can also be deduced
from the structural similarity between the isotherms and the
k-contours. The effect of the secondary flow on the temperature field is clear at the downstream location x/H=4.3. In the
upwash regions, i.e. in the area between the vortices near the
top wall and in the area of the outer region near the bottom
wall, the secondary flow is directed away from the wall and it
sweeps the hot fluid to the core of the channel. In the downwash regions, i.e. in the area between the vortices near the
bottom wall and in the area above the vortex at the outer
side, the strong downflow and lateral flow push the cold fluid
from the core of the channel to the walls.
        Figures 4-7 show that the longitudinal vortices distort
totally the usual flow and temperature field in a channel.
This will cause a considerable heat transfer enhancement.

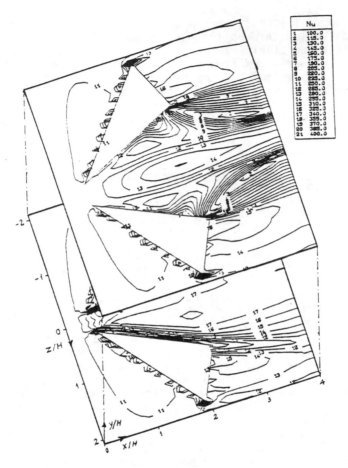

Figure 9. Local heat transfer coefficients: Contours of Nusselt numbers on the top and bottom walls from leading edge to the downstream location x/H = 4.

The Nusselt number distributions on the top and bottom
walls are shown in the form of contours in figure 9. On the
bottom wall, the heat transfer is considerably enhanced in the
region after the vortex generators and particularly in the
downwash areas near the center line. At the back side of the
vortex generator, a "dead water" zone with low Nu values is
observed. The lowest value there (line 6) directly behind the
vortex generator is only about 50% of the highest Nu value in
the downwash area. At the edge of the short perpendicular side
of the punched hole, there is a zone with relatively high Nusselt number, which is probably caused by the outflow there.
On the top wall, the heat transfer is rather smaller than
on the bottom wall. Downstream from the corner of the punched
hole, the Nusselt number becomes small due to the upwash flow.
However, near the two perpendicular sides of the hole the heat
transfer reaches a high level due to the strong flush of the
inflow.

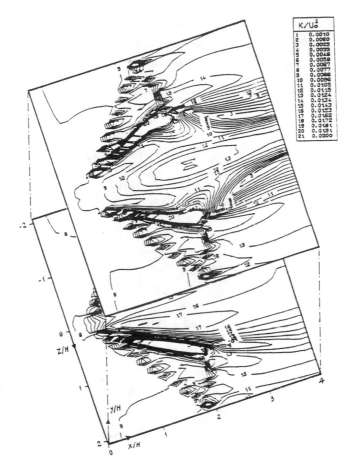

Figure 10. Distribution of the dimensionless turbulent kinetic energy near the top and bottom walls ($y^+ \simeq 90$).

   The distribution of turbulence kinetic energy near the
top and bottom walls are mapped in figure 10. A good correlation is found between the k and Nu distributions in figure 9
and 10. This similarity suggests that the level of the turbulence energy near the wall plays a key role in controlling the
heat transfer rate.

CONCLUSIONS

Present computations reveal that the longitudinal vortices,
generated by a pair of delta winglets, elevate considerably
the turbulence energy level in the whole channel. This, in
turn, leads to higher turbulent thermal diffusivities in the
flow field and thus contribute to the heat transfer augmentation in the channel. Basing on this observation, one may conclude that the heat transfer in the turbulent channel flow is
enhanced by the longitudinal vortices through two ways:

1.) Mixing of the fluids between the core and the near wall regions of the channel.
2.) Elevation of the turbulence energy level in the flow.
    The mean heat transfer augmentation over the entire channel, induced by the delta winglets in the present case, amounts to 19% for a plate area which is 30 times the area of the delta winglet.

ACKNOWLEDGEMENT

The first author (J.X. Zhu) wishes to thank the Konrad-Adenauer-Stiftung for providing a scholarship. Dr. G. Scheuerer is gratefully acknowledged for his helpful discussions.

REFERENCES

1. Edwards, F.J. and Alker, C.J.R.,
   The Improvement of Forced Convection Surface Heat Transfer Using Surface Protrusions in the Form of (A) Cubes and (B) Vortex Generators, Proc. 5th Int. Heat Transfer Conference, Tokyo, 1974

2. Fiebig, M., Kallweit, P. and Mitra, N.K.,
   Wing Type Vortex Generators for Heat Transfer Enhancement, Proc. 8th Int. Heat Transfer Conference, Vol.6, San Francisco, 1986

3. Fiebig, M., Brockmeier, U., Mitra, N.K., Güntermann, Th.,
   Structure of Velocity and Temperature Fields in Laminar Channel Flows with Longitudinal Vortex Generators, Numerical Heat Transfer, Part A, Vol.15, pp. 281-302, 1989

4. Launder, B.E. and Spalding, D.B.,
   The Numerical Computation of Turbulent Flows, Comput. Methods Appl. Mech. Eng., Vol.3, pp. 269-289, 1974

5. Hirt, C.W., Nichols, B.D. and Romero, N.C.,
   SOLA - A Numerical Solution Algorithm for Transient Fluid Flows, Los Alamos Scientific Laboratory Rept. LA-5652, Los Alamos, New Mexico, 1975

6. Pauley, W.R. and Eaton, J.K.,
   The Effect of Embedded Longitudinal Vortex Pairs on Turbulent Boundary Layer Heat Transfer, pp. 487-500, Proc. Second Int. Symposium on Transport Phenomena in Turbulent Flow, Tokyo, 1987

7. Zhu, J.X., Fiebig, M. and Mitra, N.K.,
   Numerical Simulation of a 3-D Turbulent Flow Field with Longitudinal Vortices Embedded in the Boundary Layer, submitted to the ASME Winter Annual Meeting, Dallas, 1990

Heat Transfer Between Particles and Hot Gas in Plasma Jet

K. Brodowicz, T. Dyakowski, D. Galicki
Institute of Heat Engineering, Warsaw University of Technology, 00-665 Warsaw, Poland

ABSTRACT

Heat transfer between plasma jet and solid particle was analysed. Two mechanisms of the particle heating were assumed: convective and chemical The comparison of both the heat fluxes alongtheparticle trajectory was presented.

INTRODUCTION

Heat transfer between plasma gas flowing out of a plasma torch and solid injected into this plasma jet is a problem of a practical interest in plasma spraying technology. To satisfy the layer quality requirements, particle trajectory and its thermal treatment should be controlled. On the other hand the energy of the jet should be used in an optimal way to reduce the cost of the process. So the better understanding of the interaction of particulate matter with thermal plasma is needed.

The heat transfer between a solid particle and plasma gas is the main problem of interest in this paper. For this analysis the velocity and temperature

distribution in the jet likewise the particle
trajectory were needed. The short description of
the methods used and results is given below.

The mechanism of heat transfer is assumed
to be dual - convective and chemical   due to
recombination on the negatively charged particle
surface. Both heat fluxes were estimated to find
their importance along the particle trajectory.

PLASMA GAS VELOCITY AND TEMPERETURE DISTRIBUTION

The momentum and energy equations for axisymmetrical
jets were solved. The boundary layer approach was
used. The form of the equations used meets the
GENMIX computer code requirements /1/. In von Mises
coordinates they are as following:

$$\frac{\partial u}{\partial z} = \frac{\partial}{\partial \psi} /\mu \varrho \, Ur^2 \frac{\partial U}{\partial \psi}/$$

$$\frac{\partial h}{\partial z} = \frac{\partial}{\partial \psi} /\Gamma \varrho Ur^2 \frac{\partial U}{\partial \psi}/$$

where U is velocity, h - enthalpy, r - radius,
u - effective viscosity, ϱ - density, $\Gamma = \lambda / c_p$,
λ - thermal conductivity, c_p - constant pressure,
specific heat. Equations of motion and energy were
transformed into the following form:

$$/\frac{\partial \phi}{\partial x}/ + /a+b\omega/ \frac{\partial \phi}{\partial \omega} = \frac{\partial}{\partial \omega} /c \frac{\partial \phi}{\partial \omega}/ + d$$

where ω is the nondimensional stream functions. To
solve the equations the implicit formula was used

$$D_i \phi_{i,D} = A_i \phi_{i+1,D} + B_i \phi_{i-1,D} + C_i$$

$$C_i = E_i \phi_{i,U} + F_i \phi_{i+1,U} + G_i \phi_{i-1,U} + H_i$$

The expressions for A_i, B_i, C_i and D_i were obtained by integrating the differential equation for ϕ over an appropriate control volume. The set of finite - difference equations was solved by tri-diagonal matrix algorithm.

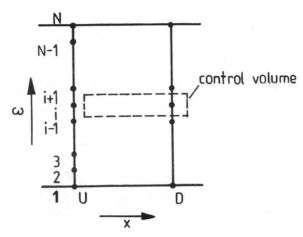

Fig.1   Illustration of a part of the $x - \omega$ grid

The velocity distribution in the nozzle plane was uniform. The initial temperature profile was assumed arbitrarily. Results were compared with the experimental data given by Fiszdon /2/.

PARTICLE TRAJECTORY

The Lagrangian formulation of the equation of motion in axial and radial direction was used.   The viscous drag force was taken into account only.

$$m \frac{dV_i}{dt} = \frac{1}{8} C_D f \rho \pi d_p^2 \mid U-V \mid /U_i-V_i/$$

V, m, d_p are the particle velocity, mass and diameter respectively; ρ is the plasma gas density. Subscript i denotes an axial or radial velocity components. The viscous drag coefficient C_D has a simple form

$$C_D = \frac{24}{Re} + \frac{6}{1+Re} + 0,4$$

Factor f is a correction due to variable property effects /3/

$$f = /q_\infty \mu_\infty / q \mu /^{-0,45}$$

Subscript ∞ denotes free stream properties and no subscript   the properties of plasma gas in the particle surface temperature.

Various initial particle position and velocity were simulated. The trajectories obtained have not showed meaningful differences.

PARTICLE HEATING

An energy balance gives rise to the following equation from which the particle temperature T_p was determined

$$\frac{dT_p}{dt} = \frac{Q}{mC_p}$$

where $Q = \propto /T - T_p/ A + Q_r$

and m is the particle mass, C_p the specific heat of the particulate matter and Q_r the heat flux released in recombination.

A particle injected in a thermal plasma is always negatively charged due to the different mobility of ions and electrons. Recombination occurs on the particle surface and the energy released is assumed to be absorbed by the particle. To calculate the recombination heat flux the molecular theory of gas was used. The composition of plasma gas with respect to the temperature was computed using the Saha eq. A limiting sphere concept assuming a spherical shell surrounding the particle /4/ was used. The negative charge of the particle causes the flux of ions I_i

$$I_i = n_i \left/ \frac{k_B T_m}{2 \pi m_i} \right/^{0,5} \exp /-x_p/$$

where $T_m = T_p + 0,78 /T - T_p/$

and $n_i = n_{i_0} \left/ \left\{ \exp/x_p/ + 1,5 \, d_p \left[\exp/x_p/-1\right] \right/ 4 \lambda_i x_p \right\}$

is the number density of ions arriving on the spherical particle /5/: n_{i_0} is the number density of ions in temperature T_m, k_B is the Boltzman constant, m_i, t_i are the ion mass and free path in T_m respectively and x_p is the dimensionless surface potential.

The recombination process releases heat flux

$$Q_r = n_i \left/ \frac{k_B T_m}{2 \pi m_i} \right/^{0,5} E \ \exp/-x_p/A$$

where E_i is the recombination energy.

The convective heat exchange coefficient was evaluated from the formula proposed by Lee /5/. Both heat fluxes to the particle along its trajectory were showed on Fig.2. The Reynolds Number varied from about 19 to 5 for atmospheric pressure and from 1,8 to 0,9 for reduced pressure /0,1 at/: the highest temperature 12500 K: particle was injected axially $d_p = 25$ mm

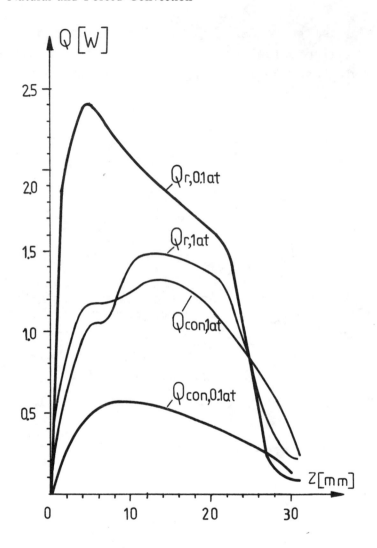

Fig.2 Character of changes of heat fluxes to the
particle along its trajectory

CONCLUSION

The calculation showed that almost all particle heating
occurs within 25 mm from the nozzle plane /for the
case analysed/. The heat flux released in recombination
seems to have at least the same meaning as the convective
one at atmospheric pressure and to be much more important
at reduced pressure. Those facts agree with the results

given in /6/ in spite of the simplifications made in the presented paper. The problem needs still experimental researches.

REFERENCES

1. Spalding, D. B., GENMIX - A Genera Computer Program for Two-Dimensional Parabolic Phenomena HMT - the science and applications of heat and mass transfer v.1 Pergamon Press, 1977
2. Fiszdon, J. Melting of Powder Grains in a Plasma Flame, Int.J.Heat Mass Transfer, Vol.22, pp. 749-761, 1979
3. Lee, Y. C. Chyou, Y. P., Pfender, E. Particle Dynamics and Particle Heat and Mass Transfer in Thermal Plasmas Part I. The Motion of a Single Particle without Thermal Effects. Plasma Chem. Plasma Proc., Vol.5, pp.211-237, 1985
4. Fuchs, N. A. On the Stationary Charge Distribution on Aeorosol Particles in a Bipolar Ionic Atmosphere, Pure Appl. Geophys, pp.185-193, 1963
5. Lee, Y. C., Chyou, Y. P., Pfender, E. Particle Dynamics and Particle Heat and Mass Transfer in Thermal Plasmas. Part II. Particle Heat and Mass Transfer in Thermal Plasmas, Plasma Chem. Plasma Process. Vol.5, pp.391-414, 1985
6. Chang, C. H., Pfender, E. Heat and Momentum Transport to Particulates Injected into Low-Pressure Plasma Jets pp.37-42, Proceedings of the 9th Int. Symposium on Plasma Chemistry, Pugnochiuso, 1989. P. d'Agostino 1989

SECTION 2: HEAT AND MASS TRANSFER

Numerical Simulation of the Screw Extrusion Process for Plastic Materials

Y. Jaluria(*), M.V. Karwe(**)
(*) Department of Mechanical and Aerospace Engineering, Rutgers University, New Brunswick, NJ 08903, USA
(**) Center for Advanced Food Technology, Cook College, Rutgers University, New Brunswick, NJ 08903, USA

INTRODUCTION

Extrusion is one of the most widely used manufacturing processes in industries dealing with materials such as plastics, polymers, and pharmaceutical products. Recently, extrusion cooking has become an important food processing operation. Most of the relevant materials in these processes are non-Newtonian, with a shear rate dependent viscosity, and therefore behave significantly different from Newtonian fluids such as water and air. In recent years, efforts have been directed toward the analysis and numerical simulation of flow in single and twin screw extruders for plastics, in order to optimize the process and improve the product quality. Experimental studies have also been carried out to understand the underlying physical phenomena [1-3]. However, many aspects are still open to further investigation, simulation, and optimization. Also, many fundamental aspects, particularly with respect to thermal transport, are not very well understood and need further work. For instance, the processing of reactive polymers and food in extruders is not well understood. The flow and heat transfer have a significant effect on the quality and the nature of the final product. Also, more work is needed on the numerical simulation of complicated flow situations, such as reactive flows and twin-screw extruders.

In one of the pioneering studies on the flow in a single-screw extruder, Griffith [4] solved for the flow of an incompressible fluid through a screw extruder taking the velocity and the temperature profiles to be essentially the same as those in a channel of infinite width and length. The effects of curvature and leakage, across the flights, were ignored. Zamodits and Pearson [5] obtained numerical solutions for a fully developed, two-dimensional, isothermal and non-Newtonian flow of polymer melts in infinitely wide rectangular screw channels.

Lidor and Tadmor [6] and, Bigg and Middleman [7] have carried out a theoretical analysis to determine the residence time distribution function, which determines the time spent by a fluid particle in the extruder, and the strain distribution in screw extruders. Tadmor and Gogos [8] and Fenner [3] have discussed the flow of a polymer in the various sections of an extruder. Elbirli and Lindt [9] have reported the results of a model in which temperature was allowed to develop along the screw channel. In these models, the screw and the barrel were assumed to be at the same fixed uniform temperature. Agur and Vlachopoulos [10] have studied the flow of polymeric materials which included a model for flow of solids in the feed hopper, a model for the solid conveying zone, and a model for the melt conveying zone.

More recently, some effort has been directed at twin-screw extruders, in which two screws rotate, in the same or opposite directions, within a barrel. The geometrical complexity of the twin-screw extruders has been the main reason for the limited amount of the published literature on the modeling of the twin-screw extruders. The present paper considers both single and twin-screw extruder systems.

Janssen [2] has discussed the flow characteristics in fully intermeshing counter-rotating twin-screw extruders. Wyman [11] has analyzed the flow in shallow channels of twin-screw extruders using a simple theoretical model, with the emphasis on the down-channel flow. Secor [12] has developed a model for a co-rotating twin-screw extruder and compared the numerical predictions with experimental results. Booy [13] has proposed a mathematical model for the isothermal flow of a Newtonian liquid in co-rotating twin-screw extruders. Denson and Hwang [14-15] have studied the performance of self-wiping co-rotating twin-screw extruders under effects of some parameters, such as the leakage and cross-channel flow, and the axial pressure gradient. Hwang [16] carried out numerical simulation for both the cross-channel and down-channel flows and discussed the mixing region of a fully-wiped co-rotating twin-screw extruder, where one screw wipes the bottom of the other screw as they turn. Bigio and Zerafati [17] numerically evaluated mixing in the region between the two screws of a non-intermeshing counter-rotating twin-screw extruder using a two-dimensional model. The rate of mixing was characterized in terms of the movement of a line of fluid particles. Nguyen and Lindt [18] modeled a counter-rotating tangential twin-screw extruder using a two-dimensional finite element model with the effect of curvature and leakage flow included.

There is very little literature published on the simulation of flow and heat transfer in a die. The shape, geometry, and size of the die at the end of the extruder screw determines the operating characteristics, residence time and shape and quality of the final product. Therefore, it is extremely important to understand the transport process inside the die.

This paper reviews the work done on the flow and heat transfer inside a single-screw extruder, a twin-screw extruder and a die. Results based on different approaches to attack the problem are presented. For example, the finite-difference method is used for the simplified geometry of a single screw extruder. Finite element methods (FEM) are employed for a self-wiping screw geometry, the intermeshing zone of a twin-screw extruder and for dies having complicated geometry. Circumstances under which three-dimensional effects are important are described. The effect of heat transfer on flow characteristics, and vice-versa, are also considered.

I. FINITE-DIFFERENCE MODEL FOR A SINGLE-SCREW EXTRUDER

The simplified geometry of a single-screw extruder and the cross section of a screw channel are shown in Fig. 1. For ease of visualization and analysis, the coordinate system is fixed to the screw root and, thus, the barrel moves in a direction opposite to the screw rotation. For steady, developing, two-dimensional flow of a homogeneous fluid in single-screw extruder with shallow and long channels, i.e., for H $\ll$ W in Fig. 1, after applying the creeping flow approximation in the x and z-directions, the equations for the conservation of momentum become

$$\frac{\partial p}{\partial x} = \frac{\partial(\tau_{yx})}{\partial y} \qquad \frac{\partial p}{\partial y} = 0 \qquad \frac{\partial p}{\partial z} = \frac{\partial(\tau_{yz})}{\partial y} \qquad (1)$$

where p is the hydrostatic pressure and τ is the shear stress. The clearance between the screw flights and the barrel is assumed to be small enough to neglect the leakage across the flights from one screw channel to the neighboring one.

Within a screw channel, the temperature as well as the velocity fields change along the length of the channel. If the barrel is maintained at a fixed temperature, the velocity and temperature profiles will ultimately approach a fully developed situation, i.e., they do not change in form along the channel length. The energy equation governing the temperature in such cases is obtained by neglecting the axial convection terms. The energy equation then becomes

$$0 = \frac{\partial}{\partial y} \left(k \frac{\partial T}{\partial y} \right) + \tau_{yx} \left(\frac{\partial u}{\partial y} \right) + \tau_{yz} \left(\frac{\partial w}{\partial y} \right) \qquad (2)$$

In the presence of strong viscous dissipation effects or heat addition from the barrel, or both, the thermal convection along the z direction (along the channel length) is significant. Therefore, the temperature field develops along the z direction. The velocities may also change with the downstream position as a result of this change in temperature, if the fluid viscosity is dependent upon its temperature. In such cases, the energy equation becomes

$$\rho C w \frac{\partial T}{\partial z} = \frac{\partial}{\partial y} \left(k \frac{\partial T}{\partial y} \right) + \tau_{yx} \left(\frac{\partial u}{\partial y} \right) + \tau_{yz} \left(\frac{\partial w}{\partial y} \right) \qquad (3)$$

where T is the local temperature, ρ the density, C the specific heat at constant pressure, and k the thermal conductivity of the fluid. The shear stress τ_{yx} and τ_{yz} are given by

$$\tau_{yx} = \mu \frac{\partial u}{\partial y} \qquad \tau_{yz} = \mu \frac{\partial w}{\partial y} \qquad (4)$$

where μ is the molecular viscosity of the fluid. The following constitutive equation is used to represent the dependence of μ on T and on the strain rate $\dot{\gamma}$ for the non-Newtonian fluids considered:

$$\mu = \mu_o \frac{\dot{\gamma}^{n-1}}{\dot{\gamma}_o} e^{-b(T-T_o)} \quad , \quad \dot{\gamma} = [(\frac{\partial u}{\partial y})^2 + (\frac{\partial w}{\partial y})^2]^{1/2}$$

$$(5)$$

where u and w are the velocity components in the x and z directions, respectively, b is the temperature coefficient of viscosity, n is the power law index, and the subscript o denotes the reference conditions. This is the power law fluid approximation, which has been extensively employed in the literature.

The boundary conditions and the flow rate constraints are shown in Fig. 2 where V_{bz} and V_{bx} are the z and x components of the barrel velocity V_b, respectively. The subscript "dev" corresponds to the fully developed case at the inlet temperature T_i. The screw has been taken as isothermal in most studies reported in the literature. However, a more practical circumstance is represented by the adiabatic condition at the screw surface. The above equations are nondimensionalized in terms of the following dimensionless variables:

$$x^* = \frac{x}{H}, \quad y^* = \frac{y}{H}, \quad z^* = \frac{z}{H}, \quad u^* = \frac{u}{V_{bz}}, \quad w^* = \frac{w}{V_{bz}}$$

$$\theta = \frac{T-T_i}{T_b-T_i} \quad , \quad p^* = \frac{p}{\bar{p}}, \quad \bar{p} = \mu \frac{V_{bz}}{H},$$

$$\beta = b(T_b-T_i), \quad \dot{\gamma}^* = \frac{\dot{\gamma}H}{V_{bz}},$$

$$\bar{\mu} = \mu_o \{\frac{V_{bz}H}{\dot{\gamma}_o}\}^{n-1} e^{-b(T_i-T_o)} \quad , \quad Pe = \frac{V_{bz}H}{\alpha}$$

$$G = \frac{\mu V_{bz}^2}{k(T_b-T_i)}$$

where α is thermal diffusivity, Pe is termed the Peclet number, and G is termed the Griffith number. The parameter β represents the dependence of viscosity on temperature.

The dimensionless equations thus obtained are

$$\frac{\partial p^*}{\partial x^*} = \frac{\partial}{\partial y^*} \left(\frac{\partial u^*}{\partial y^*} [\dot{\gamma}^*]^{(n-1)} e^{-\beta\theta} \right) \tag{6}$$

$$\frac{\partial p^*}{\partial z^*} = \frac{\partial}{\partial y^*} \left(\frac{\partial w^*}{\partial y^*} [\dot{\gamma}^*]^{(n-1)} e^{-\beta\theta} \right) \tag{7}$$

$$Pe \; w^* \frac{\partial \theta}{\partial z^*} = \frac{\partial^2 \theta}{\partial y^{*2}} + G[\dot{\gamma}^*]^{(n+1)} e^{-\beta\theta} \tag{8}$$

Similarly, the boundary conditions are also obtained in nondimensional form. The constraints on the flow are obtained, in dimensionless form, as

$$\int_0^1 u^* \; dy^* = 0, \qquad \int_0^1 w^* \; dy^* = q_v = \frac{Q/W}{HV_{bz}} \tag{9}$$

Here, the parameter q_v represents the dimensionless volumetric flow rate, generally called the throughput emerging from the extruder.

Thus for a given screw and temperature distribution at the barrel surface, the parameters that govern the numerical solution are Pe, G, n, q_v, and the viscosity temperature variation parameter β.

I.2 NUMERICAL SOLUTION

The governing dimensionless equations (6)-(9) are conveniently solved by means of finite-difference techniques for simple geometries. The computational domain is shown in Fig. 2. The computations were carried out over 101 x 41, 101 x 61, and 101 x 81 grids. The results were essentially unchanged when the grid was refined to 101 x 81 from 101 x 61 and, therefore, a 101 x 61 grid with $\Delta y^* = 0.166$ and $\Delta z^* = 2.5$ was selected. Since the energy equation (8) is parabolic in z, boundary conditions are necessary only at $z^* = 0$ to allow marching in the z direction and, thus, obtain the solution in the entire domain. The boundary conditions at $z^* = 0$ were provided in terms of the developed velocity profiles at $T = T_i$. These were obtained by solving the momentum equations (6)-(7) by means of an implicit finite-difference scheme [19]. The iterative Newton-Raphson method was used to satisfy the conditions on the total flow rates given by Eq. (9). This requires initial guesses for the values of $\partial p^*/\partial z^*$ and $\partial p^*/\partial x^*$. The iteration was terminated when the pressure gradients satisfied the following convergency criterion:

$$max \; [\Delta(\frac{\partial p^*}{\partial z^*}), \qquad \Delta(\frac{\partial p^*}{\partial x^*})] \leq \epsilon \tag{10}$$

where Δ stands for the absolute fractional change between two consecutive iterations and ϵ is a chosen small quantity. A value for ϵ was found to be adequate, by varying its value till the results were essentially independent of a further decrease. This particular convergence criterion is not useful when the values of the pressure gradients become small, i.e., close to zero. Under such circumstances, the absolute change in the values of the pressure gradient was considered for convergence.

Using the boundary conditions in terms of u^*, w^*, and θ at any upstream z location, the energy equation is solved to obtain the temperature distribution at the next downstream z location. Equation (8) is solved using both the fully implicit and the Crank-Nicolson schemes [3].

The numerical procedure was applied to a wide range of governing parameters. It was found that for small values of q_v, i.e., $q_v \leq 0.15$, and for small values of n, i.e., n < 0.1, the finite-difference algorithm based on the Newton-Raphson method did not converge. Small values of q_v represent a die of small opening, which restricts the flow considerably. Here, $q_v = 0.5$ corresponds to the no die situation and, therefore, there is no pressure rise from the hopper to the die in this case. This situation is similar to the Couette flow between two parallel plates in the absence of a pressure gradient. At small values of q_v, a significant amount of the flow is in the negative z direction. Therefore the marching scheme in the positive z direction cannot be employed. Equations (6)-(8) were also solved using the method described by Fenner [3]. This method was found to give convergent solutions at smaller values of q_v and n, for isothermal and fully developed non-isothermal circumstances. However, both methods did not yield numerical solutions for non-isothermal, temperature developing circumstances, at very low values of q_v, mainly because of significant amount of backflow. The full, elliptic, equations must be solved in such cases.

I.3 RESULTS AND DISCUSSION

The results presented here are based on the coordinate axes fixed to the rotating screw. The ratio of axial screw length L to channel height H is taken as about 70, corresponding to practical extruders. Upon obtaining the numerical solution for the velocity and temperature fields, various quantities of interest, such as the heat input from the barrel, local Nusselt number Nu_H, bulk temperature, shear stress, and pressure at various downstream locations, are calculated. For brevity, only some of the typical results are presented here. The residence time distribution (RTD) is calculated by numerically simulating the experimental procedure of injecting a die into the flow [8].

Figure 3 shows the results in terms of isotherms and isovelocity lines along the unraveled screw channel. The temperature and velocity profiles at four locations are also shown. The temperature of the flow is seen to increase above the barrel temperature. Therefore, beyond a certain downstream location, heat transfer occurs from the flow to the barrel if the barrel is maintained at a fixed temperature. This implies initial heat input at the barrel followed by heat removal further downstream. This effect is due to the viscous heating of the fluid.

Figure 4 shows the variation of the dimensionless pressure P^* and the pressure gradient $\partial P^*/\partial z^*$ along the screw channel, in comparison with the values for a Newtonian fluid. Figure 5 shows the corresponding bulk temperature rise and the local Nusselt number at the inside surface of the barrel. These trends agree with the physical behavior in actual systems.

The residence time distribution (RTD) is an important characteristic of the extrusion process. This was obtained by numerically simulating the flow of a slab of a dye as it flows from the hopper to the die [17]. The cumulative function F(t), which indicates the fraction of the total amount of die carry out of the die is plotted as a function of time in Fig. 6. It was found that the residence time distribution was only slightly affected by the barrel temperature. It was mainly affected by the throughput parameter q_v, which substantially influences the flow field.

II. FINITE ELEMENT MODEL

Finite element models were developed to simulate the flow inside

1) Single screw channel having non-rectangular, or complex profiles

2) Intermeshing zone of a tangential type twin-screw extruder.

The geometry of the tangential co-rotating twin-screw extruder is shown in Fig. 7. A section normal to the screw axes is shown in Fig. 8(a,b). The flowfield region can be divided in two regions, 1) translation region (T), and 2) mixing region (M). In the translation region, the flow is essentially the same as that in a single-screw channel with a similar geometry. A significant amount of mixing takes place in the mixing region located in the center.

In the T-region, it is reasonable to assume that the velocity field does not change significantly along the screw channel direction, i.e., z-coordinate direction, for a fully developed flow [21,22]. In the M-region, it is assumed that the velocity field does not change significantly in the axial direction, i.e., z-coordinate direction, as compared with the other two directions (x and y coordinate directions). In other words, $\partial V/\partial z \ll \partial V/\partial y$ or $\partial V/\partial x$. Based on these assumptions and using the coordinate system for the T- and M-regions sketched in Fig. 8, the velocity fields in these two regions can be represented as:

$$V = u(x,y)i + v(x,y)j + w(x,y)k \qquad (11)$$

where i, j, k are unit vectors in the three directions.

The momentum and continuity equations for the isothermal, steady-state, creeping flow are given as:

$$\frac{\partial p}{\partial x} = \frac{\partial}{\partial x}\left[2\,\mu\left(\frac{\partial u}{\partial x}\right)\right] + \frac{\partial}{\partial y}\left[\mu\left(\frac{\partial u}{\partial y} + \frac{\partial v}{\partial x}\right)\right] \qquad (12)$$

$$\frac{\partial p}{\partial y} = \frac{\partial}{\partial z}\left[\mu\left(\frac{\partial u}{\partial y} + \frac{\partial v}{\partial x}\right)\right] + \frac{\partial}{\partial y}\left[2\mu\left(\frac{\partial v}{\partial y}\right)\right] \qquad (13)$$

$$\frac{\partial p}{\partial z} = \frac{\partial}{\partial x}\left[\mu\left(\frac{\partial w}{\partial x}\right)\right] + \frac{\partial}{\partial y}\left[\mu\left(\frac{\partial w}{\partial y}\right)\right] \qquad (14)$$

$$\frac{\partial u}{\partial x} + \frac{\partial v}{\partial y} = 0 \qquad (15)$$

Thus, a three-dimensional flow analysis for the T- and M-regions can be carried out in terms of inplane and axial flows using a three-dimensional finite element model, with marching in the z-direction.

In the present study, a power-law model has been used to describe the shear rate dependent viscosity of non-Newtonian fluids, as given by

$$\mu = K(T)\dot{\gamma}^{n-1} \qquad (16)$$

where $K(T)$ is a constant which is a function of temperature T, and $\dot{\gamma}$ is the generalized shear rate, which in the present velocity field, can be reduced to:

$$\dot{\gamma} = [2(\frac{\partial u}{\partial x})^2 + (\frac{\partial u}{\partial y} + \frac{\partial v}{\partial x})^2 + 2(\frac{\partial v}{\partial y})^2$$

$$+ (\frac{\partial w}{\partial x})^2 (\frac{\partial w}{\partial y})^2]^{1/2} \qquad (17)$$

II.1   Finite Element Formulation and Solution Procedure

As far as the inplane flow is concerned, the principle of virtual velocity [21] is used to formulate the finite element method for solving the x and y momentum equations together with the continuity equation. Six-node triangular elements have been used in the present study with a quadratic and a linear interpolation for the velocity and pressure fields, respectively.

The x-y momentum equations and the z-momentum equation are coupled to each other through the viscosity function in case of a non-Newtonian fluid. However, in case of a Newtonian fluid, they are independent of each other. Due to the coupled momentum equations, an iteration scheme is required to solve these equations for u, v, and w velocity components and pressure p, which is outlined by Kwon et al. [20,23].

II.2   SIMULATION OF THE TRANSLATION REGION

The flow inside the T-region of a twin-screw extruder is very similar to the flow inside a single-screw extruder of the same screw design. In this study, the screw channel was assumed to be fully filled with the material. The computational domain of the T-region is taken as the screw channel cross-section. It may be mentioned here that screw channel with any profile can be modeled using the FEM. Mesh discretizations of two examples of screw profiles are shown in Fig. 9 for a rectangular screw profile and the so-called self-wiping screw profile, respectively. The two screw profiles have the same maximum width and depth, but the rectangular screw profile does not have gaps between the screw and the barrel. In this paper, only the results of simulation using the self-wiping screw profile are presented. Unless otherwise indicated, the material used throughout the examples in this paper was Polystyrene, with the following properties: $K(T) = 47090$ [poise sec^{n-1}], $n = 0.31$ (Kwon et al., 1986).

The boundary conditions applied at the computational domain are as follows:

- At the barrel surface, $u = V_{bx} = -V_b \sin\phi$, $v = 0$ and $w = V_{bz} = V_b \cos\phi$.

- At the inlet, traction t_x due to a certain pressure p, $v = 0$, and $\partial w/\partial x = 0$.

- At the outlet, traction $t_x = 0$, $v = 0$, and $\partial w/\partial x = 0$.

- At the screw surface, $u = v = w = 0$.

As has been mentioned earlier, the flow in the T-region is simulated in terms of the inplane and the axial flows. Figure 10 shows typical inplane and axial velocity fields, respectively, for an adverse pressure gradient in the down-channel direction. As can be seen from Fig. 10(b) there is a backflow at the bottom of the screw channel due to the pressure rise. The corresponding pressure contours for this case are shown in Fig. 10(c).

In this study, one of our interest was to study the flow mechanism of the material particles inside the screw channel which leads toa better understanding of the mixing inside the screw channel. Based on the inplane and axial velocity fields obtained from the FEM model, one can introduce particles inside the screw channel and follow the movement of these particles along the channel. A very small time interval was chosen for the numerical integration of the velocity field to obtain particle paths. The results of such particle tracings are presented in Fig. 11 in different views, for the case of an adverse pressure gradient. Four representative particles were traced for a given channel length of 1.5 screw turns. As can be seen in this figure, the particles undergo spiral movements, except particles near the barrel surface (e.g., particle A) which go straight across the flight gap into the adjacent channel. The backflow is represented by particles C and D which flow backward at the bottom of the screw channel. The spiral movement of the particles inside the screw channel clearly promotes mixing within the single screw extruder, or, the T-region in case of the twin-screw extruders.

II.4   SIMULATION OF THE MIXING REGION

In the numerical simulation of the mixing region, it is assumed that there is no gap between the screw flight and the barrel. In this simulation, the screws are rotating while the barrel is stationary. The cross-section of the M-region normal to the extruder axis is the computational domain, for which the mesh discretization is shown in Fig. 12. In this figure, the computational domain of the M-region has been taken to include part of the T-region in order to show the transition from the T-region to M-region. The boundary between the T-region and M-region was not clearly defined. However, it will be discernible from the results of the velocity and pressure fields in the following discussion.

The boundary conditions for the M-region were as follows:

- At the inlets, a chosen velocity profile was applied.

- At the outlets, a zero traction component in the x-direction was applied together with a zero velocity in the y-direction.

- At the barrel surface, the no-slip boundary condition was applied.

- At the screw root surface, the velocities in the x- and y-directions were applied for a given screw rotational speed.

Several combinations of boundary conditions have been used in this study. As for the boundary conditions at the inlet, two velocity profiles are of our interest. The first one was a linear velocity profile while the other was the velocity profile obtained in an annulus with a rotating inner surface and a stationary outer surface, at a given level of annular pressure gradient.

The typical inplane and axial velocity fields, path lines and pressure contours of the co-rotating tangential twin-screw extruder are shown in Fig. 13 (a), (b), (c) and (d), respectively, for a linear velocity profile applied at inlet. It is clearly seen that in the region away from the central part, the inplane velocity field is very uniform and the pressure contours show a constant pressure gradient. The narrow region where the velocity and pressure fields begin to deviate from beginning uniform distribution represents the boundary between the T- and M-regions. This boundary is found to be located very close to the central part.

As can be seen also from Fig. 13, the material from one screw channel is divided into two different flow paths in the M-region. Material near the screw flows into the same channel while the other part near the barrel flows into the other channel. This flow behavior represents the mixing that occurs inside the M-region of the co-rotating tangential twin-screw extruder.

Coupling of the Two Regions

To obtain a complete solution for a twin screw geometry, the finite difference model and the finite element model for the intermeshing zone were coupled together at appropriate locations. Figure 14(a) shows the variation of the pressure for the translation and mixing regions. Figure 14(b) shows the corresponding variation of bulk temperature of the fluid.

In the coupling model, the mixing zone was taken as isothermal, at the temperature corresponding to the end of the previous translation region. Figure 14(a) shows the average pressure along the channel for 1.5 screw turns. In the mixing region, the pressure goes through maximum and minimum as shown in Fig. 14(a).

II.5   3-D Finite Element Model for Simulation Flow in a Die

A three dimensional Finite-Element Model has been developed to simulate isothermal flow in a die. For conciseness, only typical results are presented here.

Figure 15 shows   geometry of a die and the corresponding pressure drop along the lie length, for different operating speeds.

Thus, knowing the die characteristics in terms of pressure drop, temperature change, one can obtain the extruder characteristics, by knowing the screw characteristics. This is shown in Fig. 16. The point of intersection of the curve corresponding screw characteristics and die characteristics represents the operating point of the extruder. This information is useful in the design of a system.

CONCLUSIONS

The numrical simulation of the fluid flow and heat transfer in the channel of a single-screw extruder has been carried out for non-Newtonian fluids. A finite-difference numerical procedure is developed in order to obtain the velocity and temperature fields in the screw channel. The heated barrel is maintained at fixed temperature. The screw is taken to be insulated. The residence time distribution is obtained numerically. The flow is mainly governed by the dimensionless flow rate parameter q_v. It is found that the temperature field is strongly affected by the dimensionless parameters G and Pe. However, the corresponding changes in the velocity field are small. It is also found that at small values of the dimensionless throughput q_v and the power law index n, a significant amount of flow is in the negative z direction. In the presence of strong viscous dissipation, heat transfer is observed to occur from the fluid to the barrel downstream in the later portion of the extruder. It is found that the residence time distribution was not significantly affected by the values of power law index n or the barrel temperature distribution for the range of parameters considered in this study. It was mainly dependent upon the flow rate parameter q_v.

A new approach towards modeling tangential twin-screw extruders has been applied. The flow inside the extruder was divided into translation and mixing regions. The results of the simulation of the translation region are directly applicable to the simulation of a single-screw extruder having the same barrel and screw channel design. The complete simulation of the tangential twin-screw extruders can be obtained by coupling the simulations of the T-region and M-region at the interface of both regions. A three-dimensional finite element model for complicated die geometries has been developed.

ACKNOWLEDGEMENT

The authors would like to thank Dr. T.H. Kwon, Dr. V. Sernas, Dr. S. Gopalakishna, M. Gupta, and T. Sastrohartono for their contribution to this work. This work was supported by a grant from Center for Advanced Food Technology, Rutgers University.

REFERENCES

[1]   Tadmor, Z. and I. Klein. 1978. Engineering Principles of Plasticating Extrusion. Huntington: Kleiger Publishing Co.

M

[2]   Janssen, L.P.B.M. 1978. Twin Screw Extrusion. Amsterdam: Elsevier
        Scientific Publishing Co.

[3]   Fenner, R.T. 1979. Principles of Polymer Processing. New York:
        Chemical Publishing.

[4]   Griffith, R.M. 1962. Fully Developed Flow in Screw Extruders. I&EC
        Fundamentals, 1(3).

[5]   Zamodits, H.J. and J.R.A. Pearson. 1969. Flow of Polymer Melts in
        Extruders, Part I. Effect of Transverse Temperature and of a
        Superimposed Steady Temperature Profile. Transactions of the
        Society of Rheology 13(3).

[6]   Lidor, G. and Z. Tadmor. 1976. Theoretical Analysis of Residence
        Time Distribution Functions and Strain Distribution Functions in
        Plasticating Screw Extruders. Polymer Engineering and Science, 16.

[7]   Bigg, D.M. and S. Middleman. 1974. Mixing in a Screw Extruder,
        Model for Residence Time Distribution and Strain. I&EC Fund, 13.

[8]   Tadmor, Z. and C. Gogos. 1979. Principles of Polymer Processing,
        New York: John Wiley & Sons.

[9]   Elbirli, B. and J.T. Lindt. 1984. A Note on the Numerical Treatment
        of the Thermally Developing Flow in Screw. Polymer Engineering
        and Science, 24(7).

[10]   Agur, E.E. and J. Vlachopoulos. 1982. Numerical Simulation of Single-
        Screw Plasticating Extruder. Polymer Engineering and Science, 22(17).

[11]   Wyman, C.E. 1975. Theoretical Model for Intermeshing Twin Screw
        Extruders: Axial Velocity Profile for Shallow Channels. Polym. Eng.
        Sci., 15:606-611.

[12]   Secor, R.M. 1986. A Mass Transfer Model for a Twin-Screw Extruder.
        Polym. Eng. Sci., 26:647-652.

[13]   Booy, M.L. 1980. Isothermal Flow of Viscous Liquids in Corotating
        Twin Screw Devices. Polym. Eng. Sci., 20:1220-1228.

[14]   Denson, C.D. and Hwang, Jr., B.K. 1980. The Effect of Leakage and
        Cross-Channel Flows on the Performance of Co-Rotating Twin Screw
        Extruders. SPE ANTEC Tech. Pap., 26:107-109.

[15]   Denson, C.D. and Hwang, Jr., B.K. 1980. The Influence of the Axial
        Pressure Gradient on Flow Rate for Newtonian Liquids in a Self
        Wiping, Co-Rotating Twin Screw Extruder. Polym. Eng. Sci.,
        20:965-971.

[16]   Hwang, Jr., B.K. 1982. Fluid Flow Studies in Twin Screw Extruders.
        Ph.D. Thesis, University of Delaware.

[17]   Bigio, D. and Zerafati, S. 1988. A Numerical Procedure for Evaluation of Mixing in the Nip Region of a Non-Intermeshing, Counter-Rotating Twin Screw Extruder. SPE ANTEC Tech. Pap., 34:85-88.

[18]   Nguyen, K. and J. Lindt. 1988. Finite Element Modeling of a Counter-Rotating, Tangential (crt) Twin Screw Extruders. SPE ANTEC Tech. Pap., 34:93-95.

[19]   Karwe, M.V. and Y. Jaluria. 1988. Numerical Simulation of Fluid Flow and Heat Transfer in a Single Screw Extruder for Non-Newtonian Fluids. ASME Winter Annual Meeting, Chicago.

[20]   Kwon, T.H., T. Sastrohartono and Y. Jaluria. 1988. Finite Element Simulation of Mixing Phenomena in the Intermeshing Zone of a Twin-Screw Extruder for Non-Newtonian Fluids. In H. Niki and M. Kawahara (Eds.), Computational Methods in Flow Analysis, Vol. 1, p. 635-642, Okayama University of Science, Okayama, Japan.

[21]   Malvern, L.E. 1969. Introduction to the Mechanics of a Continuous Medium. Englewood Cliffs, NJ:Prentice-Hall, Inc.

[22]   Dawson, P.R. and Thompson, E.G. 1978. Finite Element Analysis of Steady-State Elasto-Visco-Plastic Flow by the Initial Stress-Rate Method. Int. J. Num. Meth. Engng., 12:47-57.

[23]   Kwon, T.H., S.F. Shen and K.K. Wang. 1986. Pressure Drop of Polymeric Melts in Conical Converging Flow: Experiments and Predictions. Polym. Eng. Sci., 26:214-224.

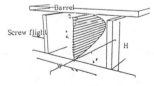

Fig. 1 Simplified Geometry of Single-Screw Extruder

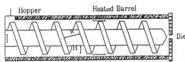

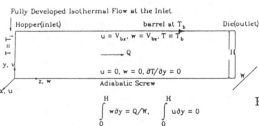

Fig. 2 Boundary Conditions for the Finite-Difference Model

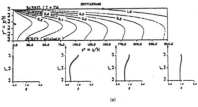

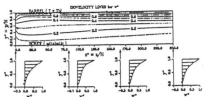

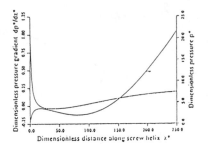

Fig. 3 Isotherms and Isovelocity lines and corresponding profiles for a non-Newtonian Fluid with n = 0.5 at ϕ = 16.54, G = 10.0, Pe = 5000, β = 1.61 and q_v = 0.30

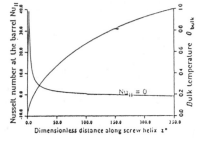

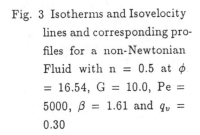

Fig. 4 Variation of the dimensionless pressure gradient $\frac{\partial p^*}{\partial z^*}$ and the dimensionless pressure p^* along z^*

Fig. 5 Downstream variation of the dimensionless bulk temperature θ_{bulk} and the Nusselt number $N u_H$ at the barrel inside surface

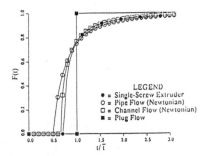

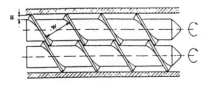

Fig. 6 Residence Time Distribution : Variation of the Cumulative Distribution Function F(t) for different flow configurations

Fig. 7 Geometry of Tangential Co-rotating Twin-Screw Extruder

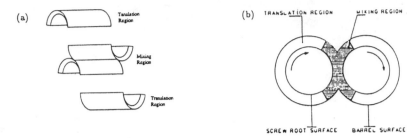

Fig. 8 Section normal to the Twin-
Screw axes

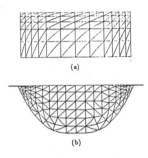

Fig. 9 Mesh Discretization of Single-
Screw channels in the FEM
model

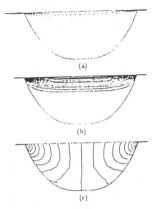

Fig. 10 In-plane and axial Velocity Fields, and Pressure
field in the screw channel
for adverse pressure gradient

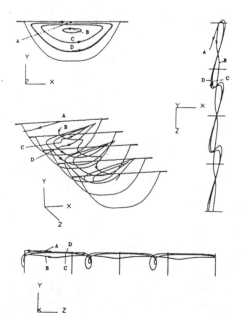

Fig. 11 Particle Tracings of Coupled Cross- and Down-Channel
Flows in the Screw Chan-
nel with an adverse dp/dz,
for a Non-Newtonian Ma-
terial

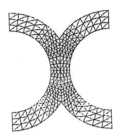

Fig. 12 Mesh Discritization of the
Mixing Region

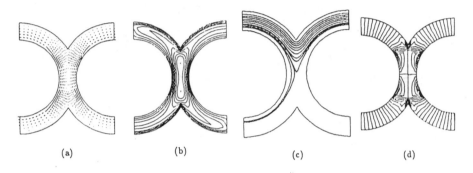

(a)   (b)   (c)   (d)

Fig. 13 Axial Velocity Field, Path Lines and Pressure Contours in the
Mixing Region

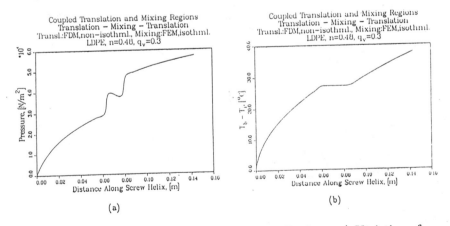

(a)   (b)

Fig. 14 Results of the Coupling of the two Regions: a) Variation of
Pressure and b) Variation of the bulk temperature along 1.5
screw turns in a tangential Twin-Screw Extruder

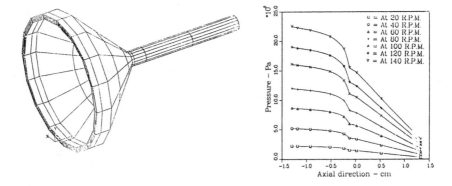

Fig. 15 Die geometry for the 3-D FEM model and corresponding pressure drop along the die

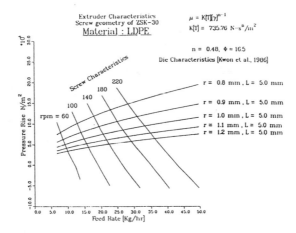

Fig. 16 Screw and Die Characteristics in terms of throughput vs pressure drop

An Examination of the Performance of a 3-Level Timestepping Algorithm - Coupled Heat and Mass Transfer Computing

H.R. Thomas, S.W. Rees

School of Engineering, University of Wales College of Cardiff, Newport Road, Cardiff, CF2 1XH, Wales

ABSTRACT

An examination of the performance of Lees' three level timestepping algorithm, in the non-linear finite element analysis of coupled ground heat and mass transfer is presented. The work performed consists of a series of numerical experiments, whereby various characteristics of the scheme are explored. The results obtained are encouraging, both in terms of the computational effort required to produce a solution and the overall stability and convergence of the scheme. Furthermore, it is postulated that the oscillatory nature of the solution response, as errors are induced, can be used to advantage in an interactive computing environment to control acceptable timestep sizes and hence minimise errors. It is concluded that the algorithm is worthy of consideration for use in computing problems of the type addressed here.

INTRODUCTION

The numerical analysis of coupled ground heat and mass transfer is a subject of considerable topical interest. Nuclear waste disposal [1] and seasonal heat storage [2] are but two typical areas of application. Extensive work is recognised to be required in all these fields and in particular the finite element method is widely used in the analysis of relevant practical engineering problems.

The application of the finite element method can be considered in two parts, spatial discretisation and timestep modelling. The question of the performance of the timestep algorithm is therefore important when considering the overall efficiency of the solution method. The work presented in this paper focuses on this question, for one particular timestepping scheme.

The algorithm chosen for investigation is a three level scheme proposed by Lees [3]. Its use is explored here in the context of a non-linear analysis, where it offers the advantage that direct evaluation of the relevant coefficients may take place, at the mid-time level. This aspect is obviously potentially attractive when compared to other schemes where iterations at each timestep are required. It also possesses the advantages of simplicity, which makes it easy to program, and second order accuracy.

Bonacina and Comini [4] investigated the local accuracy, stability and convergence of the scheme, within the context of the two-dimensional finite difference solution of non-linear heat conduction problems. They showed the scheme was unconditionally stable and convergent. The first application of the algorithm in finite elements appears to be due to Comini, Del Guidice, Lewis and Zienkiewicz [5], again for the case of non-linear heat conduction problems. They used extensive numerical experiments to examine the behaviour of the algorithm and concluded that the scheme was also unconditionally stable and convergent in the context of their new applications. Their numerical experimentation approach is continued here, within the context of coupled ground heat and mass transfer.

The work reported is based on a formulation of ground heat and mass transfer analysis described in detail in reference [6]. Full details will not therefore be repeated here; only the salient features will be highlighted. A one-dimensional application is considered, for ease of presentation and interpretation. In particular, surface drying from a 0.5 m column of loam is analysed.

THEORY AND FINITE ELEMENT FORMULATION

The governing differential equation describing the flow of moisture in a non-isothermal unsaturated soil can be written as [6]:

$$\frac{\partial \theta_\ell}{\partial t} = \nabla \cdot (D_\theta \nabla \theta_\ell) + \nabla \cdot (D_T T) + \frac{\partial K}{\partial z} \qquad (1)$$

where θ_ℓ is the volumetric liquid content, t is the time, T is the temperature, D_θ is the isothermal moisture diffusivity, D_T is the thermal moisture diffusivity, K is the unsaturated hydraulic conductivity and z is the elevation.

The governing differential equation describing the transfer of heat through the soil can be expressed in the form [6]:

$$C\frac{\partial T}{\partial t} = \nabla.(\lambda+L\rho_\ell \epsilon D_T)\nabla T + \nabla.(L\epsilon\rho_\ell D_\theta)\nabla\theta_\ell$$

$$+ L\rho_\ell \frac{\partial(\epsilon K)}{\partial z} \tag{2}$$

In the above expression C is the volumetric heat capacity, λ the thermal conductivity, L the latent heat of vaporisation of water, ϵ the phase conversion factor and ρ_ℓ the density of liquid water.

Equations (1) and (2) are solved simultaneously as a coupled set of differential equations in order to evaluate the variation of volumetric moisture content and temperature in space and time. Fixed boundary conditions of

$$T = T_c \quad \text{or} \quad \theta_\ell = \theta_{\ell c} \tag{3}$$

can be applied for the heat and mass transfer fields respectively, together with flux boundary conditions of

$$\frac{\lambda\partial T}{\partial n} + (1-\epsilon)L(q_m)_\Gamma + (q_h)_\Gamma = 0 \tag{4}$$

and

$$\frac{K\partial\Phi}{\partial n} + \frac{D_T\partial T}{\partial n} + (q_m/\rho_\ell)_\Gamma = 0 \tag{5}$$

again for the heat and mass transfer fields respectively $(q_m)_\Gamma$ and $(q_h)_\Gamma$ refer to moisture and heat fluxes at the surface, respectively, while Φ is the total potential for moisture flow, which in this case is the sum of capillary potential and the elevation.

A finite element solution of the above equations was subsequently derived through the application of Galerkin's weighted residual approach. This led to the following set of equations:

$$\underline{D}_q \underline{T} + \underline{D}_\epsilon \underline{\theta}_\ell + \underline{C}_q \underline{\dot{T}} + \underline{J}_q = 0 \tag{6}$$

$$\underline{D}_T \underline{T} + \underline{D}_\theta \underline{\theta}_\ell + \underline{C}_\theta \underline{\dot{\theta}}_\ell + \underline{J}_\theta = 0 \tag{7}$$

$$\text{i.e.} \quad \underline{D}\,\underline{\phi} + \underline{C}\,\underline{\dot{\phi}} + \underline{J} = 0 \tag{8}$$

The application of Lees' algorithm to Equation (8) yields

$$\underline{D}^n\left[\frac{\phi^{n+1}+\phi^n+\phi^{n-1}}{3}\right] + \underline{C}^n\left[\frac{\phi^{n+1}-\phi^{n-1}}{2\Delta t}\right] + \underline{J}^n = 0 \tag{9}$$

which on rearranging, produces the recurrence relationship

$$\underline{\phi}^{n+1} = \left[\frac{D^n}{3} + \frac{C^n}{2\Delta t} \right]^{-1} \times \left[\frac{D^n\underline{\phi}^n}{3} + \frac{D^n\underline{\phi}^{n-1}}{3} - \frac{C^n\underline{\phi}^{n-1}}{2\Delta t} + \frac{\underline{J}^n}{} \right] \quad (10)$$

The superscript n refers to the time level and Δt is the timestep size. It can be seen that in this approach the vector $\underline{J}$ is evaluated at time level n, in order to avoid the iterative calculations required if an average of the three fluxes over the three timesteps is used.

EXAMPLE PROBLEM CONSIDERED

The problem consists of a soil stratum subjected to moisture content losses due to an elevated temperature and a reduced moisture content at the surface. The example is therefore representative of level ground conditions subjected to uniform environmental conditions.

The thermophysical parameters used in the analysis are functions of the volumetric moisture content and the temperature of the soil. It is known, however [7], that variations with respect to temperature are less acute than variations with respect to moisture content. In the first instance, therefore, and for the sake of simplicity, only variations with respect to moisture content are considered here.

An example of drying of a loam soil has been chosen for analysis, and on the basis of an extensive literature search the parameter variations shown in Table 1 have been chosen as representative. Values at intermediate moisture contents are obtained by liner interpolation. Values at moisture contents outside the range 0.05 to 0.45 are assigned either the maximum or minimum value as appropriate. The saturated volumetric moisture content of the loam is 0.5.

An analysis of temperature and moisture content variations in the upper 0.5m of the stratum has been carried out. The zone under consideration has been discretised into five, eight node parabolic 'Serendipity' isoparametric elements as shown in Figure 1.

Initial conditions of an uniform temperature throughout of $10^\circ C$ have been assumed, together with a volumetric moisture content of 0.45 at the base of the column. The specific moisture capacity of the loam, which can be deduced from the variations of D_θ and K shown in

TABLE 1   SOIL PARAMETERS

θ $\dfrac{cm^3}{cm^3}$	λ $\dfrac{cal}{cm\ sec\ {}^{o}C}$	C $\dfrac{cal}{cm^3\ {}^{o}C}$	D_θ $\dfrac{cm^2}{sec}$	K $\dfrac{cm}{sec}$	D_T $\dfrac{cm^2}{sec\ {}^{o}C}$	ϵ
0.05	0.45E-03	0.3000	7.94E-05	8.91E-09	1.32E-07	0.7560
0.10	1.68E-03	0.3188	0.15E-03	2.11E-08	1.66E-07	0.6900
0.15	2.75E-03	0.3375	0.25E-03	4.90E-08	1.82E-07	0.5490
0.20	3.84E-03	0.3563	0.40E-03	0.10E-06	3.00E-07	0.3330
0.25	4.63E-03	0.3750	0.57E-03	0.24E-06	7.15E-07	0.1399
0.30	5.40E-03	0.3938	0.72E-03	0.33E-06	1.91E-06	0.0520
0.35	5.45E-03	0.4125	1.15E-03	0.58E-06	5.01E-06	0.0200
0.40	5.70E-03	0.4313	1.66E-03	0.84E-06	1.51E-05	0.0066
0.45	5.75E-03	0.4500	2.88E-03	1.20E-06	5.01E-05	0.0020

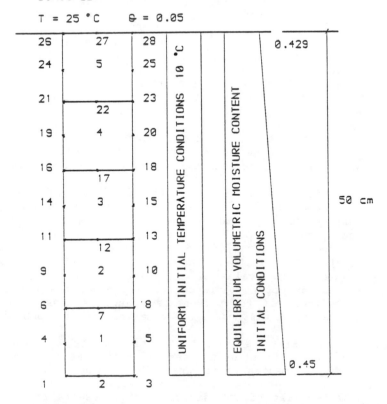

Figure 1.   Finite Element Mesh, Initial and Boundary conditions

Table 1, dictates the equilibrium variation of volumetric moisture content with elevation. This variation has been computed and applied as initial conditions of non-uniform volumetric moisture content variation with depth.

Fixed boundary conditions of a temperature of 25°C and a volumetric moisture content of 0.05 have been applied at the surface, all other boundaries being both impermeable and adiabatic. A diagrammatic summary of this information is given in Figure 1.

RESULTS

The first requirement of this exercise was the establishment of a benchmark solution against which subsequent comparisons could be made. Theoretical solutions of non-linear coupled transient heat and mass transfer problems such as this are, however, not available. The philosophy adopted here therefore is to use the results from a Backward Difference predictor-corrector scheme to provide benchmark solution values. In particular as the value of the timestep sizes is reduced and the mesh refinement increased, the results tend towards a converged solution. When this exercise was carried out here, the converged results shown in Figure 2 were obtained. These give the variations of temperature and volumetric moisture content at nodes 5, 13 and 25, representing conditions near the base, centre and surface of the column. The results presented were obtained using a constant timestep size.

The effect of mesh refinement on the accuracy of the results was also investigated. The results when compared with the benchmark solution showed very little difference.

In order to improve the efficiency of the solution algorithm and facilitate further computational work, a varying timestep size was incorporated in the analysis. In the approach adopted the timestep size is changed according to the nodal variations of temperature and volumetric moisture content that are taking place. If the percentage change in the value of a variable over a time interval exceeds an upper tolerance, at any node, the timestep size is halved. If the percentage change is less than a lower tolerance, at all nodes, the timestep size is doubled.

A first assessment of the quality of results that could be achieved using Lees algorithm was obtained using a varying timestep size analysis, using upper and lower tolerances of 5% and 3% respectively, and an initial timestep size of 300 seconds. The results produced, when compared with the benchmark solution, gave excellent correlation.

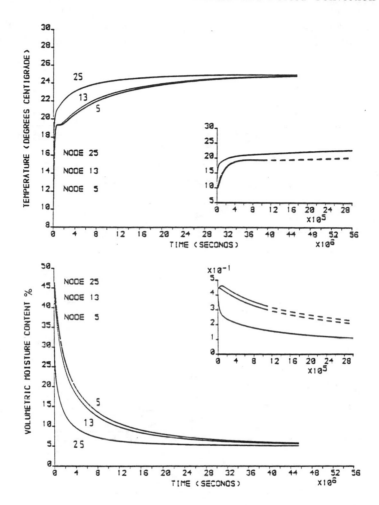

Figure 2.   Benchmark Solutions for nodes 25, 13 and 5.

In order to explore more fully the characteristics of the timestepping scheme, the effect of a number of features on the accuracy of the results and the computational performance have been considered.

Considering first the influence of initial timestep size, a series of runs were computed using progressively larger timesteps in the range 300 to 10,000 seconds. In all cases values of 5% and 3% were employed for the upper and lower tolerances respectively. It became apparent that beyond a certain timestep size Lees' algorithm produced oscillatory behaviour. The oscillations naturally became markedly more pronounced as the timestep size increased further. However, it was found that no matter how great the induced oscillations, the results ultimately converged to the correct steady state solution.

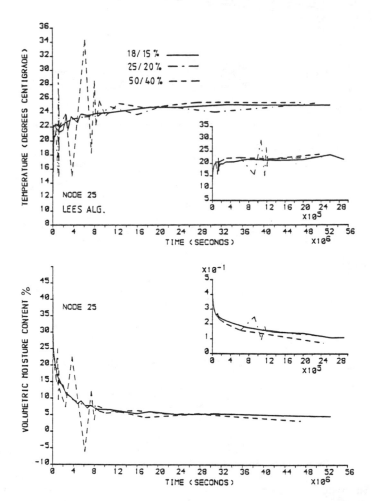

Figure 3. Lees' Algorithm - the effects of varying the timestep control tolerances.

Returning to an initial timestep size of 300 secs, the influence of relaxing the timestep size variation tolerances was investigated. Values of upper and lower tolerances of 8% and 5%, 13% and 10%, 18% and 15%, 25% and 20%, and 50% and 40% were examined. Lees' algorithm's oscillatory behaviour was again revealed as expected. Obviously the more relaxed the tolerances, the more rapid the increase in timestep size, and the greater the oscillations. This effect is illustrated clearly in Figure 3   where for clarity, the variations occurring at node 25 only are shown. Once again however the results were found to always converge to the correct steady state values.

On the basis of the above numerical experiments, together with other work on the same problem, not repeated in detail here because of the lack of space, it appears that the conclusions previously reached, regarding the use of the algorithm in the finite element analysis of non-linear heat conduction problems also apply here.   In particular unconditional stability and convergence of the scheme appears to persist.

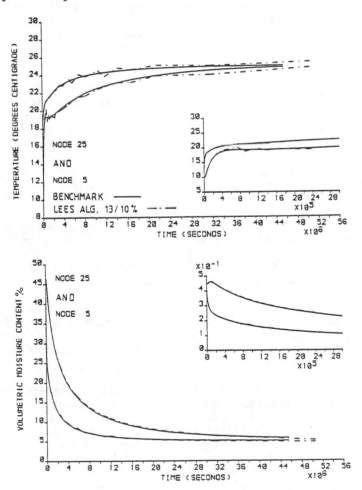

Figure 4.   Lees' Algorithm Solution and Benchmark Solution.

A further important aspect in terms of overall performance of a timestepping scheme, is the computational time taken to achieve a solution to a problem. An attempt. was therefore made to examine the computational performance of the scheme in terms of run time taken versus error level generated.

Taking as a starting point for this exercise the fact that at tolerances of 18/15%, 25/20% and 50/40%, Lees' algorithm produced unacceptably oscillatory variations in results, the solutions produced using tolerances of 13% and 10% were chosen for investigation. The values achieved are shown in Figure 4 together with the benchmark solution. For clarity of presentation only variations at nodes 5 and 25 are now illustrated. Clearly some oscillations are still visible, notably in the temperature variation at node 25. Compared to the benchmark solution maximum error levels of 4.56% in temperature and 2.94% in moisture content have been generated.

As a comparative exercise, results from the Backward Difference scheme, operating at the same timestep variation tolerances were generated. The results obtained gave maximum error levels of 3.27% and 9.40% in the temperature and moisture content fields respectively. Furthermore the CPU time was some 2.8 times greater than found necessary for the Lees' algorithm calculations with the timestep sizes employed. The number of corrections required to reach convergence with the Backward Difference scheme varied at each timestep between 2 and 5. This feature accounts for the difference in computational time.

In view of the fact that the computational time of alternative Backward Difference schemes could be reduced, by means of accelerated convergence schemes, for example, the results of the above exercise should be taken as a general guide of Lees' algorithm's performance, rather than precise comparisons. Within this context, the results obtained are however clearly encouraging and indicate the suitability of the algorithm for the work proposed.

Returning to reconsider the results presented in Figure 4, it is possible that the oscillatory nature of the results presented may proved unacceptable under some circumstances. The oscillations can of course be eliminated by reimposing tighter tolerances on the timestep variation procedure. Considering the 5% and 3% tolerances, the computational time for this work was some 90% greater than for the previous Lees' algorithm values. The total CPU time is still less therefore than the Backward Difference results previously quoted, despite the much lower error levels generated.

These results are particularly encouraging for Lees' algorithm proponents, especially when considered in conjunction with the oscillatory nature of the results obtained. It is believed that one of the inherent drawbacks to the use of a direct solution scheme is the absence of the confidence that the user obtains when using a predictor-corrector scheme, of knowing that the solution has been driven to converged values at every timestep. However, since the results obtained here clearly show that the divergences from a relatively high accuracy are accompanied by oscillatory behaviour, the onset of such oscillations can be taken as clear, visible indication of the initiation of distress in the solution. Interactive graphics facilities, which are a commonplace feature nowadays, can therefore be used to readily judge acceptable timestep sizes. Furthermore, it appears on the evidence presented here that such reductions in timestep size can be accommodated without suffering undue computational time penalty.

CONCLUSIONS

The purpose of this paper was to assess the performance of Lees' algorithm timestepping scheme in the analysis of coupled transient heat and mass transfer finite element problems. In the absence of known theoretical solutions, comparisons of the various aspects of the schemes were judged against a benchmark solution. This in turn was generated from the use of the Backward Difference scheme, driven to the limit of a converged set of results as both the timestep size was reduced and the mesh discretisation increased.

A number of observations and conclusions can be recorded after the completion of the work as follows:

(A)   The validity of Lees' algorithm as an accurate timestepping scheme for the non-linear problem under consideration has been confirmed by the correlation of results achieved when comparisons with the benchmark solution were performed.

(B)   The results of numerical experiments indicate that the conclusions previously reached, regarding the use of the algorithm in the finite element analysis of non-linear heat conduction problems also apply here. In particular unconditional stability and convergence of the scheme appears to persist.

(C)   As assessment of the computational performance of the scheme, in terms of error levels generated, compared to the benchmark solution, versus run CPU time consumed, gave encouraging results. For comparable error levels, its computational time is in the order of one third of a Backward Difference scheme examined here. Even when the error levels are reduced in Lees' algorithm runs, so that any oscillatory results are removed, the total CPU time remains competitive.

(D)   It is postulated that the oscillatory nature of the algorithm's solution response can be used to advantage in an interactive environment to control acceptable timestep sizes and hence minimise errors.

It is concluded that the algorithm is worthy of consideration for use in problems of the type addressed here.

ACKNOWLEDGEMENTS

This research work was sponsored by S.E.R.C. (Grant GR/D/77049). Their financial support is gratefully acknowledged.

REFERENCES

1.   Winograd, I.J., 1974, 'Radioactive Waste Storage in the Arid Zone'. EOS, American Geophysical Union, Vol. 55, 884-894.

2.   Nir, A., Doughty, C. Tsang, C.F., 'Seasonal Heat Storage in Unsaturated Soils: Example of Design Study'. 21st Intersociety Energy Conversion Engineering Conf., San Diego, California, 1986.

3.   Lees, M., 1966, 'A Linear Three-level Difference Scheme for Quasilinear Parabolic Equations'. Maths. Comp., 20:516-522.

4.   Bonacina, C. and Comini, G., 1973, 'On the solution of the non-linear heat conduction equation by numerical methods'. Int. J. Heat Mass Transfer, 16:581-589.

5.   Comini, G., Del Guidice, S., Lewis, R.W. and Zienkiewwicz, O.C., 1974, 'Finite element solution of non-linear heat conduction problems with special reference to phase change'. Int. J. Num. Meth. Engng, 8:613-624.

6.   Thomas, H.R., 1987, 'Nonlinear Analysis of Heat and Moisture Transfer in Unsaturated Soil'. J. Eng. Mech. ASCE, Vol. 113, No. 8, 1163-1180.

7.   Ewen, J., 1987, 'Combined Heat and Mass Transfer in Unsaturated Sand Surrounding a Heated Cylinder'. Ph.D. Thesis, University of Wales.

Computational Study of Heat and Mass Transfer in a Single Screw Extruder for Non-Newtonian Materials

S. Gopalakrishna, Y. Jaluria

Department of Mechanical and Aerospace Engineering, Rutgers University, New Brunswick, NJ 08901, USA

ABSTRACT

A numerical study of the transport phenomena arising in a single screw extruder channel is carried out. A non-Newtonian material is considered, using a power law model for the variable viscosity. Chemical reaction kinetics are also included. Finite difference computations are carried out to solve the governing set of partial differential equations for the velocity, temperature and species concentration fields, over a wide range of governing parameters for the case of a tapered screw channel.

The numerical treatment for this combined heat and mass transfer problem is outlined. A marching procedure in the down-channel direction is adopted and the validity of the scheme for practical problems is discussed. For large viscous dissipation, the material heats up considerably due to the prevailing shear field, affecting the viscosity significantly, and results in large changes in the pressure development at the end of the channel. The rate of reaction controls the mass diffusion rate which in turn affects viscosity and the flow significantly. The dimensionless throughput, q_v, is one of the most important parameters in the numerical solution. The dimensionless pressure variation is very sensitive to q_v, and orders od magnitude changes are possible for small variations in q_v. Schemes for dealing with other important effects such as back flow, heat transfer by conduction in the barrel, and the effect of the die are also outlined.

INTRODUCTION

Screw extrusion is a thermomechanical processing operation where the raw material is fed into a hopper and is forced through the passage between a rotating screw and a stationary barrel. The processed material comes out through a die of a specific shape. Single and twin screw extruders are used widely in the polymer and food processing industry. The high shear and temperature environment inside the screw channel results in mixing of the material and leads to chemical reactions that constitute the cooking process in food processing. This paper discusses the numerical simulation of the complex heat and mass transfer interactions for the simple geometry of the single screw extruder.

A number of numerical studies on the flow of polymers in extruders have been carried out [1, 2, 3, 4]. In these studies, the screw and the barrel were assumed to be at the same, uniform, temperature. Karwe and Jaluria [5] incorporate a more realistic, adiabatic boundary condition at the screw. A number of important aspects relating to the flow and heat transfer in non-Newtonian fluids are highlighted in that study. There is no available literature on the numerical simulation of mass transfer in extrusion, even though it is of practical interest.

A finite difference numerical study has been carried out here for the simulation of the transport processes in extrusion, to obtain the velocity, temperature and mass concentration variation along the length of the screw channel. The variation in channel depth due to taper is accounted for. The effects of viscous dissipation and chemical reaction are included. The governing, dimensionless parameters are obtained by appropriate scaling, and the effect of varying each of these parameters on the resulting flow, thermal and concentration fields is studied in detail parametrically. The limits of applicability of this numerical scheme are identified in terms of ranges of validity of some of the parameters. Results indicate that the throughput, q_v, determines whether the scheme converges to a solution or not. Other parameters that affect pressure rise and concentration distribution include the reaction rate S, the Peclet number Pe and the power law index n.

PROBLEM FORMULATION

The simplified geometry of a single screw extruder is shown in Fig. 1. For ease of visualization and analysis, the coordinate system is fixed to the screw root and, thus, the barrel moves in a direction opposite to the screw rotation. Such a formulation is commonly employed in the literature [1, 2].

Some of the assumptions made in deriving the governing equations include: small channel curvature, W ≫ H for the channel cross-section, lubrication approximation, no significant back flow, and negligible clearance between the screw flight and the barrel.

The governing equations are those of mass, momentum, energy and species diffusion for the flow of a non-Newtonian fluid in a channel. The constitutive equation for viscosity of a non-Newtonian material is written as:

$$\mu = \mu_0 \left(\frac{\dot{\gamma}}{\dot{\gamma}_0}\right)^{(n-1)} e^{b/T} e^{-B_m c_m} ; \tag{1}$$

where

$$\dot{\gamma} = \left[\left(\frac{\partial u}{\partial y}\right)^2 + \left(\frac{\partial w}{\partial y}\right)^2\right]^{1/2}$$

Here, u and w are the velocity components in the x and z-directions, respectively, b is the Arrhenius temperature coefficient of viscosity, B_m is the concentration coefficient, n is the power law index and the subscript "o" denotes the ambient, reference, conditions.

The equations for balances of momentum, energy and species concentration are written in dimensionless form as

$$\frac{\partial p^*}{\partial x^*} = \left(\frac{H_2}{H}\right)^{n+1} \frac{\partial}{\partial y^*}\left(\frac{\partial u^*}{\partial y^*}[\dot{\gamma}^*]^{(n-1)} e^{\beta_1/\{\theta(\beta-1)+1\}} e^{-b_m c^*}\right) \tag{2}$$

$$\frac{\partial p^*}{\partial z^*} = \left(\frac{H_2}{H}\right)^{n+1} \frac{\partial}{\partial y^*}\left(\frac{\partial w^*}{\partial y^*}[\dot{\gamma}^*]^{(n-1)} e^{\beta_1/\{\theta(\beta-1)+1\}} e^{-b_m c^*}\right) \tag{3}$$

$$Pe w^* \frac{\partial \theta}{\partial z^*} = \left(\frac{H_2}{H}\right)^2 \frac{\partial^2 \theta}{\partial y^{*2}} + \left(\frac{H_2}{H}\right)^{n+1} G[\dot{\gamma}^*]^{(n+1)} e^{\beta_1/\{\theta(\beta-1)+1\}} e^{-b_m c^*} \tag{4}$$

$$Pe w^* \frac{\partial c^*}{\partial z^*} = Le\left(\frac{H_2}{H}\right)^2 \frac{\partial^2 c^*}{\partial y^{*2}} + Sc^{*m}, \quad S = 0 \text{ for } \theta < \theta_{gel} \tag{5}$$

where lengths are scaled with H_2 (except for y which is scaled with the local height H), and velocities are scaled with V_{bz}.

$$H(z) = H_1 - (z/z_{max})(H_1 - H_2)$$

$$\theta = \frac{T - T_i}{T_b - T_i}, \quad c^* = \frac{c_m}{c_{mi}}$$

The parameters that arise from this scaling include:

$$\beta = T_b/T_i, \quad \beta_1 = b/T_i, \quad b_m = B_m c_{mi}$$

$$Pe = V_{bz}H_2/\alpha, \quad G = \frac{\bar{\mu}V_{bz}^2}{k(T_b - T_i)} \tag{6}$$

$$Le = D/\alpha, \quad S = S' H_2^2 c_{mi}^{m-1}/\alpha$$

Similarly, the boundary conditions are also obtained in the dimensionless form. The constraints on the flow are obtained, in dimensionless form, as

$$\int_0^1 u^* dy^* = 0,$$

$$\int_0^1 w^* dy^* = \frac{Q/W}{H V_{bz}} = q_v H_2/H, \tag{7}$$

where

$$q_v = \frac{Q/W}{H_2 V_{bz}}.$$

Thus, for a given screw configuration, the parameters that govern the problem are: ψ , n , β , β_1 , b_m , θ_{gel} , Pe , G , Le and q_v.

NUMERICAL SCHEME

Equations (2) - (7) are solved by means of finite difference techniques. The computations were carried out over y × z grid sizes of 41 x 101, 61 x 101, 81 x 101 and 121 x 201. A sample plot of the mesh refinement analysis is shown in Figure 3. The results were essentially unchanged when the grid was refined to 121 x 201 from 61 x 101 and therefore, a 61 x 101 grid was selected. Since the energy and mass transport equations, equations (4) and (5), are parabolic in z, boundary conditions are necessary only at z = 0 to allow marching in the z-direction and, thus, obtain the solution in the entire domain. The throughput parameter q_v determines the limits of applicability of the marching scheme. As a rough estimate, a value of q_v less than 0.2 resulted in significant back flow, for typical values of the other parameters, limiting the use of marching for the numerical solution. In physical terms, the presence of a die at the end of the extruder channel creates a restriction in the flow. This makes the problem of flow in the channel numerically elliptic in nature. Additional boundary conditions are required at the end of the channel. Thus, the characteristics of the die, which relate the pressure drop between the end of the channel and the ambient to the throughput of the extruder, are closely coupled to the extruder characteristics.

A numerical method similar to the one developed by Fenner (1979) is employed for solving the momentum equations, equations (2) and (3), at a given z location iteratively.

Integrating Equations (2) and (3) over y^*, the velocity gradients are expressed as:

$$\frac{\partial w^*}{\partial y^*} = \pi_z(y^* - y_0^*)F(y^*) \tag{8}$$

$$\frac{\partial u^*}{\partial y^*} = \pi_x(y^* - y_1^*)F(y^*) \tag{9}$$

where y_0^* and y_1^* represent y-locations in the channel (from the screw root to the barrel) where the shear stresses are zero, $\pi_z = \partial p^*/\partial z^*$, $\pi_x = \partial p^*/\partial x^*$, and

$$F(y^*) = \dot{\gamma}*^{1-n} e^{-\beta_1/\theta(\beta-1)+1} e^{b_m c^*} \left(\frac{H}{H_2}\right)^{n+1} \tag{10}$$

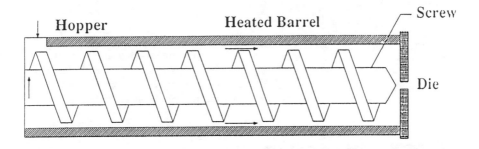

Figure 1: Simplified geometry of a single screw extruder with a rectangular screw profile

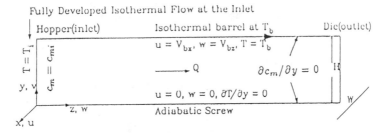

Figure 2: Boundary conditions for the numerical model, employing a co-ordinate system fixed to the screw

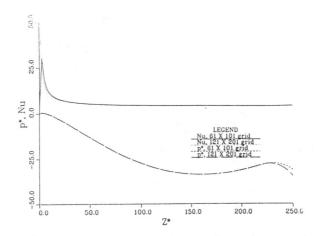

Figure 3: Variation in dimensionless pressure p^*, and Nusselt Number Nu, with down-channel distance z^* for two grid sizes; $n = 0.5$, $G = Pe = 3000$, $G = 1$, $\beta = 1.134$, $\beta_1 = 10.0$, $b_m = 4.0$, $S = -1$, $le = 0.001$, $\theta_{gel} = 0.5$, No Taper

Since $(\dot{\gamma}*)^2 = \left(\frac{\partial w^*}{\partial y^*}\right)^2 + \left(\frac{\partial u^*}{\partial y^*}\right)^2$, we get

$$F(y^*) = \left[\pi_z^2(y^* - y_0^*)^2 + \pi_x^2(y^* - y_1^*)^2\right]^{\frac{1-n}{2n}} e^{-\beta_1/[n\,(\theta(\beta-1)+1)]} e^{b_f c^*/n} \left(\frac{H}{H_2}\right)^{n+1/n} \tag{11}$$

The conditions to be satisfied by the velocities u^* and w^* are:

$$
\begin{aligned}
w^* \quad \{at \; y^* = 1\} &= \quad 1 \\
u^* \quad \{at \; y^* = 1\} &= \quad \tan\phi \\
q_v \qquad H_2/H &= \quad \int_0^1 w^* dy^* \\
0 \qquad\quad &= \quad \int_0^1 u^* dy^*
\end{aligned}
\tag{12}
$$

After some algebraic manipulations, the following four equations for the four unknowns in the LHS are obtained:

$$
\begin{aligned}
\pi_z &= (J_0 - J_1 - J_0 q_v H_2/H)/(J_0 J_2 - J_1^2) \\
\pi_z y_0^* &= (J_1 - J_2 - J_1 q_v H_2/H)/(J_0 J_2 - J_1^2) \\
\pi_x &= (J_0 - J_1)\tan\phi/(J_0 J_2 - J_1^2) \\
\pi_x y_1^* &= (J_1 - J_2)\tan\phi/(J_0 J_2 - J_1^2)
\end{aligned}
\tag{13}
$$

where

$$J_m = \int_0^1 \alpha^m F(\alpha)d\alpha. \tag{14}$$

The solution algorithm for obtaining a converged velocity solution is enumerated below.

1. Guess $\pi_z, \pi_z y_0^*, \pi_x, \pi_x y_1^*$

2. Calculate $F(y^*)$ using Equation (11)

3. Calculate the J integrals using Equation (14)

4. Solve Equations (13) simultaneously using the Gauss-Jordan algorithm

5. Obtain Converged flow solution at each time step

The iterations are complete when the pressure gradients satisfy the following convergence criterion:

$$max[\Delta(\partial p^*/\partial z^*),\; \Delta(\partial p^*/\partial x^*)] \le 10^{-4}. \tag{15}$$

where Δ stands for the absolute value of the fractional change between two consecutive iterations. This particular convergence criterion is not useful when the values of the pressure gradients become very small, i.e., close to zero. Under such circumstances, only the absolute change in the values of the pressure gradient is considered for convergence. Values of the criterion other than 10^{-4} were also tried, and it was found that satisfactory convergence was obtained with the above value within a reasonable number of iterations (typically 5-10).

Using the boundary conditions, in terms of u^*, w^*, θ and c^* at any upstream z location, the energy equation, equation (4), is solved to obtain the temperature distribution at the next downstream z location. Equation (4) is solved using the fully implicit scheme [6]. In this scheme, a tridiagonal system of equations is obtained by using the new, uncalculated, values of the dependent variable from the differencing operation in the y-direction at any nodal point in the numerical scheme. The tridiagonal system is solved using the well-known Thomas (TDMA) algorithm, which is very efficient [7]. The mass transport equation, equation (5), is solved next using, once again, the fully implicit scheme to obtain the values of concentration at the next location in the marching direction. With the temperature and concentration distributions obtained at the next downstream location, the momentum equations, equations (2) and (3), are solved iteratively, as discussed above, to obtain the velocity distribution there. This procedure is repeated until the end of the extruder channel is reached. The integration in equation (14) was carried out numerically using Simpson's one-third rule [7].

During the course of carrying out the parametric study, a number of interesting features of the numerical scheme were noted. It was found that viscosity characterization in terms of β, β_1 and b_m is very closely linked to the magnitude of the reaction rate S that is specified for the particular reaction. For given values of β, β_1 and b_m, the pressure development as a function of downstream channel distance is affected considerably by minor changes in S. In addition, the range of q_v that produces reasonable pressure development with channel distance is quite limited by back flow considerations on the one hand and unacceptable pressure drops on the other. Under the assumptions made in this analysis, a variety of physical mechanisms can be studied, although a realistic representation of the processing of high viscosity materials is not directly feasible.

RESULTS AND DISCUSSION

Results are presented in terms of distributions of velocity, temperature and moisture concentration along the screw channel as well as across the channel depth. The variation of pressure along the channel as a function of the governing parameters is examined here. Contour plots of constant velocity, temperature and moisture are also obtained. The coordinate system is fixed to the screw as described earlier.

Figure 4 shows the calculated isotherms and the temperature profiles at four downstream locations for the non-tapered case for a non-Newtonian fluid, n = 0.3. The isotherms indicate that fluid temperature can be higher than the imposed barrel temperature, i.e. θ larger than 1, due to viscous dissipation represented by the parameter G. The numerical scheme imposes no limitation on the allowable positive values of G. However, a negative value of G, corresponding to T_b smaller than T_i, results in significant back flow. For the marching scheme employed here, a typical limiting value of G below which the scheme does not converge is of the order of -1.

Figure 5 shows the moisture concentration contours. As shown here, the effect of the sink term, S, due to the reaction, is manifested in the form of a decrease in the moisture concentration due to bonding of water for gel formation as the material reaches the gelatinization temperature θ_{gel}. This loss of moisture occurs at a rate specified by the sink term S and is obviously more rapid for larger S.

The results of the simulation for a typical taper angle of 0.1432° are shown in Figure 6. The screw root is continuously tapered from the feed section to the die. This taper angle corresponds to a change in channel depth from H_1 to H_2 (typical geometry corresponding to the Brabender single screw extruder). In the figure, the taper section is shown enlarged for clarity. There is a maximum in the pressure profile at a distance of about two-thirds the total channel length.

For the non-isothermal case, as the non-Newtonian material flows in a tapered channel, it gets squeezed in the decreasing gap. However, the net flow rate has to be maintained at the same value at any channel cross-section. As shown in the figure, the spacing of the streamlines is closer towards the die. The velocity profiles at the four downstream locations show a markedly different behavior for tapered channels as compared to the case with no taper. It may be noted that the initially curved velocity profile becomes linear (corresponding to drag flow) as the section becomes shallower. The temperature profile becomes almost completely uniform across the channel depth in the regions close to the die for a tapered channel.

Figure 7 shows the effect of varying the strength of the sink on the pressure rise along the channel. As compared to the case of S = 0, the dimensionless pressure at the die is seen to be lower for higher sink strengths. There is once again, a limitation to the rate of reaction S that can be imposed on the system. For very large rates, typically S ≈ -10, the sudden, sharp decrease in the moisture level represents a blockage to the flow, and hence the scheme does not converge

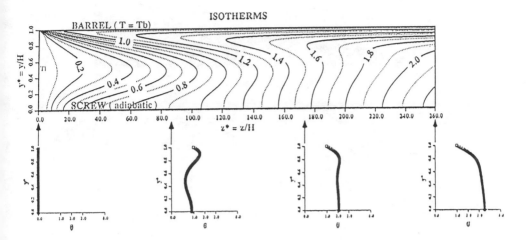

Figure 4: Isotherms, and temperature profiles at four downstream locations for a non-Newtonian fluid with n = 0.3, $\phi = 16.54\,^{\circ}$, $q_v = 0.25$, Pe = 7050, G = 0.005, $\beta = 1.134$, $\beta_1 = 10.0$, $b_m = 1.0$, S = -1000, Le = 0.001, $\theta_{gel} = 0.5$, No Taper

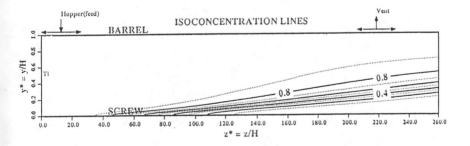

Figure 5: Lines of constant moisture concentration c^* for conditions given in Figure 4.

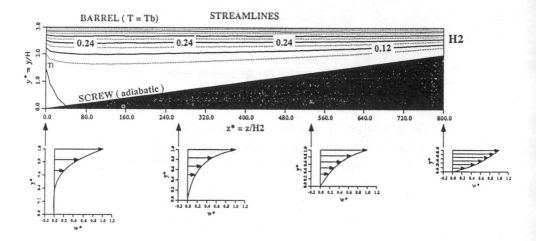

Figure 6: Streamlines, and velocity profiles at four downstream locations in a tapered channel; n = 0.5, qv = 0.2, Pe = 3000, G = 1.0, β = 1.134, β_1 = 5.0, b_m = 0.0, ψ = 0.1432°

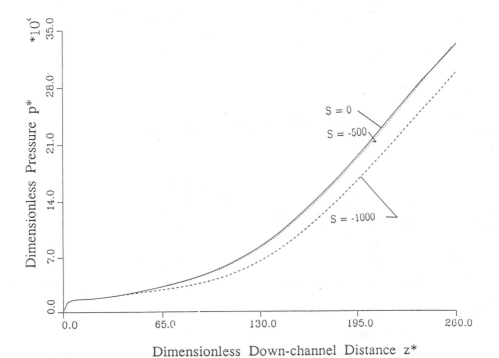

Figure 7: Effect of the strength of the moisture sink S on the variation of the dimensionless pressure p^* along the screw channel length z^* for the conditions given in Figure 4

for large magnitudes of S. It is also important to note that proper characterization of viscosity and kinetics of the chemical reaction are crucial to a successful simulation. Our experience has indicated that the solution is very sensitive to the Arrhenius and exponential dependencies of viscosity.

Figure 8 shows the effect of changing the power-law index n of the material on the pressure obtained in the extruder channel. The curves are for $G = 0.001$, $q_v = 0.3$ and $S = -1000$. The pressure rises from the hopper to the die along the screw helix, as expected. The pressure gradient is higher for increasing n. For non-Newtonian fluids, $n < 1$, and the viscosity decreases with an increase in shear rate. The Newtonian fluid, $n = 1$, therefore, gives rise to larger viscous drag which in turn leads to larger pressure gradients required to overcome it.

The effect of the throughput q_v on the pressure development at the die is investigated next. As shown in Figure 9, a smaller value of q_v, which corresponds to greater restriction to the flow at the die, results in a larger pressure rise as compared to the case with $q_v = 0.4$, which is close to the open die situation for the isothermal flow of Newtonian fluids ($q_v = 0.5$). There is a limiting value of the throughput q_v that can be fed through an extruder beyond which pressure development is affected considerably. This limiting value depends on the particular conditions governing the flow, for example, the power-law index n and the viscosity coefficients β, β_1 and b_m (0.5 for isothermal, Newtonian flow).

Further extensions to this work include a complete simulation including the elliptic nature of the general governing equations. It is expected that a coupling of die flow simulation with the extruder simulation by means of numerical models and/or empirical correlations will provide us a means of correcting the scheme in such complicated flow situations. Such an exercise is currently being carried out. Additionally, heat transfer by conduction along the screw and the barrel can play an important role in determining the shear history and temperature profiles at any cross-section. Experiments on a laboratory scale extruder have shown that the barrel temperature is not generally uniform axially. As an extension to the analysis presented here, the axial conduction in the barrel and screw need to be solved simultaneously with the convection problem in the fluid. This obviates the need to specify conditions for temperature except at the hopper and die.

CONCLUSIONS

A numerical simulation of the transport phenomena occurring in the flow of a non-Newtonian fluid through a single screw extruder has been carried out. A general methodology has been developed for the numerical treatment of various physical processes arising in such flows. Dimensionless forms of the equations have been derived and solved using finite differences, and results have been obtained

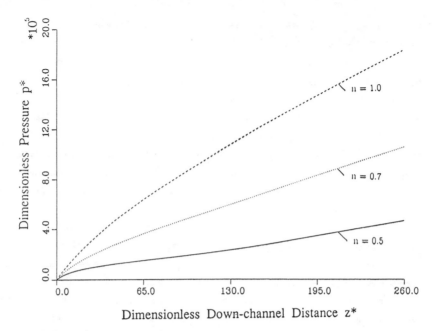

Figure 8: Effect of the power-law index n on the variation of dimensionless pressure p^* along the screw channel length z^* for $q_v = 0.3$, $G = 0.001$, other conditions as in Figure 4

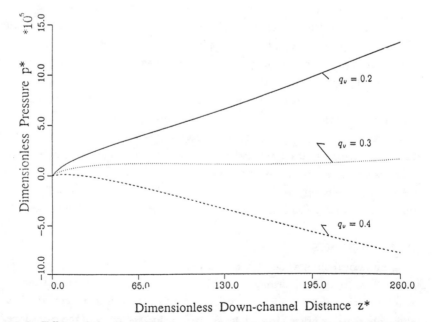

Figure 9: Effect of the throughput q_v on the variation of dimensionless pressure p^* along the screw channel length z^* for $n = 0.5$, $G = 0.0$, other conditions as in Figure 4

for a wide range of governing parameters. The limits of applicability of the marching scheme used here are discussed. Guidelines are provided that allow physical interpretation of effects based on the simulation results.

For significant viscous dissipation within the material (represented by the parameter G), the material temperature rises above the imposed barrel temperature by as much as 100 %. For negative values of G, there is significant back flow. Other parameters that significantly affect the output characteristics include the reaction rate S and the throughput q_v. For throughputs larger than a limiting value, typically 0.3, the pressure can actually decrease from the hopper to the die. Numerical experimentation has shown that the limiting values of S are closely linked to the viscosity dependence on concentration, i.e. b_m. Tapering of the screw channel is an important means of controlling the pressure developed at the die.

ACKNOWLEDGEMENTS

This work was supported by the Center for Advanced Food Technology, a New Jersey Commission on Science and Technology Center at Rutgers University. The authors would like to thank Professors V. Sernas and T. H. Kwon for discussions throughout this work.

REFERENCES

1. Tadmor, Z. and Gogos, C. Principles of Polymer Processing, Wiley, New York, 1979.

2. Fenner, R. T. Principles of Polymer Processing, Chemical, New York, 1979.

3. Fenner, R. T. Developments in the Analysis of Steady Screw Extrusion of Polymers, Polymer, Vol. 18, 1977.

4. Elbirli, B. and Lindt, J. T. A note on the Numerical Treatment of the Thermally Developing Flow in Screw Extruders, Polymer Engineering and Science, Vol. 24, No. 7, 1984.

5. Karwe, M. V. and Jaluria, Y. Numerical Simulation of Fluid Flow and Heat Transfer in a Single Screw Extruder for Non-Newtonian Fluids, in ASME Winter Annual Meeting, Chicago, 1988. Also, in Numerical Heat Transfer, to appear, 1989.

6. Jaluria, Y. and Torrance, K. E. Computational Heat Transfer, Hemisphere, New York, 1986.

7. Jaluria, Y. Computer Methods for Engineering, Allyn and Bacon, Needham Heights, Massachusetts, 1988.

N

Viscous Heating of Power Law Fluids with Temperature-Dependent Viscosity

K. Kozioł, S.G. Niwiński

Department of Technology and Chemical Engineering, Technical University of Poznań, 60-965 Poznań, Poland

ABSTRACT

The transient velocity and temperature profiles have been calculated for the plane Couette flow of power law, shear thinning fluids with an exponential dependence of the viscosity upon temperature. The influence of heat generation due to viscous dissipation has been considered. Detailed numerical results are given to show the effects of the flow index and the Prandtl number on the rate of momentum and heat transfer. The temperature increase is especially large in the boundary layer near the moving plate in the case of unsteady flow of fluids with small value of the Prandtl number, when the frictional heating is significant.

INTRODUCTION

The flow of fluids is never exactly isothermal. The enthalpy increase caused by viscous heating can significantly alter flow field and heat transfer in laminar motion of Newtonian and non-Newtonian fluids. There are many practical applications, such as viscometry, plastics extrusion, and lubrication, where viscous heating is an important problem during the flow of fluids.

The temperature rise at the moving surface in plane Couette flow is dominated by the local velocity gradient and viscosity of the fluid. Heating can lead to a dramatic decrease in viscosity. The complexities that arise in an unsteady flow are associated with the strong nonlinear viscosity-temperature dependence.

The essential feature of viscous fluid with

high shear rate is the thermal feedback occuring in
an ideal Couette flow. If the fluid flows,it becomes
hotter. This increases the flow rate, which in turn
can increase the rate of viscous heating. The viscosity decreases as the result of increased temperature, and the applied shear stress tends to increase. The shear stress is a maximum,when the shear
rate increase is offset by the decrease in viscosity.
If the rate of this heat source is small enough to
conduct the heat away to the walls,steady conditions
are established. On the other hand, the maximum temperature tends to infinity. The thermal runaway, as
the process goes on, can lead to the hydrodynamic
thermal burst.

One of the most significant advances in the
field of idealized plane Couette flow came with the
papers of Gruntfest [1,2]. The special attention was
attached to the departures from Newtonian behaviour
of the fluid as a result of thermal feedback. The
extended analysis,including viscous heating and variable viscosity effects, and applied to the same geometry of flow without inertial effects,was given by
Joseph [3,4]. It has been established that steady
state flow must develop a point of inflection at the
slit center. The existance of critical stress,double
valued solutions for flow and temperature distributions and stability characteristics are considered.

Gavis and Laurence have obtained analytical solutions for plane and circular Couette flows of Newtonian [5] and power law [6] fluids with exponential
temperature dependence of the viscosity in terms of
elementary functions. Nihoul [7] has extended their
resuls including hyperbolic dependence of heat conductivity and viscosity on the temperature. Again,as
in the case of [5,6], the ambiguity of double valued
dependence of temperature and velocity distributions
on shear stress is resolved through the Brinkman
number.

Surprisingly, simple analytical solutions were
found by Bird and Turian [8-11] for Newtonian, power
law,and Ellis models with temperature dependent viscosity and thermal conductivity,then applied to cone
and plate viscometer. Martin [12] was able to find
exact solutions for viscometric flows of power law
fluid between rotating concentric cylinders and parallel plates. A discussion of exact values of
critical parameters in viscometric flows and comparison with the upper bounds estimated by Joseph [4],
has been presented by Trowbridge and Karran [13].
Sukanek and Laurence [14] succeeded in finding ex-

perimentally the existence of maximal shear stress in plane and circular Couette flows using Newtonian fluid with strong temperature dependent viscosity.

Theoretical investigations of viscous heating in the unsteady temperature field in plane Couette flows of Newtonian [15] and power law [16] fluids have been presented by Winter. Eckert and Faghri[17] first posed and solved the coupled momentum – energy problem yielding results for the transient velocity and temperature distributions in the plane Couette starting flow of Newtonian fluids. The purpose of this paper is to present an extension of Eckert's and Faghri's analysis to non-Newtonian, power law fluids set up to the similar conditions.

FORMULATION

Governing equations and boundary conditions
Initially, the fluid remains at rest between two in-finite plane surfaces, and its temperature is uni-form and equal to that of the walls. One of the wall is set in motion with constant velocity at time t=0. The origin of co-ordinate system is placed at this surface. The temperatures of the lower and the upper surfaces are kept at their original values. It is assumed that the pressure in the flow direction is uniform.

Kinematic starting process and thermal develop-ment due to frictional heating are going on simulta-neously as a consequence of the rapid deformation of fluid at the moving wall. In that case, the inertial forces cannot be neglected. Therefore, if the direc-torial derivative terms vanish, one can obtain the equation of motion

$$\rho \frac{\partial u}{\partial t} = \frac{\partial}{\partial y} \left[\eta \frac{\partial u}{\partial y} \right] \qquad (1)$$

in which ρ, η and t are the density, apparent visco-sity and time, y is the co-ordinate normal to the flow direction, and u is the local velocity.

The conservation of energy for this geometry requires

$$\rho c \frac{\partial T}{\partial t} = k \frac{\partial^2 T}{\partial y^2} + \eta \left[\frac{du}{dy} \right]^2 \qquad (2)$$

The initial and boundary conditions for the velocity and for the temperature field are:

$$t \leqslant 0 \quad : \quad u(y,t) = 0 \quad , \quad T(y,t) = T_o$$
$$y = 0 \quad : \quad u(y) \quad = 0 \quad , \quad T(y) \quad = T_o \qquad (3)$$
$$y = h \quad : \quad u(y) \quad = U \quad , \quad T(y) \quad = T_o$$

Basic assumptions

The equations of change describe the flow and the energy transport process in the fluid layer with the following assumptions:
- the fluid is incompressible,
- no slip at the walls,
- inertia forces are compared to viscous forces,
- density, thermal conductivity and specific heat are constant,
- the temperature and the shear rate dependence of apparent viscosity is postulated to be a function

$$\eta = \left[\frac{h}{U} \frac{du}{dy} \right]^{n-1} \eta_{o,r} \exp\left[-\beta(T - T_o) \right] \qquad (4)$$

where $\eta_{o,r}$ is the viscosity evaluated at the reference shear rate U/h, and at the initial temperature of the fluid.

Dimensionless presentation

The following dimensionless variables and parameters were selected to make dimensionless the equations of change:

$$t^+ = \frac{kt}{c\rho h^2} \quad , \quad y^+ = \frac{y}{h} \quad , \quad T^+ = \beta(T - T_o)$$

$$\qquad (5)$$

$$u^+ = \frac{u}{U} \quad , \quad Pr_{o,r} = \frac{c\eta_{o,r}}{k} \quad , \quad Na_{o,r} = \frac{\beta U^2 \eta_{o,r}}{k}$$

With these notations the flow field and the temperature field can be expressed in the following form:

$$\frac{\partial u^+}{\partial t^+} = Pr_{o,r} \frac{\partial}{\partial y^+} \left\{ \left[\frac{\partial u^+}{\partial y^+} \right]^n \exp(-T^+) \right\} \qquad (6)$$

$$\frac{\partial T^+}{\partial t^+} = \frac{\partial^2 T^+}{\partial y^{+2}} + Na_{o,r} \left[\frac{du^+}{dy^+} \right]^{n+1} \exp(-T^+) \qquad (7)$$

The initial and boundary conditions take the form:

$$t^+ \leqslant 0 \quad : \quad u^+ = 0 \quad , \quad T^+ = 0$$
$$y^+ = 0 \quad : \quad u^+ = 0 \quad , \quad T^+ = 0 \qquad (8)$$
$$y^+ = 1 \quad : \quad u^+ = 1 \quad , \quad T^+ = 0$$

The equation of motion (6) and the equation of energy (7) are highly coupled through temperature dependent viscosity. The Nahme–Griffith number is a useful measure of generation effects, which compares the temperature increase due to viscous heating with changes in the viscosity. The Prandtl number is the dimensionless coefficient for momentum diffusion.

Thus, positive transient solutions of the problem satisfying the equations of change with initial and boundary conditions prescribed above can exist as a function

$$(T^+ \text{ or } u^+) = f (Na_{o,r}, Pr_{o,r}, t^+, y^+, n) \qquad (9)$$

At some stage, if the conduction balances heat generation, the final steady state arises. Then, the asymptotic temperature and velocity profiles can be performed as

$$(T^+ \text{ or } u^+) = f (Na_{o,r}, y^+, n) \qquad (10)$$

In this case, the stress is constant across the fluid layer, and thus, the flow and temperature fields are independent of Prandtl number.

Numerical method
The finite difference formulation is used to solve the set of coupled and strongly nonlinear partial differential equations (6) and (7) with boundary conditions of the first kind. The control volume approach [18] is selected for constructing a finite difference approximation. The resulting expressions are modified near the boundaries to incorporate boundary conditions.

The procedures used to solve the coupled equations employ marching with time scheme and fully implicit method, sometimes called the backward-time, central-space method, to evaluate spatial derivatives. Two initial profiles of velocity and temperature at small enough time $t^+ = 10^{-5}$ were computed for flow index $n=1$, and then resolved to any other value of n using the underrelaxation schemes with small relaxation coefficient. The spatial region of the fluid layer was divided with 250 grid points. Initial time steps of order 10^{-9} were changed dynamically to the value of 10^{-3} for fully developed velocity and temperature fields. Only results for small and large values of Prandtl number will be presented below.

RESULTS

Figures 1 and 2 perform profiles of the dimensionless velocity and dimensionless temperature as a

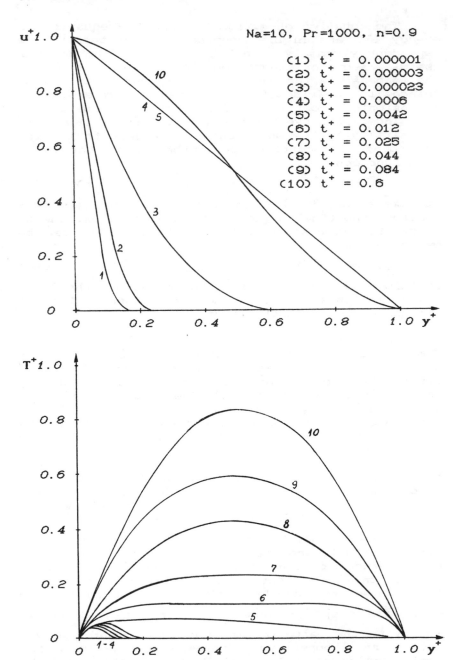

Fig.1.   Dimensionless velocity u^+ and temperature T^+
        profiles for Nahme-Griffith number Na = 10,
        Prandtl number Pr = 1000, and n = 0.9.

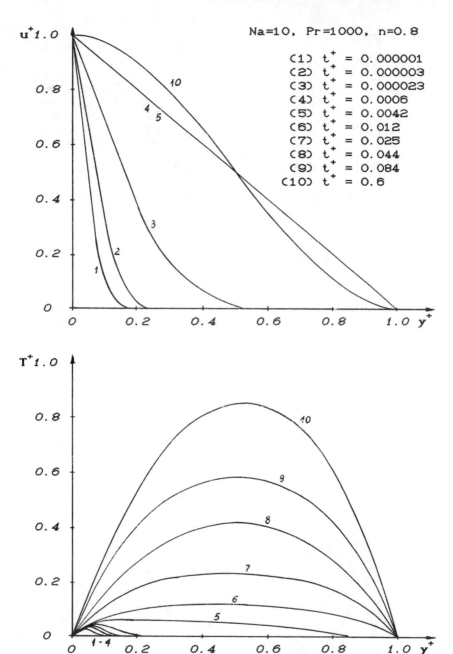

Fig. 2. Dimensionless velocity u^+ and temperature T^+
         profiles for Nahme-Griffith number Na = 10,
         Prandtl number Pr = 1000, and n = 0.8.

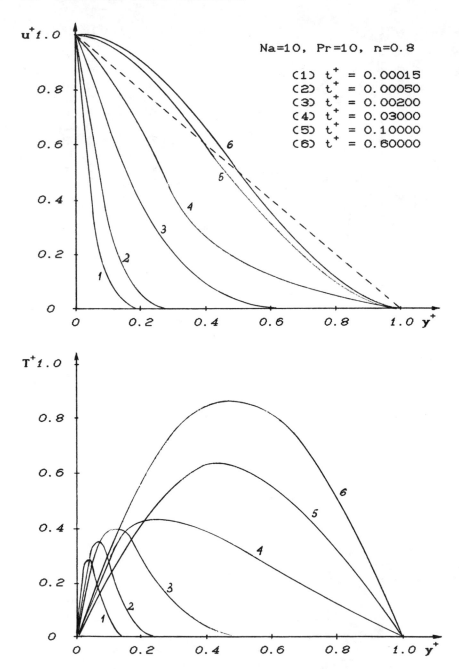

Fig. 3. Dimensionless velocity u^+ and temperature T^+
profiles for Nahme-Griffith number Na = 10,
Prandtl number Pr = 10, and n = 0.8.

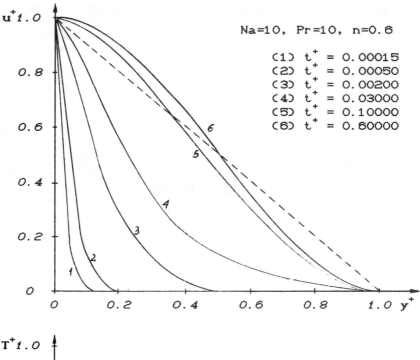

Na=10, Pr=10, n=0.6

(1) t^+ = 0.00015
(2) t^+ = 0.00050
(3) t^+ = 0.00200
(4) t^+ = 0.03000
(5) t^+ = 0.10000
(6) t^+ = 0.60000

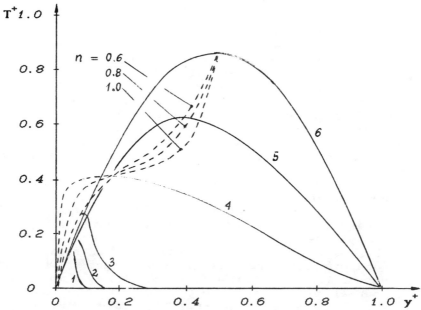

Fig. 4. Dimensionless velocity u^+ and temperature T^+
profiles for Nahme-Griffith number Na = 10,
Prandtl number Pr = 10, and n = 0.6.

function of the distance and time co-ordinates for
fluids with Prandtl number of 1000, Nahme-Griffith
number of 10, and flow index having values 0.9 and
0.8,respectively. Figures 3 and 4 contain representative results for the transient flow and temperature
fields in fluids with Prandtl number of 10, for viscous heating analogous to the previous situation,and
for flow indices of values 0.8 and 0.6.

It can be observed in Figs.1 and 2 that initial
value of temperature near the moving wall is smaller
for lower values of the flow index. The same is true
for fluids with small Prandtl number from Figs.3 and
4. The asymptotes for maximum temperature values,
ploted as a function of dimensionless distance in
Fig.4 can be convenient to describe the influence of
flow index on the instantaneous temperature distribution.

CONCLUSIONS

The plane Couette flow forced by the sudden start of
the wall with constant velocity can be considered as
a limiting case of the flow between two infinite
concentric cylinders with the thickness of the fluid
layer to be small compared to the radius. Except for
the rheological model, there is a full analogy to
Eckert's and Faghri's analysis [17], with respect to
the co-ordinate system chosen to perform the equations of change, the initial and boundary conditions
and the temperature dependence of viscosity. Thus,
the results obtained for Newtonian flow in [17] were
treated as the references to these, presented now.
Most of the basic features described by Eckert
and Faghri are observed in our investigations. This
is true, particularly, for the following similar-
ities:
a) asymptotic velocity profiles are S-shaped for
   both, small and large values of Prandtl number,
b) there is, for large values of Prandtl number, the
   straight line velocity distribution (nearly the
   fully developed flow), and - up to that moment -
   there are very small changes in the initial temperature distribution (this is agreed with the
   assumption that kinematic starting process and
   thermal development can be treated as separable),
c) for fluids with small Prandtl number, the initial
   temperature increase in a thin layer near the
   moving wall is of the order comparable to the
   maximum steady state value - the assumption mentioned above does not apply in this case,
d) on the other hand, when the Prandtl number is
   small,the transient velocity profile is disturbed

in this manner that it will never assume the re-
ctilinear distribution.

Generally, for flows of fluids with dependent
physical properties and viscous dissipation, it is
difficult and time-consuming to solve the equations
of change in full detail. The complexity of computational methods and procedures arises from the nature
of the governing equations. Numerical simulation of
the system is often the only way to give the useful
understanding of the coupled momentum-energy transport phenomena, when the hydrodynamic development
region is not considerably shorter than the thermal
entrance region.

NOMENCLATURE

c	specific heat	[J/(kg K)]
h	width of the flat slot	[m]
k	thermal conductivity	[W/(m K)]
n	power law flow index	[-]
t	time	[s]
u	local velocity	[m/s]
y	spatial co-ordinate	[m]
T	temperature	[K]
U	constant wall velocity	[m/s]
β	temperature coefficient of viscosity	[1/K]
η	apparent viscosity	[N s/m^2]
ρ	density	[kg/m^3]
Na	Nahme-Griffith number	[-]
Pr	Prandtl number	[-]

Indices
+ denotes the dimensionless variable
o refers to the initil value
r denotes the reference value

REFERENCES

1. Gruntfest, I.J. Thermal feedback in liquid flow;
   plane shear at constant stress, Trans.Soc.Rheol.,
   Vol.7, pp 195-207, 1963.
2. Gruntfest, I.J. Apparent departures from Newtonian behavior in liquids caused by viscous heating
   Trans.Soc.Rheol.,Vol.9, pp 425-441, 1965.
3. Joseph, D.D. Variable viscosity effects on the
   flow and stability of flow in channels and pipes,
   Phys.Fluids,Vol.7, pp 1761-1771, 1964.
4. Joseph, D.D. Stability of frictionally-heated
   flow, Phys.Fluids,Vol.8, pp 2195-2200, 1965.
5. Gavis, J. and Laurence, R.L. Viscous heating in
   plane and circular flow between moving surfaces,
   Ind.Eng.Chem.Fundam.,Vol.7, pp 232-239, 1968.

6. Gavis, J. and Laurence, R.L. Viscous heating of power-law liquid in plane flow, Ind. Eng. Chem. Fundam., Vol.7, pp 525-527, 1968.

7. Nihoul, J.C.J. Nonlinear Couette flows with temperature-dependent viscosity, Phys. Fluids, Vol.13, pp 203-204, 1970.

8. Bird, R.B. and Turian, R.M. Viscous heating effects in a cone and plate viscometer, Chem. Eng. Sci., Vol.17, pp 331-334, 1962.

9. Turian, R.M. and Bird, R.B. Viscous heating in cone-and-plate viscometer - II. Newtonian fluids with temperature-dependent viscosity and thermal conductivity, Chem. Eng. Sci., Vol.18, pp 689-696, 1963.

10. Turian, R.M. Viscous heating in the cone-and--plate viscometer - III. Non-Newtonian fluids with temperature-dependent viscosity and thermal conductivity, Chem. Eng. Sci., Vol.20, pp 771-781, 1965.

11. Turian, R.M. The critical stress in frictionally heated non-Newtonian plane Couette flow, Chem. Eng. Sci., Vol.24, pp 1581-1587, 1969.

12. Martin, B. Some analytic solutions for viscometric flows of a power law fluid with heat generation and temperature dependent viscosity, Intern. J. Nonlinear Mech., Vol.2, pp 285-301, 1967.

13. Trowbridge, E.A. and Karran, J.H. A discussion of critical parameters which can occur in frictionally heated non-Newtonian fluid flows, Int. J. Heat Mass Transfer. Vol.16, pp 1833-1848, 1973

14. Sukanek, P.C. and Laurence, R.L. An experimental investigation of viscous heating in some simple flows, AIChE Journal, Vol.20, pp 474-484, 1974.

15. Winter, H.H. The unsteady temperature field in plane Couette flow, Int. J. Heat Mass Transfer, Vol.14, pp 1203-1212, 1971.

16. Winter, H.H. Warmedissipation in Polymerschmelzen bei ebener Schleppstromung, thermischer Anfahrvorgang und Gleichgewichtszustand, Rheol. Acta, Vol.11, pp 216-223, 1972.

17. Eckert, E.R.G. and Faghri, T.A. Viscous heating of high Prandtl number fluids with temperature--dependent viscosity, Int. J. Heat Mass Transfer, Vol.29, pp 1177-1183, 1986.

18. Patankar, S.V. Numerical heat transfer and fluid flow, Hemisphere Publishing Corporation, Washington, 1980.

Heat and Mass Transfer in the Absorber of Neutral Gas Absorption Refrigeration Units

D.A. Kouremenos, A. Stegou-Sagia, E.D. Rogdakis, K.A. Antonopoulos

Mechanical Engineering Department, National Technical University of Athens, 42 Patission Street, 106 82 Athens, Greece

ABSTRACT

In this study a simulation method is presented for the composite process of the absorption of ammonia in the absorber of a small inert gas absorption refrigeration unit. The momentum, heat and mass transfer differential equations and the continuity equation are solved within the absorber by using finite-difference techniques. The physical situation considered is a vertical two phase annular ammonia-water-hydrogen flow. A thin annular film of ammonia-water solution flows downwards on the interior surface of a tube and absorbs a quantity of ammonia from the ammonia/hydrogen gas mixture flowing upwards. The computer algorithm developed uses the thermodynamic and transport properties of the two streams presented in the form of analytical expressions in terms of pressure, temperature and concentration. The results include detailed information of the absorption process i.e. the velocity, pressure, concentration, enthalpy and entropy fields. The presented computer-based technique may be used in the design of related equipment.

1. INTRODUCTION

In previous studies [1,2], numerical predictions of heat and mass transfer in the evaporator of neutral gas absorption refrigeration units have been presented. Here a numerical algorithm is developed for calculating heat and mass transfer in the absorber of such units. The situation considered has been simplified as shown in Figure 1: A thin annular film of liquid NH_3/H_2O mixture (low mass fraction of

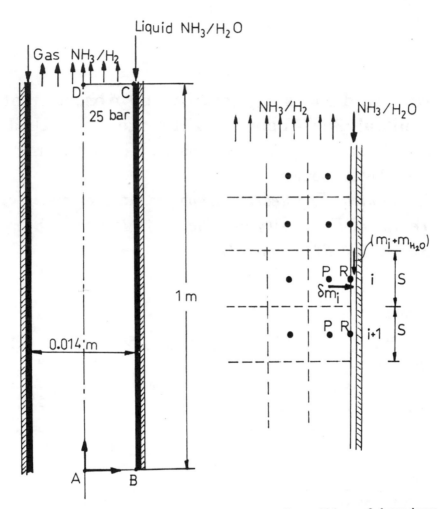

Fig.1   Absorber and solution
domain (ABCD)

Fig.2   Imposition of boundary
conditions along the
film surface.

NH$_3$ in the mixture) flows downwards on the interior surface of a vertical tube and absorbs NH$_3$ from an upwards flowing gaseous NH$_3$/H$_2$ mixture. The numerical algorithm developed solves the momentum, energy and mass transfer differential equations together with the continuity equation using finite-difference techniques and provides the complete fields of velocity, temperature and mass fraction of NH$_3$-vapour within the NH$_3$/H$_2$ mixture. The entropy and enthalpy fields are also calculated by use of relations developed [3], which provide these quantities in terms of pressure, temperature and mass fraction. The mass fraction of NH$_3$ along the film of liquid NH$_3$/H$_2$O mixture is also calculated.

2. GOVERNING EQUATIONS

The transport equations governing steady, two-dimensional, viscous flow with simultaneous heat and mass transfer, may be written in terms of cylindrical polar coordinates x, r within the solution domain ABCD (Figure 1) as:

$$\frac{1}{r}\left[\frac{\partial}{\partial x}(\rho u r \Phi) + \frac{\partial}{\partial r}(\rho v r \Phi) - \frac{\partial}{\partial x}(r\,\Gamma_\varphi\,\frac{\partial \Phi}{\partial x}) - \frac{\partial}{\partial r}(r\,\Gamma_\varphi\,\frac{\partial \Phi}{\partial r})\right] = S_\varphi$$

(1)

where ρ is the density and Φ is a general dependent variable standing for any of the following: the velocity component u in the axial direction x; the velocity component v in the radial direction r; the mass fraction f of NH$_3$ within the NH$_3$/H$_2$ gas mixture; the temperature T.   The exchange coefficient Γ_φ and the source term S_φ are given, for each value of Φ, in Table 1.   The symbols μ, Pr, Sch and p appearing in Table 1 denote the viscosity, the Prandtl number, the Schmidt number and the pressure, respectively.

The thermodynamic and transport properties of the gaseous NH$_3$/H$_2$ mixture and the liquid NH$_3$/H$_2$O mixture are available in the form of analytic functions in terms of the

Table 1. Values of the variable Φ and the terms Γ_φ and
S_φ in equation (1)

TRANSPORT EQUATION	Φ	Γ_φ	S_φ
x-momentum	u	μ	$-\partial p/\partial x$
r-momentum	v	μ	$\mu\,v/r^2 - \partial p/\partial r$
energy	T	μ/Pr	0
mass	f	μ/Sch	0
continuity	1	0	0

temperature, the pressure and the NH_3 mass fraction. Details about the derivation of these functions may be found in [3,4]. As an example, the expressions developed for the enthalpy h (in KJ/Kg) and the entropy s (in KJ/Kg K) of the NH_3/H_2 gas mixture are given below:

$$h = \sum_{i=1}^{3} A_i \, f^{i-1} \tag{2}$$

$$s = \sum_{i=1}^{3} C_i \, f^{i-1} \tag{3}$$

where:

$$A_1 = \sum_{i=1}^{3} \left[\sum_{j=1}^{4} a_{ij} \, p^{j-1} \right] T^{i-1} \tag{4}$$

$$A_2 = \sum_{i=1}^{4} \left[\sum_{j=1}^{4} a'_{ij} \, p^{j-1} \right] T^{i-1} \tag{5}$$

$$A_3 = \sum_{i=1}^{4} \left[\sum_{j=1}^{4} a''_{ij} \, p^{j-1} \right] T^{i-1} \tag{6}$$

$$C_i = \sum_{j=1}^{3} c_{ij} \, f^{j-1}, \quad i=1,2,3 \text{ (for p=20 bar)} \tag{7}$$

In the above equations f (in Kg NH_3/Kg mixture) is the mass fraction, T (in °C) is the temperature, p (in bar) is the total pressure of the mixture, and a_{ij}, a'_{ij}, a''_{ij}, c_{ij} are coefficient given in [1].

3. BOUNDARY CONDITIONS

The following boundary conditions have been imposed on the boundaries of the solution domain ABCD, shown in Figure 1:

Boundary AB (inlet):   At the inflow boundary AB the velocity, temperature and mass fraction profile is prescribed.

Boundary BC (liquid NH_3/H_2O film): The interface BC between the NH_3/H_2 gas mixture and the liquid NH_3/H_2O film is treated as a moving wall, i.e. the axial velocity u is prescribed along this boundary, while the radial velocity v is taken equal to zero. The temperature T or the heat flux is prescribed along boundary BC (as a first approximation a constant temperature T=30 °C has been imposed along BC). With reference to Figure 2, the mass δm_i (in Kg/s) of NH_3 absorbed at location i is:

$$\delta m_i = \left[\rho DS \ \frac{f_P - f_R}{(PR)} \right]_i \tag{8}$$

where ρ and D are the density and the diffusion coefficient of the NH_3/H_2 mixture, S (in m^2) is the side of the boundary cell and f_P, f_R are the NH_3 mass fractions in the NH_3/H_2 mixture at nodes P and R, respectively. As the NH_3 mass δm_i is added to the film, the NH_3 mass fraction, F, in the film at the next location $i+1$ becomes:

$$F_{i+1} = \frac{m_{i+1}}{m_{i+1} + m_{H_2O}} = \frac{(m_i + \delta m_i)}{(m_i + \delta m_i) + m_{H_2O}} \tag{9}$$

The partial pressure P_{NH_3} (in bar) of the NH_3-vapour (saturation condition) on the film surface at location $i+1$, is approximated (for $T = 30$ °C) in terms of F_{i+1} by relation:

$$\left(P_{NH_3} \right)_{i+1} = 1.7 + 9 \ F_{i+1} + F_{i+1}^2 \tag{10}$$

The above partial pressure (assuming approximately equilibrium conditions) defines the mass fraction $\left(f_R \right)_{i+1}$ of NH_3 in the NH_3/H_2 gaseous mixture at the boundary point R (for $T = 30$ °C) as:

$$\left(f_R \right)_{i+1} = 0.019619 + 0.265058 \left(P_{NH_3} \right)_{i+1} - 0.0289935 \left(P_{NH_3} \right)_{i+1}^2 \tag{11}$$

According to the above procedure all boundary values of f can be calculated in terms of the corresponding values at the previous location along the film. Therefore along the interface boundary BC, either f can be prescribed or the flux of absorbed NH_3 can be calculated and inserted as a sink term S_φ in equation (1) for the boundary cells.

Boundary CD (outlet): At the outflow boundary CD only the distribution of the normal velocity u has to be specified. It is calculated during the numerical solution procedure by adding to the adjacent interior nodal values an increment such that the overall continuity is maintained.

Boundary DA (axis of symmetry): On the axis of symmetry DA, the velocity component normal to the axis is zero as are the normal gradients of the other variables.

4. METHOD OF SOLUTION

The methodology described in [5-10] has been employed for solving the set of differential equations (1), i.e. a computational grid of coordinate lines is generated within the solution domain ABCD (Figure 1) and the differential equations are intergrated over the control volumes of this grid to yield finite-difference equations of the form:

$$A_P \Phi_P = \sum_i A_i \Phi_i + S, \qquad i = N, S, E, W \qquad (12)$$

where Φ stands for any of the variables u, v, T, f and the coefficients A express the combined effect of convection and diffusion. The summation is over the North, South, East and West neighbouring nodes N, S, E, W of the typical grid node P.

5. RESULTS AND CONCLUSION

Results are shown in Figures 3 to 6 (discussed lated) which correspond to the following conditions:

(a)   Tube length: 1 m
(b)   Inner tube diameter: 0.014 m
(c)   Mean film thickness: 0.002 m
(d)   Total pressure of the NH_3/H_2 gas mixture: 25 bar
(e) H_2 inlet mass flow rate: 100 g/h
(f)   Inlet mass fraction of the gaseous mixture:

0.65 Kg NH_3/Kg mixture

(g)   Inlet velocity profile: laminar, fully-developed
(h)   Constant temperature of the NH_3/H_2 mixture at inlet and

along the film: T=30 °C
(i)   Film mass flow rate: 41 g NH_3/h, 364 gr H_2O/h

(NH_3 mass fraction 0.1 g NH_3/g mixture)

The above mentioned Figures 3 to 6 show the radial variation of u, f, s and h at three axial locations, i.e. inlet, middle, outlet. In Figure 3 it is observed that axial velocity increases owing to the decrease of density as the NH_3/H_2 mixture becomes poorer in NH_3 along the flow.   The decrease of NH_3 mass fraction of the gaseous NH_3/H_2 mixture along the flow, owing to absorption of NH_3 by the film, is clearly shown in Figure 4.   The expected entropy increase in the axial and radial directions is shown in Figure 5. Lastly, Figure 6 shows the increasing enthalpy in the radial direction and along the flow.   The enthalpy increase is due:

(a)   to the heat produced by the absorption of NH_3, and

(b)   to the compression of H_2 (the partial pressure of H_2
increases along the flow because the partial pressure of NH_3 decreases while the total pressure remains constant)

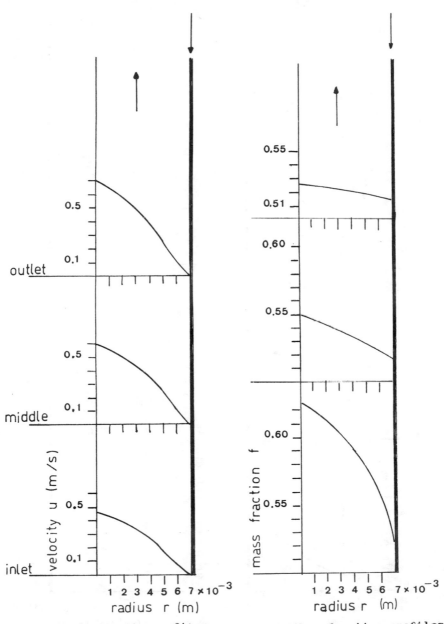

Fig. 3   u-velocity profiles
along the absorber

Fig. 4   Mass fraction profiles
along the absorber

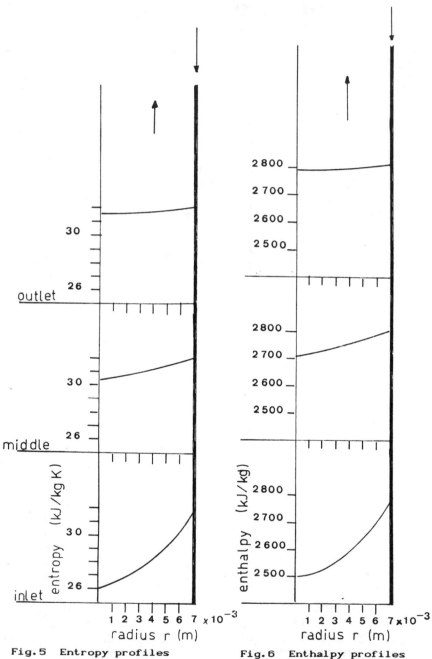

Fig. 5   Entropy profiles
along the absorber

Fig. 6   Enthalpy profiles
along the absorber

It is believed that the procedure developed, after validation by comparison with experimental data, may be employed in the design of absorbers of neutral gas absorption refrigeration units.

REFERENCES

1.   D.A. Kouremenos, K.A. Antonopoulos, A. Stegou-Sagia. "Entropy Productions during Evaporation in annular NH_3 Flow", Proceedings of the Winter Annual Meeting of the American Society of Mechanical Engineers, AES-Vol. 10-1, pp. 29-34, San Fransisco, California, 1989.

2.   D.A. Kouremenos, A. Stegou-Sagia, K.A. Antonopoulos. "The Prediction of Heat and Mass Transfer in the Evaporator of Neutral Gas Absorption Refrigeration Units". Proceedings of the Second World Congress on Heating, Ventilating, Refrigerating and Air-conditioning, CLIMA 2000, Volume "General Problems of Energy Refrigeration Components and Systems", pp. 85-90, Sarajevo, Yugoslavia, 1989.

3.   D.A. Kouremenos, A. Stegou-Sagia. "The psychrometric problem for the evaporation of NH_3 in NH_3/H_2 atmosphere in neutral gas absorption refrigeration units for pressures 17.5 bar to 27.5 bar. Waerme-und Stoffeubertragung, 1988, Vol. 22, 373-378.

4.   D.A. Kouremenos and E.D. Rogdakis. "Device modeling and simulation of NH_3/H_2O absorption units for refrigeration, heat pumps and heat transformers", ASME Winter Annual meeting, Anaheim, California, Vol. 2, pp. 97-106 (1986).

5.   S.V. Patankar, D.B. Spalding. "A calculation procedure for heat, mass and momentum transfer on three-dimentional parabolic flows", Int. J. Heat Mass Transfer, 1972, Vol. 15, 1987.

6.   L.S. Caretto, A.D. Gosman, S.V. Patankar, D.B. Spalding. "Two calculation procedures for steady, three-dimentional flows with recirculation", Proc. of the Third Int. Conf. on Numerical Methods in Fluid Mechanics, Springer Verlang, 1972, Vol. 2, 60-68.

7.   K.A. Antonopoulos. "Heat Transfer in Tube Banks under Conditions of Turbulent Inclined Flow", Int. Journal of Heat and Mass transfer, 1985, Vol. 28, No.9, 1645-1656.

8.   K.A. Antonopoulos. "Heat Transfer in Tube assemblies under conditions of laminar axial, transverse and inclined flow", Int. Journal Heat and Fluid flow, 1985, No.3, 193-204.

9.   K.A. Antonopoulos. "The Prediction of Turbulent Inclined Flow in Rod Bundles", Computers and Fluids, Vol. 14, No.4, pp. 361-378, 1986.

10. K.A. Antonopoulos and A.D. Gosman. "The Prediction of Laminar Inclined Flow in Tube Banks", Computers and Fluids, Vol. 14, No.2, pp. 171-180, 1986